P9-CFT-910

A SHEARWATER BOOK

The Dominant Animal

The Dominant Animal

Human Evolution
and the Environment

Best wishes –

Paul R. Ehrlich and Anne H. Ehrlich

ISLANDPRESS / Shearwater Books
Washington · Covelo · London

A Shearwater Book
Published by Island Press

Illustrations on pages 28, 37, 48 (top row), 60, 125, 179, 180, 188, 272 by John and Judy Waller.

Library of Congress Cataloging-in-Publication Data
Ehrlich, Paul R.
The dominant animal : human evolution and the environment / Paul R. Ehrlich and Anne H. Ehrlich.
p. cm.
"A Shearwater book."
Includes bibliographical references and index.
ISBN-13: 978-1-59726-096-1 (cloth : alk. paper)
ISBN-10: 1-59726-096-7 (cloth : alk. paper)
1. Human beings—Effect of environment on—History.
2. Human evolution. 3. Human-animal relationships—History.
I. Ehrlich, Anne H. II. Title.
GF13.E47 2008
304.2—dc22 2007050706

British Cataloguing-in-Publication data available.

Printed on recycled, acid-free paper ♻

Design by David Bullen

Manufactured in the United States of America

10 9 8 7 6 5 4 3 2

Keywords: genes, human origins, cultural development, Darwin, population, climate change, biodiversity, conservation, energy, disease

To Ruth and William Ehrlich
and Virginia and Winston Howland,
gone but sorely missed,
and
to Sally, Penny, and Lisa,
with love.

Contents

The Dominant Animal

Prologue

HUMAN BEINGS live in a world of change and always have. But in recent decades, the world has been changing faster than ever before, largely because of human modifications of our planet, and the pace is accelerating. Those modifications have, at least temporarily, enabled the power and consumption patterns of a billion or so people to be enormously enhanced and allowed a couple of billion to be doing all right, while leaving a few billion others living in poverty and, often, hopelessness. The acceleration of change traces to the rapid expansion in the human population since World War II, and to an explosive flowering of science and technology that has greatly increased the ability of our species to manipulate resources and the natural world.

The remarkable technological accomplishments of modern human beings have had unfortunate, if unintended, consequences. They have not been matched by comparable advances in how wisely we treat one another and our environment. As a result, the weight of great human numbers coupled with our unprecedented technological capacity now threatens to overwhelm Earth's ability to sustain what has become a global civilization. Those unintended consequences—civilization's threats to its own ability to persist—are often called the "human predicament."

How one species, *Homo sapiens*, has become so powerful that it can significantly undermine the ability of Earth's environment to support much of life—including our own—is a central theme of this book. Humanity's rise to dominance is a result of both genetic and cultural evolution, both of which led to scientific advances that have spawned ever more powerful technologies. Both kinds of evolution have occurred largely in response to changing environments, and both, in turn, have been responsible for dramatic environmental alterations.

Knowledge of these reciprocal evolution-environment interactions is critical to our ability to make wise decisions affecting the long-term success of our species and of the natural world upon which it is utterly dependent.

Besides providing the means to transform most of Earth's land surface, disrupt the life of its oceans, and significantly alter its atmosphere, science and technology have greatly enhanced our understanding of how the world works. Armed with computers, satellites, chemistry labs, electron microscopes, binoculars, butterfly nets, and theories, scientists have gained a reasonably comprehensive understanding of how Earth and its myriad inhabitants—including ourselves—interact and how they have changed over time. Having anything close to this level of knowledge is something new in the more than 100,000-year history of our species. In theory, we could use that knowledge to create a sustainable civilization—one in which human beings live happy, productive lives into the indefinite future. Whether we can manage that in practice remains to be seen.

Relatively recently, people's understanding of the world was quite different. An educated Englishman in the seventeenth century believed the Creation, including humanity, to be organized into a great "chain of being" that stretched "from the foot of God's throne to the meanest of inanimate objects."[1] Everything, living or not, was assigned its place in an unchanging order: angels ranked between God and kings, kings above commoners, people above other animals, lions over mice, mice above plants, plants above rocks, and so on. It was a *chain*, not a ladder; one couldn't climb or descend it. The basic characteristics of human beings weren't thought to change, nor usually did their status in society; butterflies didn't change, mountains didn't change, the air people breathed didn't change, and poor girls didn't marry royalty. Indeed, nothing fundamental was thought to have been altered since the world was created, on October 23, 4004 BC, as Archbishop of Armagh James Ussher calculated in 1650. In modern terms, seventeenth-century savants were ignorant about the most basic aspects of the world.

But by Archbishop Ussher's time, ironically, change was already in the wind. Despite having been (in theory) put on the throne by God, Charles I of England was executed for "treason" on January 30, 1649, by human beings. This act of regicide defied a millennium of custom and sermons and indicated a weakening of belief in the great chain. In fact, it heralded a fundamental shift. Previously, most scholars had focused on received wisdom from texts. In the West, these included the Bible as well as the classical works of scholars, especially Aristotle, who wrote extensively on physics, philosophy, natural history, logic, and psychology three and a half centuries before Christ; and Thomas Aquinas, with the five proofs of the existence of God he developed in the late thirteenth century. Both Aristotle and Aquinas themselves had an empiricist bent, meaning they tried to make sense of the natural world through observation and experience (though not experiment), as most scientists do today. But in the Middle Ages most people were not trying to acquire new knowledge from nature; they had been taught to believe what their putative superiors said and to do as they were instructed.

After Ussher's time, the focus on received wisdom was gradually replaced in the West by a new spirit of independent inquiry and discovery. Galileo (1564–1642), rolling marbles down inclined planes and making careful measurements to investigate gravity, began to undermine the vague Aristotelian notion that each object sought its "natural place." Science was being born. A "question-and-test" school of intellectual discourse was weakening the old "believe-and-obey" school. In the Age of Enlightenment that followed, spurred by scientific advances and growing discontent with oppressive monarchies, ideas of change and progress were appearing everywhere in Europe, along with belief in the power of reason to explain the universe and to improve people's lot. In part, ascendance of that view was due to the work of the extraordinary mathematician and physicist Isaac Newton (1643–1727), who, among many other things, showed how the mathematical laws that described the motion of objects on Earth also governed the movements of celestial bodies. Newton was followed a century and a half later by that greatest of biologists Charles Darwin (1809–82), who explained how the vast diversity of living creatures was generated. His ideas gave the coup de grâce to the prevalent static view of the world—he "Newtonized" biology by showing that myriad seemingly disparate facts could be explained by a set of unifying rules of change.

Increased understanding of physical science laid the groundwork for the industrial revolution, which later mass-produced wonders ranging from pistols with replaceable parts to automobiles, jet aircraft, digital computers, and nuclear missiles. At the same time, the discoveries of biological science started to end plagues and improve health, and thus lower death rates, and, by so doing, encouraged unprecedented human population growth. Those biological discoveries also began to explain where human beings had come from, how we fit into nature, and how we got smart enough to create and apply science, become the dominant animal on the planet, and even contemplate our possible destinies.

By substantially increasing the power of human beings to modify their environments, the industrial revolution and the population explosion laid the groundwork for a nineteenth- and twentieth-century human conquest of nature on a scale hitherto undreamed of. Societies around the globe cleared vast areas of forest to raise crops and build cities, lacing the world with railroads and then highways, filling the skies with jet aircraft, and creating a vast array of plastics and other chemical products never seen in nature. If at first this seemed a triumphal march, by the middle of the twentieth century a growing minority of scientists and others had begun to realize that the "conquest" also amounted to a vast assault on the global environment that had increasingly serious implications for the future of humanity.

Questions of where people came from—that is, how we gradually changed from tiny, mouselike creatures 60 million years ago into the planet's dominant animal—and of how we have both altered and been influenced by our physical and biological environments are inextricably intertwined. This book deals with what scientists have discovered about the origins of people, what is known of our diverse cultures, and how our environments shaped those origins and cultures and now shape the human future. And it explains the equally important obverse of the coin: how we are reshaping our global environment, helping to steer our species' trajectory. It is a story about scientific discovery in the human realm. It describes both *what* scientists have found out about us, our surroundings, and the dramatic consequences of our activities, and *how* science, the human activity that gave us the power to dominate Earth, can help us better understand the predicament we have created for ourselves and thereby avoid its worst consequences.

The Dominant Animal is thus intended to be a concise account of human beings' interactions with one another and with the biophysical world in which we evolved, and how we came to dominate land and water, atmosphere, microbes (maybe!), plants, and animals. To understand the roles we play in our social and biophysical environments, we need to look at what science can tell us about topics as seemingly disparate as climate change, genes, sex, religion, epidemics, ethics, education, politics, and nuclear war. This volume attempts to explain what human dominance means for the functioning of our planet and therefore for our future. Amazing as it may seem in an era of increased environmental awareness, this story is rarely told in its entirety.

Traditional books that include an evolutionary perspective focus mainly on *genetic* evolution—change in the hereditary endowment of organisms. In this approach, the environment, the physical and biological surroundings with which each individual interacts daily, is normally viewed as a background factor. It is discussed primarily as a cause of genetic change—something to which populations of organisms "adapt" over the course of many generations as their genetic endowments are altered. In practice, however, the two sides of the gene-environment interaction must be viewed together. That bacteria can evolve genetically to be resistant to the antibiotics people add to their environments is important for us to know. It is equally important to understand the exact ways in which the bacterial environment is being altered—the kinds and amounts of antibiotics that are being added. Decisions about that will have a profound effect on the duration of an antibiotic's efficacy and its health-enhancing effects. It is thus important to understand the basic mechanisms of genetic evolution, which provide a fundamental background to the entire science of life, as well as to inform people about such issues as pesticide use, antibiotic resistance, and the threat of emergent diseases.

Most books that focus on evolution largely ignore *cultural* evolution—change in the information people possess that is not found in their genes—which is even more important than genetic evolution in understanding the sources of current events. It was genetic evolution that produced brilliant, behaviorally flexible, highly social apes—ourselves. It was cultural evolution, building on those accomplishments, that determined most aspects of our environment-modifying behavior. The discovery of antibiotics and of the ways to apply them to our greatest benefit is a small example; the development and proliferation of science and science-based technology is a big one.

If books on evolution, in our view, ought to pay more attention to the environment and to cultural evolution, so books on environmental science, we believe, should pay more attention to genetic and cultural evolution. Understanding genetic-evolutionary interactions is critical to tasks such as preventing epidemics and designing optimal fishing strategies. If people concentrate on catching the big fishes, for instance, fishes whose genes allow them to breed at smaller sizes will soon predominate, and they will produce fewer young than big individuals. Use a genetically uninformed fishing strategy and you'll discover the fish in your nets growing smaller and fewer.

But cultural evolution is critical as well; pay no attention to how cultural evolution works and you may be unable to fix genetic-evolutionary problems even if the solution is clear. For instance, with appropriate interventions the culture of fishers could be redirected to catching a range of sizes and throwing some of the big ones back. Beyond that, questions about cultural evolution are crucial to understanding the origins of human power relationships that bear on the environment: how, for example, states evolved from tribal chiefdoms, allowing societies to organize and specialize to the point that they could menace the global environment, and how political systems might be modified to make achieving sustainability, rather than suffering catastrophe, more likely.

Knowing about evolution and human origins helps us to understand what makes us human and informs us about our possible destinies, the choice of fates largely to be determined by how we treat one another and our planetary home. As a triumphant species, we are risking our ability to sustain that triumph by, for instance, threatening the systems that provide us with food and water and maintain a satisfactory climate. That most people lack essential knowledge about our relationship to the natural world, which is humanity's life-support system, is a major contributing factor to *Homo sapiens'* deepening predicament.

This book is an attempt to address that widespread lack of awareness, and in particular to show how human society is a product of continuous evolutionary change. That continuous change is a mixture of genetic evolution in populations of people and other organisms, cultural evolution within and between societies, and evolution of the planet as its physical and chemical characteristics change in response to natural and human-generated forces. It's a remarkable and ongoing story. By knowing our evolutionary past and understanding the forces that have shaped our present, we will be better positioned to fashion a more sustainable future.

CHAPTER 1

Darwin's Legacy and Mendel's Mechanism

"Nothing in biology makes sense except in the light of evolution."

THEODOSIUS DOBZHANSKY, 1973[1]

HURRICANE KATRINA was a new kind of experience for most Americans—a huge natural disaster that demonstrated how poorly prepared the United States was, in 2005, for natural disasters. It also underlined that something strange is going on with the weather, even if Katrina's destructiveness itself might just have been a rare event in normal variability in the size and paths of hurricanes. If you regularly watch TV, read newspapers, or go to the movies, you can hardly have missed news about unusual weather. Global temperatures are going up, glaciers are melting, storms seem more frequently intense and so do droughts, and sea level is gradually rising.

Other animals are noticing too, at some level. Polar bears are finding it more difficult to make a living as the sea ice from which they hunt seals disappears. Coral animals in some places are dying as seawater warms, threatening the existence of coral reefs. If you are a bird-watcher, you might have noticed that some migratory birds from Latin America are arriving earlier each spring on their North American breeding grounds. In the Yukon of Canada, red squirrels are having their young earlier because of the great quantities of spruce seeds made available by a warming climate. In Europe, flowers are blooming about a week earlier in spring than they did in 1975—and in 2006 they were still blooming in Moscow in November.

Populations Evolving

That populations of organisms do not remain static in the face of environmental change is a recurring theme in the study of the natural world. Today, numerous animals and plants are changing their lives in response to the human-caused alteration of climatic regimes—the annual sequence of changes in temperature and rainfall—in their environments. While the average temperature increase of Earth has been less than a degree Celsius (1°C is 1.8°F) over the past fifty years, in the same period at the high-altitude laboratory in Colorado where we've worked since 1960, magpies have arrived from the lowlands for the first time and many flowers are blooming earlier, all apparently in response to earlier spring melting of the alpine snow.

Pitcher-plant mosquitoes (*Wyeomyia smithii*) are a good example of such climate-related changes. They inhabit eastern North America, and their early stages (eggs, larvae, and pupae) all live in water trapped in the "pitchers" of a carnivorous plant, *Sarracenia purpurea*. The plants grow primarily in nitrogen-poor soil and supplement their "diet" by digesting insects, mostly ants and flies, that die when trapped by downward-pointing hairs in their tubular leaf, which the small, hovering mosquitoes easily avoid. Southern populations of the mosquitoes produce five or more generations per year; northern populations, just one.

The larvae (wigglers), which hibernate in the leaves of the host plant, use the length of the day to determine when to enter that dormant state and when to resume activity. The critical day lengths at which the mosquitoes carry out these functions are quite rigidly controlled by their genes. But over the past thirty years, the genetically controlled clocks of more northerly populations have shifted their response so that hibernation does not begin until the day length becomes even shorter. As a result, the northern populations now behave much as the southern populations do, waiting until later in the fall before hibernating. Without that shift, as a warming climate in North America lengthens the growing season, the larvae would hibernate too soon. They would need to survive on the fat they had stored—not just survive through the winter as previously, but also without feeding through a warm period at the end of summer. This would lessen their chances of being alive when warm weather returned in the spring. With the shift, their necessary hibernation is shorter, helping them to survive without eating.

Other animals have not been able to change their behavior in response to a warming climate. Some populations of pied flycatchers (*Ficedula hypoleuca*), an attractive brown, black, and white insect-eating bird in Europe, have declined in size by 90 percent. The reason? The peak of insect abundance is occurring ever earlier as the environment warms, and the birds' customary breeding time is now too late for their offspring to have an optimal food supply. They have not successfully adjusted to the challenge of climate change, but neither have their populations remained static; they, along with other slow adjusters, are blinking out.

Darwin and Wallace's Great Idea

Why do some organisms adjust successfully to environmental change while others do not? For a biologist, there is no more basic or fascinating question, and it now has a pretty good answer—thanks in large part to scientific foundations laid by two Victorian Englishmen. In 1859, Charles Darwin (1809–82) and the brilliant polymath Alfred Russel Wallace (1823–1913) simultaneously proposed the first basically correct model for the mechanism that causes shifts such as that in mosquito hibernation. Historically, Darwin has received most of the credit for this world-changing idea—justifiably, since he supported his conjecture with an abundantly documented book, *On the Origin of Species*. He had formed many of his ideas on the basis of a five-year trip around the world as a naturalist on the British naval survey ship HMS *Beagle* (1831–36) and had subsequently corresponded about them with numerous colleagues.

It came as a shock to him when, in 1858, Wallace sent him a paper outlining Darwin's basic idea. Darwin, at the urging of friends, had his and Wallace's idea presented jointly at a meeting of a scientific society. But he followed this in 1859 with *Origin*, which sold out in a day and cemented his reputation as the greatest of all biologists. Like many great ideas, the Darwin-Wallace theory, which has come down to us identified by the term "natural selection," was disarmingly simple. Basically, it recognized that variation ordinarily exists among individuals within a natural population and as a result their interactions with their environments are likely to differ. Such variation was also long recognized in the characteristics of domesticated plants and animals—wheat with grain that was hard or easy to harvest, cows that gave more or less milk, dogs that could herd sheep or were hopeless at it, hogs that were fatter or thinner, and so on.

Farmers (who, from the standpoint of wheat and cows, are part of the environment) selected individuals with desired characteristics and encouraged their breeding, thereby over time creating strains that were easier to reap in the case of wheat or that gave more milk in the case of cows. Eventually, wheat was produced that had seed heads so heavy their stems could barely hold them up; cows appeared that could give fifteen gallons of milk per day. Darwin himself drew much of his view of selection in nature from observing the results of selective breeding of domestic animals, including the production of fancy plumage varieties by pigeon fanciers.

Darwin and Wallace were both inspired by an observation that had been made earlier by pioneering political economist Thomas Malthus, now most famous for warning that increases in numbers could cause the human population to outstrip its food supply. Malthus noted that animals commonly produced many more offspring than could survive. Darwin and Wallace drew the conclusion that those organisms that could take best advantage of their environments would be the ones most likely to survive; and by their survival and reproductive output, they were the ones "selected" by nature to be the parents of—and pass on their biological characteristics to—the next generation. Darwin recalled in his autobiography: "In October 1838, that is, fifteen months after I had begun my systematic inquiry, I happened to read for amusement Malthus on *Population*, and being well prepared to appreciate the struggle for existence which everywhere goes on from long-continued observation of the habits of animals and plants, it at once struck me that under these circumstances favourable variations would tend to be preserved, and unfavourable ones to be destroyed. . . . Here, then, I had at last got a theory by which to work."[2] "Favourable variations" here, of course, means those animals and plants most likely to survive and have many offspring.

Thus appeared the idea of natural selection, which caused a paradigm shift in the biological sciences. The widely held conceptual worldview at the time, that the diversity of organisms was created all at once and would be forever the same, gave way to another, one in which new kinds (species) were being continuously produced by a gradual process: natural selection. Furthermore (and equally revolutionary at the time), the theory included the complementary idea that species would also be going extinct.

For our purposes here, natural selection can be viewed as the differential reproduction of individuals with varying genetic endowments that are members of the same population—the individuals of a species in a given area at the same time.[3] While natural selection is not the only process that causes organisms to change generation after generation, it is the only one that makes them appear to be "designed" to survive and thrive in a given environment. There is, we emphasize, no actual design—it just looks that way to a naïve observer. It is important to remember that those "given environments" also change in their characteristics in time and space. The construction of a housing project may dramatically alter the environment of the plants and animals in a meadow, farm field, or woodland, for example. And differences in climate from place to place—at the extremes, from the heat of the tropics to the cold of the poles—affect where particular species may thrive. Selection modifies each population of an organism to adapt it to its environment, but environments usually differ in many respects from one place to another and from one time to another. Thus the selection pressures on different populations of the same organism or on the same population at different times may vary considerably, which, as we will see later, is of great importance in natural selection's role in producing life's diversity.

Islands in Time

Islands, it turns out, have been particularly good places to study the effects of natural selection, sometimes with surprising results. Observations of island organisms during his historic 1831–36 voyage on the *Beagle* subsequently helped Darwin perceive the results of natural selection and grasp its power to cause organisms to diverge from one another, first by a little bit, and then by more and more. What we call "evolution" today, he called in the first edition of *Origin* a combination of "descent with modification" and "much extinction." He saw, for example, that island dwellers normally do not share characteristics with one another that make them especially suited for life on islands, nor are they most similar to one another and different from mainland creatures as a whole. Instead, they usually appear to have diverged from similar species that live on the mainland closest to their island home.

But one curious aspect *is* shared by many island organisms. One would expect them to be very mobile, since they had to reach the island in the first place. But once they have been on the island awhile, organisms commonly have *less* physical ability to disperse than do their relatives that live on continents. This comparative inability to disperse means they have adjusted genetically (or, as evolutionists say, adapted) to isolated island conditions. Evolutionists interpret this sedentary tendency as the result of natural selection against anything that promotes long-distance movement. The ability to disperse 100 miles carries no reproductive advantage for a bird isolated on a 5-square-mile island 300 miles from the mainland. On the contrary, if the bird leaves the island, it is likely to drop into the sea exhausted and drown.

Species of rails (birds in the coot family) that once inhabited many Pacific islands had lost the ability to fly. By contrast, continental species of rails, the kinds that originally colonized the islands, can fly very well. Once on the islands, the rails faced no bird-eating predators such as foxes or cats, so flying was not necessary to escape being eaten. Island rails that didn't put as much energy into growing big, well-muscled wings had more energy to allocate to producing chicks. So rails that had smaller wings were "favored by selection"—just a way of saying they had more surviving offspring than rails with normal-size wings. The flightless rails survived very successfully on the islands until their environment changed dramatically a few thousand years ago—when people arrived. Then the rails were like the slow-adjusting pied flycatchers. Hungry people and the rats human beings transported around the world with them found the flightless rails easy prey and tasty—and exterminated most of them.

Recently it has been shown that selection to reduce the ability to disperse can operate over a surprisingly short time. Biologists studied annual plants of the sunflower family on about 200 small islands near Vancouver Island along the Pacific Coast in British Columbia. Populations frequently went extinct, and then the islands were recolonized from the readily dispersed mainland populations. The seeds of these plants are blown over the water attached to fluffy "parachutes" (like those of dandelions). Within a decade after plants returned to the islands, the parachutes of two species began to decrease in size over subsequent generations, and the seeds of one increased in weight—in each instance reducing dispersal ability.

Selection thus can be strong enough in some cases that dramatic results can be observed in short periods of time. That can be very good news for organisms at a time when environmental change from such factors as human modification of the climate and civilization's release of poisons into the environment threatens to exterminate many of our living companions on Earth. Unfortunately, there's an underside to the story from the human perspective; many of the organisms that today are best able to evolve rapidly are disease-causing organisms that want to make meals of us, or pests that transport those pathogens and parasites, or other pests that attack our crops.

Birds can evolve fast too, if not as quickly as pathogens. Recent work by evolutionists Peter and Rosemary Grant has revealed that the bills of Darwin's (Galápagos) finches, the birds made famous by Darwin's 1835 visit to the Galápagos islands on the *Beagle*, evolve in size and strength as rapid changes in climate require changes in diet. For instance, during a drought in 1977, a population of the medium ground finch (*Geospiza fortis*) on Daphne Major Island was subjected to intense selection pressure that favored large individuals with big bills. The drought had reduced the supply of smaller fruits and seeds, and only big-billed birds could crack the large, tough fruits that remained. Females also tended to mate with larger males, and as a result there was a detectable increase in bill size in a single generation.

In 1982 a population of large ground finches (*G. magnirostris*), almost twice as big as *G. fortis*, invaded Daphne Major. They presented little competitive problem for the medium finches until the drought of 2003 created strong competition for food. The large finches with giant bills could consume the big, tough fruits three times faster, and they physically excluded the medium finches (*G. fortis*) from the places where those fruits could be found. Among the medium finches, it was now those with the smallest beaks, which were best for dealing with small fruits and seeds, that had higher survival rates, not the big-beaked ones. As a result, the average beak size of the medium finches declined (figure 1-1). The Grants in this case were able to take advantage of a "natural experiment" (produced by an extended drought) to understand how the finches were evolving.

An example of even more rapid response to selection pressure was recently demonstrated by ecologist Jonathan Losos and his colleagues. They conducted a "field experiment" in which they manipulated one element of the environment in studying *Anolis sagrei* lizards on six small

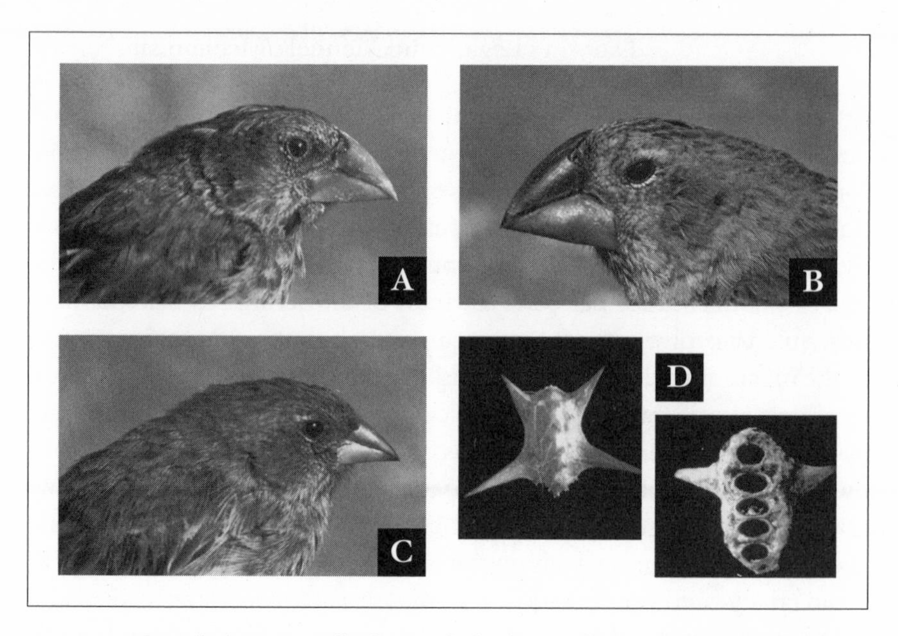

FIGURE 1-1. Natural selection rapidly changes the beak sizes of Darwin's (Galápagos) finches. Large-beaked *Geospiza fortis* (A) and *G. magnirostris* (B) can crack or tear the woody tissues of large, tough fruits (D), while small-beaked *G. fortis* (C) cannot. In (D), the left-hand fruit is intact. The right-hand fruit, viewed from the other side, has been opened by a finch, exposing five pits from which seeds have been extracted. Fruits are shown at twice the magnification of the finches. Photographs by B. R. Grant and P. R. Grant.

Bahamian islands. When they introduced a predatory lizard to the islands, the anoles in just six months evolved longer legs; those with shorter limbs were slower and more likely to get eaten. Then the anoles began to climb bushes to evade the predator. Selection then reversed, since shorter legs were a benefit in climbing; in another six months the legs were back to the original condition!

Rapid evolution doesn't occur just on small islands, of course. Perhaps the most famous example of rapid selection detected in nature was a change in the English insect known as the peppered moth. Among biologists, this case, worked out half a century ago, was a great instance of catching evolution in action, a visually compelling demonstration of Darwin's mechanism at work outside the laboratory. It also occurred when biologists, after almost a century of research and debate subsequent to Darwin, had just produced the "modern synthesis" of evolution, a unified picture of how organisms changed, and the moth seemed an ideal example.

The Modern Synthesis

Darwin had realized that the variations he observed had to be inheritable, but for decades following the publication of *Origin* almost no one had any idea about the actual mechanism of heredity. In fact, the field of genetics itself had been started by the monk Gregor Mendel in 1865, but his work wasn't recognized until early in the twentieth century, after Darwin's death. Mendel showed that the units of heredity (which were later named genes) were basically particulate. Previously it had been assumed that inheritance was a process of blending, as would occur when ink and water are mixed to make a gray liquid, so that offspring would be more or less an average of the attributes of the parents. Instead, Mendel's genes, which occurred in different versions (later called alleles) that could produce different characteristics, retained their individual attributes from generation to generation. Mendel showed they could be passed on independently and could occur in different combinations in different individuals. Unlike what would be expected in a "blending" inheritance, a child could exhibit a trait produced by a combination of genes not present in a parent but present in a grandparent—a trait that skipped a generation. That would allow, for example, a blue-eyed man to have a brown-eyed child, who in turn might give him a blue-eyed grandson.

What made the example of the peppered moth a signature story of the new synthesis was that rapid genetic changes could be observed in response to changing environmental conditions in nature—not in laboratories or hog herds. The peppered moth exists in two different forms: a "speckled" form (white with black mottling) that is camouflaged against lichen-covered tree trunks, and a "melanic" (blackened) form that is camouflaged against sooty, lichen-free trunks (figure 1-2). In 1848 the speckled form made up more than 99 percent of the peppered moth populations in the area around Manchester, England. Fifty years later, however, more than 99 percent of the Manchester-area moths were the melanic form. Genetic studies showed that the difference was caused by substitution in the population of one form (allele) of a gene by another form of the same gene. Each moth has a store of genetic information, which can be thought of as instructions coded into its genes, that interacts with environmental conditions to produce the moth we see. All the genes possessed by all the individuals in the Manchester population of peppered moths are what biologists call their gene pool. The gene pool of peppered moths underwent a dramatic change with respect to a melanic form as selection adapted the population to changing conditions in its environment.

The spread of the gene associated with the melanic form was highly correlated with soot pollution caused by the Manchester area's industrialization—hence the term "industrial melanism." Apparently, as the tree trunks became sooty, the lichens were killed, and the previously camouflaged speckled forms of the moth became conspicuous and were eaten by birds and other predators that hunt by sight. Thus the genes that produced the speckled pattern became less frequent in the population, the speckled type of the peppered moth became rare, and the now-camouflaged melanic forms proliferated. In common evolutionary parlance, melanism was an adaptation to the polluted environment.

In unpolluted areas, the melanic individuals apparently are still eaten disproportionately by birds, as they always have been, because in those areas they are easily seen against the pale, speckled background of lichens growing on tree trunks. The genes producing the speckled form confer an advantage on their owners in clean woods, and speckled moths predominate in populations occupying such habitats. This explanation was supported by the geographic coincidence of pollution and melanism and by studies of bird predation on the moths, including films that showed birds eating easily seen moths while overlooking camouflaged moths on the same tree trunks. And it recently received strong support when abatement of air pollution led to reestablishment of lichens and a subsequent decline in the frequency of melanic moths.

Were it not for the human-induced changes in the environment, the strong selection favoring the speckled form would have kept the blackened form as a rare variant, produced by an occasional random change (mutation) in the key gene. A tiny change in the moth's genetic material would lead to individuals easily spotted by birds and thus with little chance of surviving to reproduce. But much evidence also shows that natural selection is operating even where there are no fast or dramatic changes in the environment similar to rapid industrialization, but just the gradual change that has always occurred all over Earth's surface.

In the mosquito, island, and peppered moth examples, the basic evolutionary process consisted of individuals with some combination of genes reproducing more than those with other combinations. That

FIGURE 1-2. *Top*: a "speckled" form of the peppered moth (*Biston betularia*) is nearly invisible on a lichen-covered tree trunk, while above it and to the left a "melanic" (blackened) form stands out. *Bottom*: the reverse is true where the melanic form is camouflaged against a sooty, lichen-free trunk, while a moth of the "speckled" form to its right is obvious. Drawing by Anne H. Ehrlich, after photographs by H. B. D. Kettlewell.

raised the frequency of some genes in the population and, in the process, modified the kinds of individuals present. Today, evidence is increasingly appearing that climate change is causing genetic changes that parallel the story of industrial melanism and that will alter the world in which we live.

When organisms evolve, they also change the environments of other organisms. Birds in polluted woods surrounding Manchester had more trouble finding moths to eat, for example, when the moths evolved to became melanic. The environments of virtually all plants and animals (including people) have been changed by recent human activities, the result of *Homo sapiens* having evolved big enough brains to invent world-modifying technologies. Changing environments change organisms; changing organisms in turn change environments: that's pretty much the central story of life on Earth. But some organisms are not able to change fast enough to keep up with environmental change—whether they are pied flycatchers faced with the need to breed earlier in spring, or dinosaurs unable to adjust quickly to the horrendous climatic conditions following a comet's collision with Earth 65 million years ago. The road to the "much extinction" that Darwin noted is an inability to evolve in the face of environmental change.

Artificial Selection

Selection can be observed and studied not only in natural settings, but also in the unnatural, carefully controlled environments it is possible to create in laboratories. When Paul was a graduate student, he worked in the laboratory of a famous evolutionist, Robert Sokal. They raised fruit flies (*Drosophila*) (figure 1-3) in glass vials in which the environment was contaminated with DDT. Most of the fruit flies died, but those whose genes just happened to have made them somewhat resistant to DDT were more likely to survive and reproduce than were those lacking such a built-in advantage. In a few generations (each of which takes just twelve days or so), it was easy to produce a strain of flies that were resistant to that pesticide. Unnatural selection, so to speak, was at work. People (the lab workers) had substituted for nature, created a new environment, and used what is technically known as "artificial selection" to evolve that strain. Evolution took place as Paul watched, in just a matter of months. In the jargon of biologists, the resistant flies that evolved in the Sokal lab

FIGURE 1-3. This is what one of those tiny (less than one-eighth-inch-long) fruit flies (*Drosophila melanogaster*) that have played such a large part in our understanding of genetics looks like under a microscope. Photograph courtesy of iStockphoto.

had "higher fitness" or were more "fit" than the flies that succumbed to the DDT (Darwin would have said the resistant flies were "more favourable" variations).

"Fitness" (or "favourability") in the evolutionary sense does not necessarily refer to any physical characteristic, such as keen sight in a fruit fly or great flight ability. Fitness in this sense is purely a measure of relative reproductive contribution: an ugly ninety-pound human male weakling who fathers a dozen children and then drops dead at 35 is much more fit in these terms than is a tall, handsome, muscular man who lives to be 100 but is childless. This underscores that fitness does not necessarily refer to the ability to stay alive, as in the old slogan "survival of the fittest," although obviously an organism that dies before it reproduces is not very fit.

Would it be possible to make flies that were genetically susceptible to DDT be more fit? At first it seems impossible, because after all, to get a living population of DDT-susceptible flies, you can't breed the *Drosophila* that die most easily! But in fact it's simple, and it was done in the Sokal lab. All that was necessary was to divide the offspring of each parental pair of flies into two separate vials to be raised. In one set of vials DDT was introduced, in the other no DDT. Then the breeding stock used in each generation consisted of the brothers and sisters of the flies that had shown the highest mortality in the DDT vials. This produced flies that dropped dead at the sight of a bottle of DDT (okay, actually when exposed to tiny doses). The critical point that made this experiment successful is that brothers and sisters share, on average, half of their genes. Selection by breeding siblings of individuals with the desired trait works, but more slowly than direct selection. In this case, what biologists call kin selection was used to produce a change that was impossible to achieve by direct selection.

Another critical aspect of the evolutionary process was also vividly demonstrated by the Sokal lab's work with *Drosophila*. As geneticists know very well, selection operating on one trait will also change others—something that is frequently lost on non-geneticists writing about human evolution. Fruit fly pupae (the resting state during which the wormlike larvae—"maggots"—transform into flies) are normally scattered at random over the surface of the gooey medium on which the maggots grow. The lab's *Drosophila* populations that were selected for DDT resistance, however, behaved differently. As in the normal strains, the maggots as they fed crisscrossed the surface of the nutritional goo in the bottoms of the vials. But more and more, they formed their pupae only around the edges of the goo or on the glass vial wall above it. The fruit flies had been selected just for DDT resistance and got a different behavioral pattern in the bargain: peripheral pupation.

The Sokal group investigated this two-for-one result by trying a reverse experiment. They started with normal flies and created no selection pressure for DDT resistance (that is, no DDT was mixed into the medium). Instead, they selected individuals that pupated near the edge of the medium as the parents of each generation. In this case, edge-pupating strains evolved, as one might expect, but those strains, unexpectedly, also turned out to be DDT-resistant. What the Sokal lab found in fruit flies has been found in selection experiments using many different organisms: *it is usually very difficult to select for just one characteristic*.

Evolution in Human Beings

Genetic evolution—change in the hereditary composition of populations—has gone on everywhere, and it continues everywhere. Generation by generation, disease-causing bacteria, oak trees, butterflies, frogs, guppies—indeed, virtually all organisms—are perpetually changing in their genetic makeup. We are surrounded by evolution, and just like every other organism, we evolved. As Darwin himself recognized in his classic book *The Descent of Man* (1871), human beings are not just changers of environments, and thus drivers of some aspects of evolution; we ourselves have also been changed. Darwin started asking in evolutionary terms the questions that fascinate us to this day: Where did people come from? What are our origins? The key to answering these questions lies in understanding how the Darwinian process that changes mosquitoes, fruit flies, peppered moths, dandelions, rails, and Darwin's finches also changes us.

The best-known example today of differential reproduction of human beings with different genetic endowments (genotypes) concerns variation in our red blood cells. Some Africans and people of African descent living outside Africa have an unusual gene that causes the production of a special form of the hemoglobin molecule (hemoglobin makes red blood cells red and plays the crucial role of transporting oxygen to our other cells). Those who have just one copy of that unusual gene paired with a normal gene produce both normal and abnormal

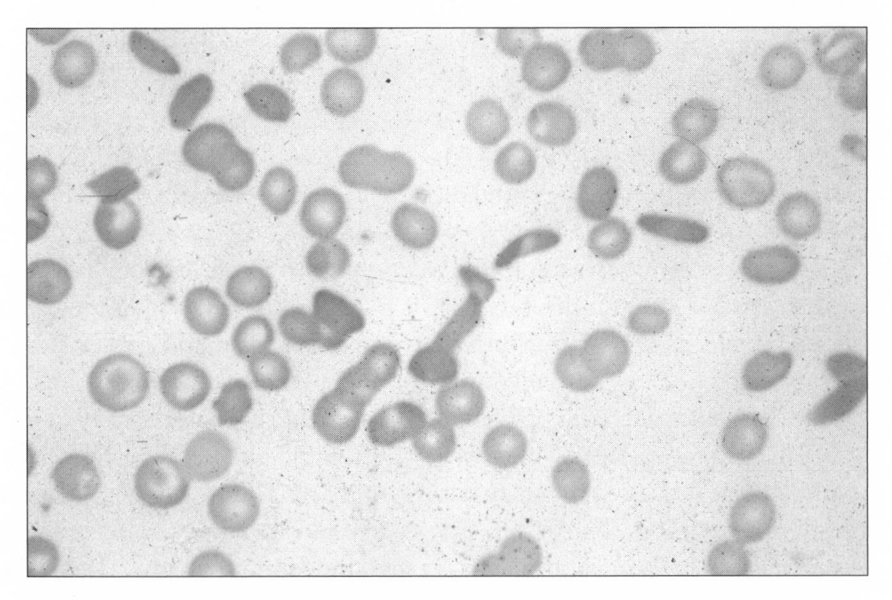

FIGURE 1-4. Microscopic view of red blood cells with the sickle-cell trait. Normal red blood cells are disk shaped; the elongated ones here are "sickled." Photograph courtesy of Dr. J. H. Crookston, Dr. R. Hasselback, and the University of Toronto.

hemoglobin and usually lead normal lives, though at high altitude or in some other situations, they can be subject to sickness or even death. Individuals who have both of the sickling genes (both alleles) have an inherited condition that causes their red blood cells to change from normal disks into forms that often look under a microscope like crescent moons ("sickle" cells; figure 1-4). Those people produce only that special kind of hemoglobin and are at grave risk of dying from sickle-cell anemia.

Why doesn't selection remove the "sickling allele" from the population? It happens that individuals whose red blood cells contain both kinds of hemoglobin are relatively resistant to the most lethal species of malarial parasite (*Plasmodium falciparum*), which, when it enters the human body, spends part of its life cycle in the red blood cells. Africans have been exposed to deadly *P. falciparum* malaria for many hundreds of generations, probably since agriculture was introduced there some 4,000 to 5,000 years ago. Agriculture provided local mosquitoes with an ideal habitat—many small water bodies in which to breed, close to large, newly sedentary human populations. Then people who carried both versions of the gene (the two different alleles) would have been at a selective advantage. Those people suffer less from malaria and are not anemic in most situations, and therefore they outreproduce individuals who carry identical alleles. The latter have only one kind of hemoglobin, either the normal one (no malaria resistance) or the one that causes sickling (likely to die from anemia). In this case, selection, instead of changing the genetic composition of the population, held it constant with both kinds of hemoglobin present.

The sickle-cell trait is still found in about 2.5 million African Americans, but in the absence of malaria in the United States it will gradually disappear over many generations. Selection operating against the genes that produce sickling hemoglobin can be seen whenever people move into areas where malarial parasites do not occur. In the relatively few generations since the slave trade was finally halted, the frequency of the sickling allele is already much lower in populations of people of African origin now living in the malaria-free Caribbean island of Curaçao, for example, than in those living in nearby malarial Suriname.

The sickling situation illuminates the difficulty natural selection has too in influencing just one thing; resistance to *P. falciparum* malaria came at the price of death in some individuals and potential sickness in others. That suggests, among other things, just how unlikely it is that many human behavioral traits (rape, seeking spouses with big bank accounts) claimed to be the direct result of natural selection actually are. Suppose, for example, a gene occurred in a population of our ancestors that produced a tendency to prefer mates with big noses. Even though bigger noses would allow slightly better detection of both food and enemies by smell, selection might never increase the frequency of such a gene if it had the additional effect of partially blocking binocular vision to each side, which could expose individuals to a risk of being eaten by predators that otherwise would be detected visually. Good side vision might yield benefits much greater than those accruing to a behavioral preference for big-nosed mates.

Even if it is difficult for selection to do only one thing, this does not mean that human beings are now beyond the reach of natural selection. Recent molecular studies indicate that *Homo sapiens* has been (and doubtless still is) showing differential reproduction related to skin color and a range of genes associated with medically relevant conditions. For example, earlier selection favoring very efficient digestion (making individuals better able to survive on small amounts of food when there are shortages) may in some areas of the world now be being reversed in response to food abundance. The previous advantage in resistance to famine now is a disadvantage because it increases the likelihood of obesity and diabetes—both of which are now epidemic in affluent societies. We are changing our environments ever faster, altering the climate, adding novel chemicals to our air, food, and water, and exposing our children to ever more electronic toys and television shows unknown when we were kids. For the foreseeable future, natural selection will be responding to some of these changes (for instance, it is not impossible that certain kinds of foods, environmental toxins, or even TV exposure may be related to fertility or reproductive behavior). Technology may greatly reduce what otherwise would be powerful selection pressures (as it does in many societies where eyeglasses can correct what in earlier times would have been fatal nearsightedness), but it will not eliminate many selection pressures and may even exacerbate some.

What Genes Do

Investigation of how other attributes may be influenced by selection that is focused on just one attribute turns out to be extremely important in comprehending the way genes and environments influence human behavior—and the overall process of evolution. There are two reasons for the tendency of selection to affect such other attributes. First, genes that are physically close together on the filaments in the nuclei of cells that carry the genes—the chromosomes—tend to be passed on together to offspring. Thus selection favoring a trait associated with one gene will often carry along the trait or traits associated with others. Technically, that tendency is called linkage. In addition, the action of one gene often may be altered or suppressed by the action of another gene at a different place on the same or different chromosome. Second, the same genes frequently (perhaps almost always—we can't be sure) influence more than one trait. We've seen several instances of this already, in the case of fruit flies whose resistance to DDT may also influence choice of pupation site, for example, and in the single gene change that produces sickling in the human sickle-cell trait, which may lead to some protection from malaria but may also lead to a suite of other changes, from spleen enlargement to susceptibility to pneumonia.

The interaction of genes with the environment produces a complex, observable phenotype (defined as the appearance, structure, and behavior of an organism). Phenotypes are the observable characteristics of individuals: blue-eyed or brown, fat or thin, early hibernator or late hibernator, melanic or speckled, docile or vicious, beautiful or ugly. No matter what organism such terms are applied to, they describe phenotypes.

Many phenotypic differences are produced entirely by environmental factors and do not reflect any underlying genetic variation. A difference in skin color in identical twins (twins that originate by division of the same fertilized egg in the womb and thus share identical genetic endowments), one sunning in Savannah and the other freezing in Fargo, would be an example. On the other hand, some individual phenotypes may be indistinguishable despite having distinct genotypes; for example, two different genotypes can produce people with brown eyes—either two identical alleles code for brown, or one allele codes for brown and the other for blue, but the latter is not expressed (to get blue eyes you need both "blue alleles"). That's why some brown-eyed couples can have a blue-eyed child and some other brown-eyed couples can't; it's because of differences in their genotypes that are not expressed in their own eye-color phenotypes.

In most organisms, phenotypes have many more features than there are genes to affect them, if each gene influences only one part of the phenotype. For instance, the *Plasmodium falciparum* malarial parasite, which once almost killed Paul, has slightly more than 5,000 genes. Paul, on the other hand, has about 25,000 genes—roughly the same number as a fruit fly. Rice may have almost twice as many genes as *Drosophila* or people, so the total number of genes is not closely tied to the complexity of an organism. But although rice lacks a brain, it does have a very complicated apparatus for carrying out photosynthesis, the process by which plants bind light energy to run their metabolism. In any case, Paul and you and a rice plant have many more than 25,000 phenotypic characteristics (in your case and Paul's, for example, hundreds of trillions of different connections among nerve cells in your brains—all elements of your phenotype). So influencing more than one phenotypic character is a necessary feature of the action of the vast majority of genes.

Behind the explanations just given are the fine-scale workings of the genes themselves, mechanisms that (directly or indirectly) are referred to almost daily on TV and in the popular press. Scientists have worked intensively ever since Mendel's research became known, and especially over the past half-century, to uncover the detailed mechanisms of heredity—of exactly how evolution takes place at the chemical level and life perpetuates itself. They have had great success in discovering how, for instance, tiny sperm and eggs can transfer massive amounts of information from parent to offspring. Much of their research has been and still is done on the same *Drosophila* fruit flies used in the Sokal lab a half-century ago and in the labs of pioneering geneticists almost fifty years before that. And much of what is known about fine-scale details of evolution has been produced by research on the human gut bacterium *Escherichia coli*, an organism that reproduces even faster than fruit flies. Its generation time can be less than a minute.

With intricate experiments, biologists have shown that genes, the physical embodiments of genetic information, are segments of molecules of deoxyribonucleic acid, the famed DNA. The gene-bearing DNA molecules in the chromosomes of animals and plants have the frequently illustrated "double helix" structure (figure 1-5). They are like twisted chemical ladders, made up of two complementary strands wrapped around each other. The subunits of each strand are called nucleotides, which linked together make identical chemical "uprights" connected by four different kinds of chemical "rungs." These rungs are

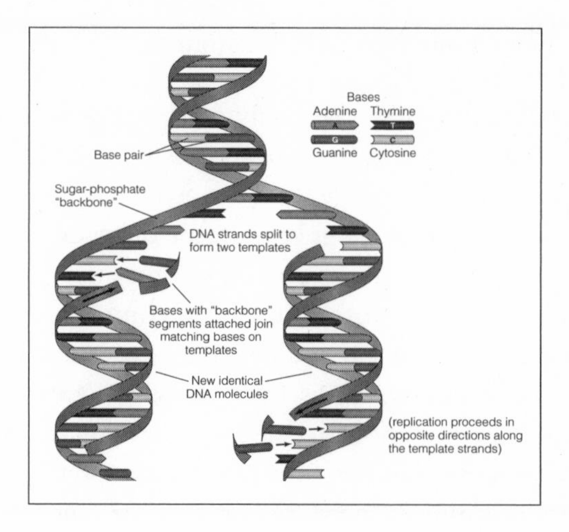

FIGURE 1-5. When DNA is replicated, the "rungs" split apart, forming two "backbone" templates, against which matching bases form two new backbones. They then unite into new DNA molecules. That is the basic biochemical mechanism by which genetic information that is coded into the "rung" sequence is reproduced and passed on.

pairs of four different chemical constituents of the nucleotides, called bases, abbreviated A (adenine), T (thymine), C (cytosine), and G (guanine). The bases always pair with the same other base: A with T, and C with G. It is the sequence of those nucleotide base pair rungs that encodes the information.

A gene consists of a series of nucleotides with different bases that determines the sequence of chemical subunits (amino acids) that will link together to make up a single giant chemical molecule called a protein. Proteins are essential parts of all organisms and participate in all living processes. A complex cellular apparatus, made up in part by various forms of DNA's close molecular relative RNA (ribonucleic acid), uses the DNA "instructions" to bring the proper amino acids together and link them in the correct sequence. Changes in just a few of those sequences may determine whether a finch has a big bill, whether a fruit fly is resistant to DDT, or even whether an organism turns out to be a chimpanzee or a human being. A complex apparatus allows the DNA to be duplicated (replicated) and the genetic information coded into it to be passed from cell to cell and adults to offspring.[4]

Proteins play many roles; they become parts of the body's physical structure, operate in its defensive immune system, and often act as catalysts—substances that speed up (or, more rarely, slow down) a chemical reaction. As catalysts, proteins control the development of an organism and its functioning. In the genetic system, proteins assembled by one gene can instruct other genes to produce their proteins or block them from assembling their proteins. Genes thus can regulate the actions of other genes. Genes can also produce a protein that inhibits their own further production. These complex regulatory interactions often occur in response to environmental cues, and scientists are discovering that the regulatory process is much more complex than they thought even a decade ago.

The whole field of genome evolution (evolution of the entire complex of DNA in an organism) is in an exciting era of change—especially in relation to huge amounts of DNA that do *not* code for proteins. Once called "junk DNA," much of it, it now appears, is involved in development—interacting with environments, turning genes on and off, determining how frequently genes can be altered (mutation rates), and directing how information in the genotype is used to create the phenotype. Indeed, the seemingly ever-increasing complexity of the functioning of the genome is even calling into question the concept of "gene" itself, since genes don't have the composition typically attributed to them and aren't quite the discrete entities once thought. And it now appears that once apparently well-settled pictures of exactly how natural selection operates may need reexamination, focusing on that issue of under what circumstances it might be possible for selection to influence the evolution of just one trait.

When we wrote about natural selection in fruit flies, then, we actually were writing about complicated processes at the level of molecules. We were writing about differentials in reproduction between individual flies that were traceable to differences in the sequence of DNA nucleotides that code for the production of different proteins in their cells. Similarly, in our distant ancestors, natural selection for upright posture or for eyes that form detailed images occurred in a complex of processes basically similar to what took place in our selection experiments with fruit flies, but it went on for many, many more generations.

Environment: Evolutionary Partner of the Genes

The importance of genes in evolution is pretty clear; genetic evolution itself can be defined as changes in the genes in a population or, perhaps more accurately now, in the genomes in a population. But the direction genetic evolution takes, whether among our primate ancestors, other animals, or plants, is determined primarily by the environment. Thus a key to understanding evolution is understanding that role of the environment and how it affects phenotypes, the raw material on which the environment, as the agent of selection, operates. Just as air pollution favored genes for melanic phenotypes in the peppered moth, the frequent evolution of pesticide and antibiotic resistance in many organisms follows the same pattern. Human beings "pollute" the environment of pests and bacteria by purposely exposing them to poisons, which at first kill many or most of them. That changes their environment to favor genes that produce phenotypes capable of avoiding death by detoxifying or excreting the poison, as we are discovering to our distress in an increasing number of instances. Since constant exposure to poisons has the same effect in agricultural fields as it did in fruit fly vials and other laboratory selection experiments, the number of insect and mite pest species now resistant to one or more pesticides is pushing 500.

Similarly, overuse of antibiotics has caused rampant antibiotic resistance in bacteria, so much so that many scientists believe that the most powerful weapons we have against these deadly little enemies may soon be nearly useless. A virulent new strain of *Clostridium difficile* (a relative of the bacterium that causes botulism), for example, is spreading rapidly and is resistant to many of the antibiotics previously used against it. One element in the new *Clostridium* menace is the impact on the normal gut environment of the overuse of antibiotics. When employed in combination with certain antacid drugs, they kill off harmless or helpful bacteria in the intestine and, by eliminating its competition, give *C. difficile* an easier time establishing and building large populations. And overuse of penicillin and its relatives has led to the evolution of methicillin-resistant *Staphylococcus aureus* (MRSA), which has become a big and deadly problem in hospitals, nursing homes, and increasingly the general public. Continually using one or a few biocides to poison pests or pathogens amounts to running a selection experiment with a nearly certain result—the evolution of resistance and the loss of effectiveness of the pesticide or antibiotic.

The environment can influence evolution in other ways. It can, for example, change the proportions of genes in a population purely by accident. If you are the only blue-eyed member of your family, and you are struck and killed by lightning before the end of what would have been your reproductive life, the proportion of eye-color genes in what your family passes on will change. Furthermore, the very process of reproduction leads to random changes in the proportion of particular genes—

a phenomenon known as genetic drift—just by the accident of which particular gene is in the sperm that fertilizes a particular egg. A key thing to remember is that genetic drift has a much larger effect on evolution in a small population than in a large one (all else being equal). While that lightning strike may profoundly affect the proportion of eye-color genes in your four-person family, its impact on the proportions of the same genes in your 300 million–person national population will be negligible.

Another kind of environmental change—individuals immigrating or emigrating—can also shift the proportions of genes in a population. If all the blue-eyed people leave town, the assortment of genes still represented in the residents (the town's gene pool) will be changed. Like selection and drift, migration is an evolutionary force. And so are factors that change a nucleotide (or multiple nucleotides) in a gene and turn it into a different one, producing a mutation. A mutation can be caused by, say, an X-ray knocking out a nucleotide in the DNA, or by an error in the complex mechanism that copies the DNA, in either case changing the DNA nucleotide sequence and changing the functioning of the gene. A mutation that changes one gene's functioning can have cascading effects in development, resulting in quite dramatic phenotypic changes. But the key point is, as we indicated above, that with possible minor exceptions, mutations are random. They normally occur without regard to the environmental situation or "needs" of the organism.

Migration, mutation, and a reshuffling (recombination) of genes that occur in the process of reproduction are evolutionary forces that produce the genetic variation among individuals in a population. That genetic variation is the raw material of evolution, providing the different genes on which, through the environment, natural selection can act. For it is only natural selection that gives *direction* to evolution, over generations improving the reproductive capacity of individuals in the environment in which they are evolving. Selection is the creative, systematic force in evolution.

But if it's so creative, how do we explain those pied flycatchers that now breed too late to feed their young adequately because global warming caused the abundance of their insect prey to be shifted to an earlier time? Easy: selection is a powerful force, but what it can do and the speed with which it operates depend on characteristics of both the environment and the genotypes of organisms. Often, as is possibly the case here, selection in response to a change may lower fitness by affecting unrelated attributes; remember, selection seldom influences just one characteristic at a time. As a result, extinction is the fate of many populations and species. Indeed, the fossil record is littered with the remains of organisms that couldn't evolve rapidly enough to keep up with certain environmental changes.

This is especially so when that change is *fast*—such as the catastrophic consequences of a strike by an extraterrestrial body. One such collision apparently shrouded the entire Earth with sunlight-blocking dust for a long period of time and wiped out almost all the dinosaurs and innumerable other organisms 65 million years ago. Another sudden change was the invasion of the Western Hemisphere by human hunters thousands of years ago; they were the primary (or perhaps only) cause of the rapid extinction of mammoths, camels, giant ground sloths, giant beavers, and other members of the "Pleistocene megafauna" that used to roam the Americas. Humanity seems to be entraining yet another such mass extinction episode now by rapidly altering environments all over the world, and the pied flycatchers and pitcher plants may be among the victims.

The Evolution Explosion

Since the pioneering work done on peppered moths and a few other organisms around the middle of the twentieth century, there has been an explosion of studies of selection and other forces of evolution, an explosion fueled in part by the laboratory study of microbial organisms with very short generation times and the availability of the new techniques of molecular genetics.

Molecular research has, among other things, allowed clear demonstrations of how selection can "create" highly intricate mechanisms. For example, some people have claimed that selection could not possibly have produced the biochemical complexity seen in cellular processes. Especially difficult, they claim, are "lock-and-key" mechanisms such as occur with hormones (chemical messengers such as testosterone) and the hormone receptors of cells. After all, if there were a change in a lock, for the system to remain functional would require a matching change in the key. But it turns out that complex lock-and-key systems *can* evolve through the action of natural selection, as molecular evolutionist Joe Thornton and his colleagues recently demonstrated. They were able to reconstruct the exact stepwise molecular sequence by which a functional interaction between a steroid hormone and its partner receptor evolved through natural selection.[5] Such work helps to put an end to any scientific debate over whether complexity can be produced by selection.

The recent advances in understanding evolution have been so dramatic that in 2005, *Science* magazine, the premier scientific journal in the United States, declared "evolution in action" (increased understanding of how the process works) the breakthrough of the year. *Science* stated in part, "Today evolution is the foundation of all biology, so basic and all-pervasive that scientists sometimes take its importance for granted."[6] In this respect, things have not really changed since the greatest evolutionary geneticist of the twentieth century, Theodosius Dobzhansky, made the 1973 statement that serves as the epigraph for this chapter.

In the face of this recognition, why then do scientists talk of the "theory" of evolution? Isn't evolution a "fact"? Well, no; in science, the only facts are things that a group of people assembled in a room can agree upon; say, that the professor has one eye on each side of her nose. Is evolution then just a hypothesis—an assumption that can be tested (the teacher is wearing a wig)? No, it isn't "just" a hypothesis, because evolution has been thoroughly tested. In science, that makes it a theory—an explanatory framework that has been exhaustively tested and can be used for making predictions. Indeed, a scientific theory is almost the direct opposite of a theory in common usage, which is a speculation ("I have a theory on why my girlfriend dumped me"). Evolution is one of the most completely explored theories we have in science, comparable to the "theory" that Earth travels around the sun. But details of the mechanisms of evolution are still under intense investigation, just as are cosmological mechanisms half a millennium after Nicolaus Copernicus proposed a heliocentric universe.

Charles Darwin did not use the word "evolution" in *Origin*'s first edition, but a century and a half ago, before genes or DNA were discovered, he captured the basic theme of this chapter—and the next—in a beautiful and oft-quoted terminal passage:

> It is interesting to contemplate an entangled bank, clothed with many plants of many kinds, with birds singing on the bushes, with various insects flitting about, and with worms crawling through the damp earth, and to reflect that these elaborately constructed forms, so different from each other, and dependent on each other in so complex a manner, have all

been produced by laws acting around us. These laws, taken in the largest sense, being Growth with Reproduction; Inheritance which is almost implied by reproduction; Variability from the indirect and direct action of the external conditions of life, and from use and disuse; a Ratio of Increase so high as to lead to a Struggle for Life, and as a consequence to Natural Selection, entailing Divergence of Character and the Extinction of less-improved forms. Thus, from the war of nature, from famine and death, the most exalted object which we are capable of conceiving, namely, the production of the higher animals, directly follows. There is grandeur in this view of life, with its several powers, having been originally breathed into a few forms or into one; and that, whilst this planet has gone cycling on according to the fixed law of gravity, from so simple a beginning endless forms most beautiful and most wonderful have been, and are being, evolved.[7]

CHAPTER 2 The Entangled Bank

"All organisms alive today trace their ancestry back through time to the origin of life some 3.8 billion years ago. Between then and now millions—if not billions—of branching events have occurred as populations split and diverged to become separate species."

SCOTT FREEMAN AND JON C. HERRON, 2001[1]

"We have moved from a view of coevolution as a stately, long-term process that molds species over eons to one in which coevolution constantly reshapes interacting species across highly dynamic landscapes."

JOHN N. THOMPSON, 2005[2]

DARWIN'S entangled bank contained a lot of different kinds of organisms: various species of plants, insects, birds, and so on. We have seen how populations of a species can evolve through time, but Darwin had an interest also in how new and different species arose, including, of course, our own. After all, the title of his magnum opus was *On the Origin of Species*.

Why, then, are there both dogs and wolves; horses and zebras; red-shouldered and red-tailed hawks; brown, rainbow, and cutthroat trout; tiger and black swallowtail butterflies? How do the different kinds arise? And how are they linked to one another so there can be such diversity in that entangled bank? The answers to these questions will begin to tell us why scientists are so concerned about what happens to other species, how we evolve with them, and the role the rise of our own species to planetary dominance has come to play in their fates.

The short answer to the "speciation" question in the modern view of evolution is actually signaled by one of the most common phenomena in biology: geographic variation in the characteristics of organisms. Geographic variation simply means that populations of the same species living in one place are different from populations of the same species living somewhere else.

Bird-watchers are well acquainted with geographic variation. For instance, in populations of the bird today known as the yellow-rumped warbler (*Dendroica coronata*) in eastern North America, males have white throats and two white bars in the folded wings. In the American West, males have yellow throats and a large white patch in the folded wings (figure 2-1). The birds are so distinctively different in their characteristic appearance that they were long considered two different species, myrtle warblers in the East and Audubon's warblers in the West. But populations in the northern Rocky Mountains of Alberta and British Columbia show intermediate characteristics, and taxonomists (scientists charged with the formal classification of organisms) now consider myrtle warblers and Audubon's warblers part of the same species. It's not just in colors that birds of the same species display geographic variation. Nine populations of the white-crowned sparrow (*Zonotrichia leucophrys*) in different parts of the San Francisco Bay Area sing distinctly different songs. And within some bird species, populations have very different migratory patterns.

Geographic variation also occurs ubiquitously in characteristics that are not at all obvious to the casual observer. For example, the frequencies of genes analyzed by molecular biological techniques vary from population to population in virtually every one of hundreds of plants and animals that have been studied. Such geographic variations are important clues to the puzzle of species origination that Darwin embodied in the title of his great work.

The very same process that causes evolutionary changes within populations—natural selection working on the results of mutation, recombination, migration, and genetic drift—also causes populations to become different from one another, forming diverse kinds of organisms. Geography, where conditions usually change with distance, is key to differentiation by being the source of most diverging selection pressures

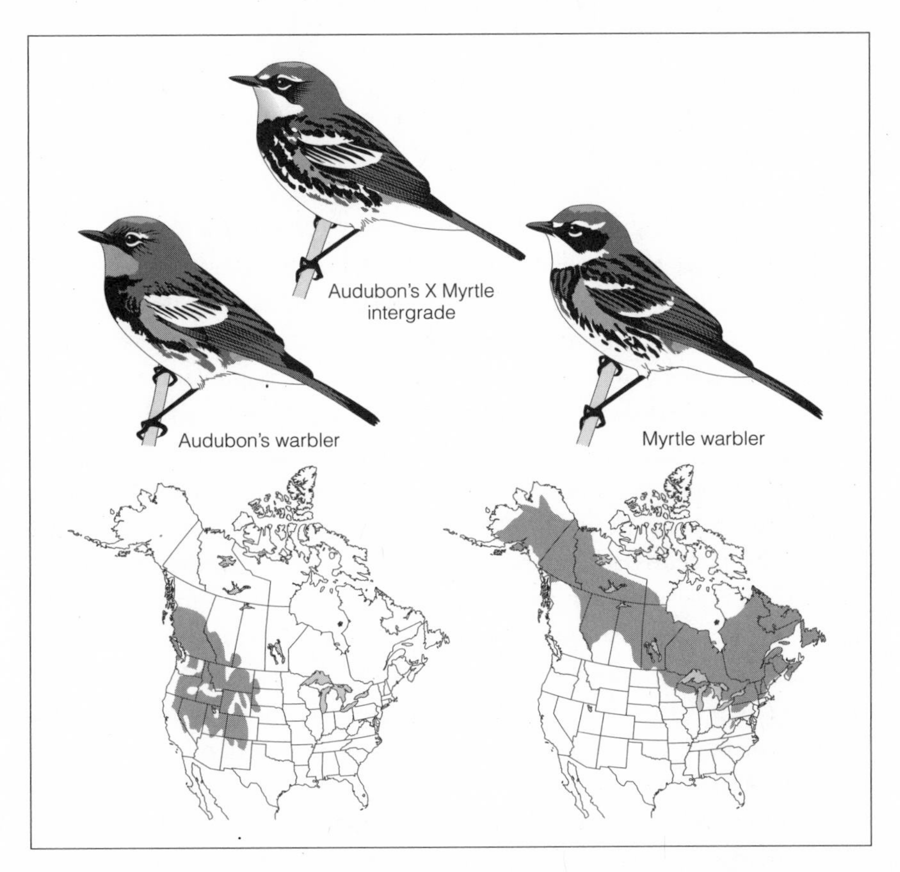

FIGURE 2-1. Speciation in process. The yellow-rumped warbler has different-appearing western (Audubon's warbler, *left*) and eastern (myrtle warbler, *right*) populations with somewhat different songs. In the Canadian Rocky Mountains—where, as the maps show, the populations overlap—intergrades (*center*) are found. Whether or not to consider these a single species or two daughter species is largely a matter of taste, since population differentiation is a continuous process. Redrawn from D. Sibley, *The Sibley Guide to Birds*. (Knopf, New York, 2000, pp. 436–437.)

and often by providing barriers that prevent interbreeding between populations. Differentiation of populations, when it goes far enough, is called speciation. Very generally, in sexually reproducing plants and animals, being different enough to be considered separate species means that two similar populations of an organism may be living together in nature, but individuals of one are not usually mating with those of the other and producing viable and fertile offspring. Collies, Labradors, pit bulls, and mutts of mixed ancestry are not different enough—we consider them all members of the same species (with the Latinized name

Canis familiaris). But wolves and coyotes are different enough to be considered separate species because they do not interbreed when they encounter one another in nature.

Speciation, along with evolutionary changes within species, is the mechanism that over millions of years generates major evolutionary patterns such as the diversification of vertebrates into fishes, amphibians, reptiles, birds, and mammals. Thus intermediate fossils show that amphibians in the distant past were diverging from fishes, and later, birds diverged from crocodiles. And natural selection has been the driving force behind it all.

The awesome process of generating the grand diversity of life, stretching back to the first appearance of living proto-organisms in Earth's early oceans or deep rock crevices more than 3 billion years ago, began with variation in environments. No two environments were identical even when very simple life was first generated in the ocean depths (or reached the oceans from a birthplace deep below ground, as some scientists hypothesize). Tides and currents, salt concentrations, ocean bottom composition, temperatures, depths, and other factors varied from place to place, just as no two environments are identical today. No environment was constant then or is constant today. Seasonal weather cycles, changes in climate, changes in topography through such mechanisms as weathering, erosion, seabed spreading as continents formed and gradually moved over the planet's face, mountain building, glaciation, and streams altering course—these are just a few of the physical processes that promote environmental change. In response, different species evolved as natural selection adjusted each population to its unique and changing environment. In turn, changes in individual species or populations usually alter the environment of other species that interact with them in the same locality—species that eat them or that they feed on; species that provide shelter, compete with them for food; and so on.

There are many examples of changes in populations of one kind of organism causing changes in other kinds. DDT is especially deadly to insects that normally eat mites, tiny relatives of ticks. When DDT was applied to a field, it greatly reduced the populations of insects that ate mites and thus caused the mites to undergo population explosions. That in turn altered the environments of many other populations of plants and animals. For instance, before World War II, spider mites were a group of minor pests of crops. After widespread use of DDT and other synthetic organic insecticides had greatly reduced populations of many of the spider mites' natural enemies, spider mites became, in the words of one group of scientists, "the most serious arthropod pests affecting agriculture world-wide."[3]

Similarly, the extermination of the passenger pigeon nearly a century ago by hunting may have been an important factor in promoting outbreaks of Lyme disease, a nasty tick-borne bacterial infection first identified in 1975 from a cluster of cases around Lyme, Connecticut. Passenger pigeons occurred in flocks of billions and fed on acorns, beechnuts, and other nuts of forest trees—collectively known as "mast." With the birds' extinction, more of the mast was available to deer mice, and populations of the rodents flourished. That in turn made the environment more favorable for mouse ticks, which transmit the bacteria (closely related to the spiral-shaped spirochaete germs of syphilis) that cause Lyme disease. Commercial hunters, unwittingly making conditions more favorable for the mice, ticks, and spirochaetes by driving the passenger pigeon to extinction, made the environment less favorable for *Homo sapiens*.

Such changes in the living environment can be as important to the diversification of organisms—to the process of creating new species—as changes in the physical environment. All populations, human and non-human, are exposed to different combinations of selection pressures, depending on the species around them and changes in the populations of those neighboring species, as well as those imposed by different rainfall regimes, geologic formations, and the like.

Geographic Speciation

The late, great evolutionist Ernst Mayr (1904–2005) first clearly set forth what has become the standard view of how new species arise, a diversification process that is sometimes called "geographic speciation." It answers the basic question of why two similar kinds of organisms (species) often live together in the same region without interbreeding, and it seems to explain the origin of most (but far from all) species.

Swainson's and hermit thrushes around the Rocky Mountain Biological Laboratory (RMBL) in Colorado fit the description of similar but non-interbreeding groups in a region. Though these two kinds of thrushes don't interbreed, they closely resemble each other (they are most easily distinguished by the different songs of the males). Presumably both are descendants of a common thrush ancestor, but how could such differentiation into two distinct species have happened?

Mayr's view, unlike Darwin's, emphasized differentiation in isolation. Under Mayr's model of geographic speciation, at some point in the past, two populations of the single ancestral species became spatially isolated from each other. Perhaps a period of climatic drying produced a grassland barrier to movement of the ancestral thrush in what was once a continuous stretch of suitable woodland. Or perhaps a chance group of thrush colonists was blown across a mountain barrier in a storm and occupied previously vacant habitat. Whatever happened, two ancestral populations were isolated. Their environments were different, so selection pressures would have been different, however subtly. Over time, different selection regimes as well as chance mutations and the random effects of genetic drift would have influenced the two populations differently.

As a result, the two thrush populations began to diverge, each on a separate evolutionary trajectory. Thousands (or millions) of years later, environmental conditions changed, and one or both populations expanded in range, bringing the two sister lineages, descended from the same ancestral population, to live in close proximity again. If sufficient divergence had occurred in the time intervening, the reunited groups would be unable to breed with each other and would remain distinct. Perhaps they had evolved different songs of the males, and sing as they might, those of one group could not attract females from the other group. Perhaps some intergroup mating took place, but the genetic endowments of the two groups had become so different that hybrids were sterile. Whatever the case, speciation had occurred, and today we have both Swainson's and hermit thrushes around our mountain cabin. If, on the other hand, the populations had rejoined but differentiation had not gone far enough, we would have only one kind of thrush around our cabin. Partially differentiated populations doubtless re-merge frequently, leaving few or no traces of historical separation.

This description of the Mayr model is necessarily simplified, but a key thing to keep in mind is that it predicts the existence at any given time of a spectrum of degrees of population differentiation among organisms. It indicates that populations are always differentiating, disappearing, and fusing. In each case, the amount of difference between populations would be related to the other organisms each population lived and interacted with, and the strength of physical and chemical environmental differences across geographic gradients that the populations occupied—all of which would cause differences in selection pressures.

If populations are continually differentiating, what keeps a species comprising multiple populations together as a unit? Many scientists believe that the evolutionary force that binds populations into one species is migration—often expressed as "gene flow." Gene flow is simply the movement of DNA from population to population, carried by individuals that transfer and breed. Our own view is somewhat heterodox. Although gene flow certainly can play a role, we think similar selection pressures often make more or less isolated populations remain genetically similar. Genotypes that, in a given habitat, produce successful phenotypes—"winning" genetic combinations, if you will—might tend to persist unless their environments diverge from one another in some dramatic way that is very important to the organisms.

Despite the complexities of species generation, biologists generally accept Mayr's relatively simplified explanation. One reason is that it fits contemporary observation. The "snapshot" of organic diversity that one sees today closely matches the geographic speciation picture; it shows every conceivable level of diversification. At one end of the spectrum is a multitude of geographically isolated but extremely similar populations that are even difficult to distinguish statistically. At the other end are many thousands of clearly distinct but similar species living together without interbreeding, where every individual can be indisputably assigned to one species or another (as with wolves and coyotes, mallard and pintail ducks, or tiger swallowtail and black swallowtail butterflies). And, when we look at changing bill sizes in Darwin's finches, or changing frequencies of melanics in peppered moth populations, we can see in action the processes that cause the differentiation of populations.

In many cases, one sees evidence that a single ancestral kind of an organism has penetrated a new geographic area, dispersed to form what became largely isolated populations subject to different selection pressures, and diversified into numerous species. Each of these species often will have evolved a different way of obtaining resources—that is, each occupying a different ecological niche. An organism's niche can be thought of as its "occupation" and its habitat as its "address." Darwin's finches are a classic case of adaptive radiation—a situation in which an evolutionary lineage radiates into a series of niches as populations adapt through selection to the diverse conditions. An ancestral finch species sometime in the past reached the Galápagos archipelago and diversified on the various islands in isolation, and then the products of that diversification periodically invaded new islands, as the large ground finch did at Daphne Major. Large and medium ground finches had the same address (habitat) there, but they were in different niches because their occupations were eating seeds of different sizes.

Nevertheless, a growing body of information suggests that geographic isolation is not always required for speciation to occur; in some cases preferences by insects for different food plants or pressures on fishes to exploit different diets in small lakes can lead to increasingly different populations and, eventually, to reproductively isolated species. There are also rather common circumstances in plants, and increasing evidence in animals, that new species can form as a result of hybridization (crossbreeding of individuals) between already differentiated species. Perhaps the best example is a species of longwing butterfly (genus *Heliconius*) that is produced by hybrids between individuals of two other species, where the hybrid individuals prefer to mate with each other rather than with individuals of the parental species. Indeed, in sunflowers it has been shown that the "same" new species can arise multiple times by hybridization of the same parental species. That is, species A and B hybridizing can give rise more than once to species C. So, while geographic speciation is the dominant mechanism, especially in animals, it is not the sole mechanism of speciation.

It is sometimes claimed that, while one can observe genetic changes within populations, the formation of new species is never observed and must be the result of special creation. In fact, even skipping cases such as those of the *Heliconius* butterflies and sunflowers just mentioned, the process of speciation has been observed in tens of thousands of cases. To deny it is the equivalent of denying that giant sequoia trees grow from seeds, since no one has ever observed one grow from seed to 300-foot-tall giant. But since their growth rates are measurable, and one can detect growth on a monthly time scale and can find trees of all sizes, it is agreed that the seeds end up as giant sequoias. Similarly, as indicated above, intermediate steps on the path to speciation are abundantly documented.

Bird-watchers are especially familiar with the intermediate stages of speciation. They are perpetually annoyed as taxonomists working on birds keep changing the names as they seek places to cut the continuum of evolutionary differentiation, trying, for example, to decide whether Audubon's and myrtle warblers should be lumped into continent-wide yellow-rumped warblers, or whether the red-shafted, yellow-shafted, and gilded flickers should be considered separate species or combined in taxonomic classification into the northern flicker. Indeed, in our somewhat heterodox view, taxonomists waste huge amounts of time trying to make up rules for cutting the continua of differentiation. "Species" is a useful term (like "mountain," "blue," "religion," and "happiness"), even though, like the others, it represents a segment of a continuum and does not have a clear, agreed-upon definition (we like "distinct kind" for species, for want of better).

No Species Is an Island: Coevolution

Selection pressures on a population are generated not only by physical alterations of the environment but also by the community (the group of other organisms) with which it interacts—what we called the living environment earlier in the chapter. Think of the birds on Darwin's entangled bank as putting selection pressures on the insects they eat, pressures for the insects to be less conspicuous, or to taste bad, or to develop other characteristics that make them less likely to be featured at an avian banquet. In turn, the birds are under selection pressure to be able to spot and catch their insect prey, to be resistant to any nasty chemicals the insects have evolved to make themselves poisonous, and generally to retain the ability to nourish themselves and their chicks with six-legged delicacies. Other organisms are almost always a source of important selection pressure on plants, animals, or microbes. Indeed, organisms are all subject to the simultaneous influences of their living companions and their physical environments—the complex in which they are embedded as members. All organisms, each of us included, are members of those complexes, which go under the name of ecosystems, among the most critical concepts in understanding life on Earth.

Today *Homo sapiens* is, for most organisms, the world's most powerful agent of selection. Human beings have changed the chemical environment for virtually all of life, for instance. Tens of thousands of human-made chemicals and an array of brand-new radioactive elements from our nuclear weapons now pollute the globe from pole to pole and to the depths of the oceans. We humans also have modified almost the entire surface of the planet, demolishing forests, converting to agriculture (or paving over) a major fraction of the land, damming and diverting rivers, overfishing and raking ocean floors with bottom trawls, and so on, thereby changing or, in effect, destroying the habitats of innumerable organisms. Through selective breeding we have deliberately altered the gene pools of the many species we domesticate. Furthermore, we have unintentionally altered the gene pools of most of the species we hunt or gather when we have caught the largest fishes and shot the elephants with the biggest tusks. The results are smaller fishes and elephants with shorter tusks.

And now people are changing the world's climate—one of the most important sources of selection pressure (and thus of geographic variation). Remember how climate-change selection has shifted the timing of the life cycle of pitcher-plant mosquitoes. Similarly, the melting of the Himalayan glaciers will dramatically change selection pressures in many Asian rivers and the lands they water.

We have also introduced exotic plants and animals (inadvertently or as crops, ornamentals, or domesticates) that have sometimes had enormous effects on other species. For instance, in central California, virtually every herb that one sees is a European weed, largely introduced along with fodder for the horses of Spanish settlers. Those weeds, encouraged by nitrogen fertilization from automobile exhaust, are threatening the beautiful native flora and fauna of once nitrogen-poor serpentine soils in the San Francisco Bay Area. In a more dramatic case, the kudzu vine, a plant imported into the United States from eastern Asia, has been described as the "weed that ate the Southeast." It is a climber that covers the ground, buildings, power lines, virtually any surface; shades out native herb species; kills mature trees; and can grow as much as a foot a day. Clearing it away costs many millions of dollars annually. As humanity changes the distribution of plants and animals over the face of our planet, it is altering the evolutionary future of many organisms.

When two species are ecologically intimate, closely influencing each other's lives as do grazers and grasses, predators and prey, hosts and parasites, cooperators and competitors, each normally becomes a major source of selection operating on the other. Grazing bison are a selective force favoring individual grasses that grow close to the ground and are thus less likely to be eaten by such large animals. Other defensive tactics also serve grasses well. They have evolved the ability to sequester silica crystals in their leaves, making them tough to eat. On the Serengeti Plain of Tanzania, grasses from heavily grazed areas of the ecosystem accumulate more silica in their leaves than do grasses from areas less subject to grazing pressure. The silica in turn tends to favor the growth and reproduction of large grazers such as bison with big, wear-resistant teeth—that is, silica-rich grasses "select for" such teeth.

Many people enjoy snorkeling or scuba diving as a hobby. If they know about coevolution, they can observe its results on coral reefs by watching cooperators (technically "mutualists"). Small cleaner fishes (species of the wrasse genus *Labroides*) can be observed picking parasites from the skin and jaws of large predators such as coral cod (*Cephalopholis miniata*). The predators have evolved recognition of the cleaners and have learned to open their mouths to allow the cleaners to enter and work, instead of snapping them up as hors d'oeuvres.

The grass-grazer and cleaner-predator relationships are examples of coevolution—the evolutionary interactions of ecologically intimate organisms. As a species, humanity is ecologically intimate with lots of other organisms, from cows and crop pests to mackerel and malaria parasites, and our coevolution with them affects us—and them—in many ways. For instance, people domesticated cattle and selected for cows that produced abundant milk. The milk in turn put a selection pressure on people. In non-milk-drinking populations, the ability to digest lactose (milk sugar) is limited to childhood. But in dairying populations, there has been selection for the ability to digest it right through adulthood—a phenomenon known as "lactase persistence" (lactase is the enzyme that breaks down lactose so it can be absorbed). Northern Europeans appear to have evolved this ability only in the past 7,000 years or so, since dairying started there. Selection to be able to use a parasite-free, protein- and calcium-rich food must have been intense, strongly favoring individuals who could tolerate it past infancy.[4] And you'll recall from the sickle-cell case that selection pressure from malarial parasites caused the evolution of resistance in human red blood cells; reciprocally, human use of antimalarial drugs such as chloroquine has caused the parasites to evolve resistance to them—a case of human-*Plasmodium* coevolution.

The original study of coevolution was of the interactions between plants and a group of animals that feed on them. It wasn't the relationship between bison and grasses but that between butterflies and the plants their caterpillars eat, studied by plant evolutionist Peter Raven and Paul in the early 1960s. Since plants cannot run away from herbivores as gazelles can flee from lions, plants have evolved static defenses. The spines cacti grow are the most obvious example; the silica crystals in grass leaves are perhaps more obscure. Even less apparent, but overall more important, are the chemical defenses that plants have evolved to poison, disorient, intoxicate, starve, or trap (in goo-like pine resin) the creatures that try to eat them.

That seems obvious today, but a half-century ago the wonderful array of complex chemicals produced by plants were thought to be "excretory products." For some reason, scientists didn't ask why evolution would favor putting the effort into making energy-rich excretory products that, rather than being excreted, would be loaded into leaves, stems, and flowers. But the main role of most of these chemicals was revealed by the way their distribution among different plants largely determined which groups of butterflies could or could not feed on those plants. Those patterns suggested that a plant would evolve a certain chemical that would deter feeding by caterpillars. In response, one population of butterflies might evolve an enzyme that detoxified the chemical, allowing those butterflies to enjoy a relatively competition-free diet—until the plant evolved improved defenses, in what might be called an "arms race." That race might go on for a long time, or one participant in the coevolutionary interaction might be unable to evolve fast enough, losing the race by going extinct.

At first botanists doubted that little insects could place sufficient selection pressure on plants to cause them to evolve defensive chemicals, but a field experiment in 1968 suggested that the selection pressure could be considerable. A tiny blue butterfly (*Glaucopsyche lygdamus*) with wings barely half an inch long laid its eggs on the flower stalks of lupine plants about three feet tall. By designating some flower stalks as "experimentals" and removing all the eggs laid on them, and designating matched flower stalks as "controls" on which the eggs were left undisturbed, plant biologist Dennis Breedlove and Paul were able to measure the impact of the tiny herbivore on the big plant. It was striking: the butterflies destroyed almost half of the potential seed production—a powerful selective force. Subsequent experiments and observations by many scientists effectively undermined the idea that herbivores did not place heavy selection pressure on plants.

In retrospect it all looks simple. Plants have responded to millions of generations of assault from herbivores by evolving an array of poisons to defend themselves. But coevolution is, by definition, a two-way street. While the plants were evolving chemical defenses, the enemies—from viruses to mammals—were evolving ways to avoid or detoxify the chemicals. Some insects that became able to detoxify some of the plants' defensive chemicals adopted those chemicals to defend themselves. Monarch butterflies (*Danaus plexippus*) are a familiar example. These beautiful creatures have black and orange color patterns that warn avian predators that they often have acquired heart poisons as a defense against being eaten (birds that try to eat them end up vomiting and learn to avoid such color patterns in the future).

In some cases, harmless insects mimic the appearance of poisonous insects in order to fool mutual predators. By simply looking like the poisonous insect, they take coevolutionary advantage of the predators' having learned to avoid dangerous or distasteful food. Some flies have evolved to look like wasps and obtained some of the benefits of having a nasty sting without actually having one. Experiments have shown that toads exposed to bumblebees actually wouldn't, after having their tongues stung, try to eat harmless drone flies that look like the bees. If successful, the harmless species gets the benefit of not being devoured, without having to spend the energy on manufacturing poisons and developing protections against them or developing a stinging apparatus. You can think of the mimicking strategy as the evolutionary equivalent of bluffing in poker, and the mimics, in economic terms, as being "free riders." But while it is obviously advantageous for the flies to fool predators into thinking they can sting, it is disadvantageous for the bees to have harmless insects impersonate them (a naïve toad having just snapped up a tasty bee-like fly would be temped to eat the next passing bee). But presumably there would be selection pressure on the bees to make them appear different from the flies. Thus a seemingly simple picture turns into a coevolutionary arms race, the flies "trying" to keep their appearance up with the bees, and the bees fleeing from looking like the flies.

In other cases, several poisonous butterfly species have coevolved

FIGURE 2-2. The monarch butterfly (*top left*) is a "model" species because its black and orange coloration announces that it is distasteful—it contains heart poisons that cause vomiting in birds that eat it. In North America the viceroy (*top right*) is a "co-model," which usually is also poisonous. The two have coevolved similar black and orange color patterns to announce their nastiness to potential predators—experience with either is likely to make the predator avoid both. Notice that the mimicry is very imperfect in detail; many tropical mimicry pairs are much more difficult to sort out as are the New Guinea butterflies (*middle and lower left*) and moths (*middle and lower right*). Lower four figures redrawn by Anne Ehrlich from R.C. Punnett, *Mimicry in Butterflies* (Cambridge University Press, New York, 1915.)

the same color pattern, making it easier for predators such as birds to learn to avoid them, as have the monarch and the viceroy, *Limenitis archippus* (figure 2-2). Such mimicry among poisonous or dangerous creatures is one of the most common products of coevolution. Other mimicry can be thought of as aggressive—such as hawks masquerading as non-predatory vultures. The eggs of cuckoos (birds that lay their eggs in the nests of other birds) have evolved to look like the eggs of their host species, lessening the chances that the hosts will toss the eggs of their enemies out of the nest.

Human beings also have adapted those defensive chemicals of plants to their own purposes, as spices, medicines, pesticides, and recreational drugs. Some plant chemicals function to disorient animals. Thus if a zebra eats parts of a plant with a hallucinogenic defensive chemical and then goes off and tries to mate with a lion, it will never molest the plant again. The same chemical, when ingested by a human being in the right quantity, can provide a "disorientation" that some find pleasant.

The coevolutionary arms races between plants and herbivores have helped create a problem for our species. Those millions of generations of evolutionary experience that herbivores have had with plant poisons almost certainly make it easier for the pests among them to evolve resistance to the poisons people have invented to kill them. Crop-destroying insects often have already developed mechanisms to detoxify the defensive poisons of certain plants, systems that with some evolutionary modification also will lessen the impact of synthetic poisons such as DDT. Carnivores, on the other hand, have less experience with plant poisons and are normally more susceptible to pesticides—as we saw in the disappearance of the tiny carnivores that attack spider mites. Presumably, they have been typically less adept at evolving detoxifying systems of their own (many more kinds of plants are toxic than are herbivores), and carnivores usually have smaller populations than the herbivores.

Our having ignored the lessons of coevolution is one reason why synthetic organic pesticides have been less successful in controlling the pests that attack crops than they might have been. Since the poisons tend to have more severe impacts on the predators that attack pests than on the pests themselves, more care should have been taken to prevent pesticides from *worsening* pest problems, as they often have. Pesticide-created plagues, such as those of the spider mites mentioned earlier, are frequent. For instance, in the Cañete Valley of Peru in the early 1950s, entirely new pests were created. Overspraying of cotton with DDT and its relatives killed off the insect predators of insect herbivores that could eat cotton but that had such small populations that they were not an economic problem. The new pests and the pesticide-resistant strains of the old ones, which quickly evolved, in eight years reduced cotton yields to levels *below* those that prevailed before pesticides were used.

Gradually, though, many farmers around the world are learning to avoid overuse of pesticides and replacing them with "integrated pest management." In IPM pesticides are used as scalpels rather than bludgeons, applied only when there is a clear problem (rather than sprayed regularly, problem or not), and other measures are taken to slow the evolution of resistance. IPM leans heavily on non-chemical methods to control pests: when appropriate, destroying places where pests pass the winter, manipulating the genetics of pests, improving the ecological circumstances of predators, and introducing new predators. The idea is not to exterminate the pests (which is almost never possible) but to manage them to keep damage at an economically acceptable level. Meanwhile, the environmental and health hazards of the broadcast use of poisons are eliminated or at least sharply reduced.

The story of antibiotic resistance is similar. We human beings, along with other vertebrate animals, have evolved an immune system to protect ourselves from tiny enemies such as viruses, bacteria, and various other parasites, and our discovery of antibiotics gave us an extremely valuable additional weapon in our coevolutionary war against bacteria, saving many millions of lives. But bacteria have a long history of exposure to antibacterial (antibiotic) poisons—many of them produced by other competing bacteria. Not only do bacteria often rapidly develop ways of detoxifying antibiotics; they also have evolved techniques for sharing their evolutionary novelties via plasmids, elements of DNA that can replicate and pass genes between bacterial species. This is an important method, for example, for the spread of antibiotic resistance. Sadly, a lack of understanding of biological evolution on the part of the medical community and the general public has led to overuse of those correctly named "wonder drugs," which has allowed those weapons to be blunted by the widespread evolution of resistance to them. Since the late 1980s some bacteria have evolved resistance even to what had become the most powerful antibiotic of last resort, vancomycin.

Coevolution is involved in many kinds of interactions among species, such as between plants and herbivores, parasites and hosts, predators and prey, competitors (species that compete for the same resources, such as food plants or prey animals), and mutualists (species that benefit from association with one another). Understanding coevolution has been the focus of a major effort among biologists in recent decades, and very substantial progress has been made.

For instance, we now realize that many individual organisms are actually the composite result of ancient mutualistic relationships, situations in which individuals of two (or sometimes more) species interact to the benefit of each. Thus once-independent energy-producing bacterial cells were incorporated into larger bacterial cells that were early ancestors of plants and animals. Those energy-producing cells evolved into energy-mobilizing mechanisms, termed mitochondria, which are found in our cells and those of all animals and plants today. Mitochondria still have separate genomes that are very similar to bacterial genomes. Other cellular organs (organelles), such as the chloroplasts that carry out photosynthesis in plant cells, are thought also to have originally evolved as free-living bacteria. Under selection pressures we only can imagine, they gave up their independence and were incorporated into other organisms, becoming organelles. Many insects (and other invertebrates) have bacterial partners that are passed between generations. The bacteria supply nutrients lacking in the insects' diets and help to explain the great diversity and evolutionary success of insects. And, it turns out, almost any plant you look at has fungi living within it. Some fungi are necessary to enable the plant to take up vital nutrients from the soil; some may even produce toxic chemicals that help protect plants from herbivores. And others are outright parasites.

New methods of reconstructing evolutionary history have begun to reveal the importance of such coevolutionary interactions over geologic time. For example, there is now growing evidence that, as was speculated in the original paper on the topic, plant-herbivore coevolution has been a major factor in the diversification over tens of millions of years of both flowering plants and insects. And experiments using bacteria and viruses that attack bacteria (bacteriophages)—the microscopic equivalent of the plant-herbivore system—show that, like evolution in response to physical and chemical changes in environments, coevolution can also occur very rapidly. It is becoming ever clearer that vast coevolutionary networks have played a central role over the eons in the generation, organization, and preservation of Earth's biodiversity—its rich panoply of populations, species, and ecosystems. The world is, and has long been, truly a dynamic coevolutionary wonderland.

So it is not surprising to find that *Homo sapiens* is deeply embedded in coevolutionary webs. For instance, the body louse depends almost entirely on human beings for its habitat and resources. It carries diseases such as typhus, and the selection pressure it puts on people may at times be severe. In return, humanity, by losing most of its body hair, selected the lice (unlike head lice) to live not on the body but in clothes. In contrast, people depend on multitudes of other organisms, perhaps most prominently rice, wheat, and maize (corn) plants, which supply the bulk of humanity's food. But we are linked to some degree to thousands of others, from bee species that pollinate crops (a mutualistic relationship) to fishes that we harvest from the sea, fungi that produce antibiotics, trees that supply timber, and on and on. In most cases (especially where there is deliberate cultivation, as of crops, domesticated mammals, honeybees, tree plantations, etc.) people put strong selection pressure on those organisms, and, through influencing our diets, forms of shelter, and energy use, the organisms return the favor. Thus European honeybees have been selected for docility (as opposed to the African "killer" strains) and rice has been selected to produce big seed heads, while many of our populations have been selected for lactase persistence and for genes that confer resistance to celiac disease, a nasty reaction to the gluten (a mix of proteins and starch) in wheat.

The original definition of coevolution has subsequently been stretched to include interactions between any two evolving systems, an extension that sometimes seems quite useful. One such extension that is important to the themes of this book is the interaction between Earth's biota (plants, animals, and microorganisms) and Earth's climate, which shares many characteristics with coevolution. Organisms influence the climate by changing the reflectivity of the planet's surface. Where forests grow, more solar energy is absorbed than in deserts, from which more of the energy is reflected back to space. Organisms also influence climate by adding greenhouse gases (ones that influence Earth's surface temperature) to, and subtracting them from, the atmosphere. When green plants grow, they take carbon dioxide from the atmosphere and release oxygen. When plants and animals run their life processes or die and decay, they release carbon dioxide. Organisms also alter the hydrologic (water) cycle: for instance, by plants holding soil in place with their roots and slowing water's flow, and by respiration, whereby plants take up water from the soil, use it in their life processes, and release it through their leaves.

Climate changes in turn alter selection pressures on virtually all organisms. For example, a change in climate may well have played a key role in the splitting of our lineage from that of chimpanzees. It seems likely that our last common ancestor with the chimps lived in woodlands in a gradually drying world, and that the chimp line followed retreating forests while our ancestors ventured out onto the spreading savannas.

We exist and are able to speculate about our evolutionary origins only because conditions in our planet's past permitted life to evolve. There is no evidence that the universe was somehow especially suitable to the development of intelligent life. The latter notion is something scientists call the anthropic principle. That principle was neatly exposed by the late author Douglas Adams (*Hitchhiker's Guide to the Galaxy*) in his parable about a thinking puddle: ". . . imagine a puddle waking up one morning and thinking, 'This is an interesting world I find myself in—an interesting HOLE I find myself in—fits me rather neatly, doesn't it? In fact, it fits me staggeringly well, must have been made to have me in it!'"

Darwin's entangled bank is full of intricate relationships, as we've seen, even though it wasn't "designed." Rather, via processes of genetic change—of speciation and coevolution—in populations in response to varying environments, and in turn via populations altering those environments, the community of organisms on the bank have evolved. Changing organisms and changing environments, changing people and changing human environments—these are enduring themes that run through the life of our species.

CHAPTER 3

Our Distant Past

"We thus learn that man is descended from a hairy quadruped, furnished with a tail and pointed ears, probably arboreal in its habits, and an inhabitant of the Old World."

CHARLES DARWIN, 1871[1]

IMAGINE the surprise of workers at a quarry near Solnhofen, Bavaria, in 1861 who were mining a form of limestone rock ideal for engraving to make lithographs, illustrations, or text for books printed from a flat stone. When one slab of the limestone was split in two, a fine fossil was revealed: the hardened remnants of a magpie-size creature that, we now know, lived some 150 million years earlier. The headless fossil had feathers and reptilian body parts, and was named *Archaeopteryx* (ancient wing). It was the first of some ten specimens. The very first find had been a single feather, and in 1875 a more complete fossil was found, which showed that the creature had a reptilian head with a toothed mouth (figure 3-1).

Archaeopteryx was the first "missing link"—a fossil showing clearly intermediate characteristics between what, in modern times, are distinct groups of organisms. It looked as if the feathers and wings of a bird had been grafted onto the body of a lizard. Its discovery was in essence the beginning of the documentation of evolution by a fossil record. That record now shows evolution's course in some detail, including changes within populations and among populations in numerous taxonomic groups.

Not the least of these is *Homo sapiens*. When most dinosaurs were eradicated in the wake of a collision of an extraterrestrial body with

FIGURE 3-1. *Archaeopteryx*, the famous bird-reptile "missing link" discovered in the nineteenth century. This fanciful picture shows the organism rising out of an accurate presentation of the beautifully preserved fossil in a slab of slate. Pencil drawing © Darryl Wheye, Science Art-Birds, based on a photograph of the Berlin *Archaeopteryx* taken by Hermann Jaeger.

Earth about 65 million years ago, many ecological niches, such as egg-eating predator or large plant-eater, were left for mammals to evolve into. One group of mammals that arose just after the decimation of the dinosaurs was characterized by grasping hands, binocular vision with depth perception, a complex brain to help them deal with complicated social arrangements, and predominantly single births followed by the infant's prolonged dependence on the mother. That group, the primates, began an adaptive radiation some 50 million years ago, and human beings are on one branch of that radiation.

On that branch, the discovery of "missing links" galore has shown that the human family tree, once thought to be a rather simple evolutionary sequence, was actually a complex evolutionary "bush" with much speciation in the past. And in recent years, traditional fossil evidence of these events has been magnificently supported by the use of new molecular techniques that get at the very basis of genetic change and are even allowing us to determine the genetic differences between living people and their close fossil relatives.

The Origin of Life

If the line from early primates to humans is the most discussed period of our evolution, it is not the earliest: our very first ancestors—the first living organisms—arose more than 3 billion years earlier. At its most basic, the entire process of genetic evolution is relatively straightforward. It involves a combination of a genetic system (a set of interacting elements that replicates itself, with some variation) enclosed by a membrane that separates it from its environment (thus making it an organism) with a mechanism, natural selection, that leads to some variants replicating more than others. Such an enclosed genetic system subject to differing environments has made possible the now-vast diversity of life (biodiversity) that evolved once life had begun, and the eventual appearance of our planet-dominating species.

Yet one part of the story is not simple. A fundamental question remains unresolved—that of where and how life itself began. There are few problems in understanding the creation, in a lifeless world, of the chemical building blocks of life; scientists have made many of them in laboratories under conditions that simulated those hypothesized for the early Earth. But how those building blocks became concentrated and assembled into metabolizing, self-replicating entities with capsules separating them from their environments is still unknown. Biochemists, geologists, and other scientists have been working on pieces of the puzzle, but they can't even rule out the possibility that life reached Earth on a meteorite originating on Mars. There are still two basic schools of thought. One is that life would naturally evolve on any Earth-like planet in the universe. The other is that it was a rare event that, about 3.8 billion years ago, simply happened on our planet. Life may have begun as a primitive chemical cycle on the surfaces of rocks, with the development of mechanisms for replication and of cells with membranes that could provide isolation from the physical environment occurring later as a result of selection. Or it might have started with the appearance of a self-replicating molecule such as RNA, a single-stranded close relative of DNA, which still functions in modern cells and which can serve both as a carrier of information and as a catalyst (the latter role being played by proteins in today's organisms). Or it might have been some combination of the two.

After learning all this, should we say we "believe in evolution"? In one common meaning of "belief," having a firm persuasion based on faith, the answer is a clear no. No scientific conclusion is ever final. But we can say we believe that the present theory of evolution is, on the basis of massive accumulated evidence, at the moment by far the best scientific explanation of the diversity of life that surrounds us. Remember, theories are the ultimate product of science: explanatory frameworks that help us to understand the world, frameworks that can be tested by observation and/or experiment and that (at their best) allow solid predictions to be made about future findings or events.

As we explained at the end of chapter 1, evolution fits that highest accolade as well as does the theory that Earth circles the sun. Evolution, like other scientific theories, is closer in meaning to "fact" than it is to the common lay meaning of "theory"—"speculation" or "conjecture." Nonetheless, if the two of us could come up with a better explanation for the rich and changing diversity of organisms, we'd publish it immediately. Science is an adversary game. It works not because scientists (try as they might) are generally more "objective" than other people, but because scientists can advance their careers by showing that the standard views of other scientists are wrong. We'd be happy to play the role relative to Darwin that Einstein played to Newton, showing how limited was the latter's domain. For a biologist, it would be exciting to show that natural selection was not the driving force of evolution, or that selection had a goal. But we're not holding our breath; no observation has been made that calls into question that life evolved, and many millions of observations show that it wasn't created in a single "act."

One of the strongest sources of evidence that evolution has occurred and that it accounts for the diversity of life consists of myriad pieces of evidence that are literally rock solid.

The Fossil Record

What most scientists consider the main evidence for evolution's occurrence, the fossil record, is so rich now that it would take many volumes to describe it (indeed, many have been written). The validity of scientific interpretation of that record is supported in many ways, but perhaps the most interesting is the repeated finding of animals and plants that had been described on the basis of fossil specimens but that later turned up alive and just as described from ancient fossils.

For instance, lobe-finned fishes were once known only as fossils from 80 million years or more ago, yet in 1938 a living specimen of a lobe-finned species (*Latimeria chalumnae*) was caught by a fisherman in the Comoros Islands off southeastern Africa. More recently, in 1994, a small population of a "fossil" tree species, the Wollemi pine, was found growing in a remote canyon in the Blue Mountains west of Sydney, Australia. It is the only known survivor of a closely related group of tree species that were common 175 to 50 million years ago; dinosaurs doubtless sheltered in its groves. The species persisted over a huge time span and was broadly distributed; the most recent fossil of the Wollemi pine was from 2 million years ago, and it was thought to have been extinct since then. The pine is now being successfully cultivated for sale as an ornamental plant, and friends of ours are growing one to use as a Christmas tree.

In the early days of evolutionary science, much attention was paid to the question of "missing links." If people were really descended from monkeys, where were the fossils of intermediate forms? Why hadn't people found traces of the links between birds and reptiles if the former really were derived from the latter? In the century and a half since Charles Darwin, scientists have learned some answers. Certain kinds of organisms don't easily become converted to fossils, and birds are among them. Some habitats don't allow good preservation, and lots of birds live in those places. And in some cases scientists just hadn't looked hard enough. But in the case of birds, they now have. In addition to the famous feathered reptile *Archaeopteryx* discovered just after Darwin published *On the Origin of Species*, a wealth of fossils has been discovered connecting dinosaurs with birds. And, in general, the other principal gaps in the fossil record, such as the ones between fishes and amphibians, reptiles and mammals, and chimpanzees and human beings, are now occupied by fossils of organisms with intermediate characteristics.

For a long time there were few fossils to connect terrestrial mammals with whales. Now several fine transitional fossils have been found, and one can trace the line from a beast with four legs to one with forelimbs transformed into flippers and hind limbs reduced to vestiges, and then to one with vestigial hind limbs disappearing as the animal became fishlike in form. Here we can see the evolution of convergence: organisms from totally different groups becoming similar—such as whales and fishes—when they are subject to similar selection pressures by the need to move relatively rapidly through water.

Doubters of evolutionary processes often claim that the absence of known intermediates makes it clear that modern life-forms cannot have evolved from early, very simple organisms. There are many creationist Web sites claiming that the fossil record does not show a continuum of forms but is shot full of gaps where there are missing links. One still stated late in 2007 that "not a single fossil with part fins and part feet has been found."[2] But discovery of just such a missing-link fossil was reported in April 2006—the finding of a fossil fish, 375 million years old, that makes a wonderful transitional form to the amphibians, which are descendants of fishes and distant ancestors of ours. The new fossil, *Tiktaalik* ("large shallow-water fish" in an Inuit language), found in the Canadian Arctic, had front fins that were partly feet. They were in the process of evolving into legs, showing signs of digits, a wrist, elbows, and shoulders. Morphologically, in many ways they fit nearly perfectly between numerous fossil fishes that show signs of evolving into tetrapods (four-legged animals) and actual tetrapods (figure 3-2). The fact is that the continuous evolution of organisms from only relatively simple ones in the distant past to the huge variety today, including many much more complex forms, is amply documented in the fossil record. More telling perhaps is the absence of, say, rabbit fossils mixed in with fossil trilobites (ancient insect relatives) in rocks from the Cambrian period, 500 million years ago. Rabbits evolved after the extinction of the dinosaurs, some 65 million years in the past, and a documented finding of a rabbit fossil 500 million years old would be a severe challenge to modern evolutionary theory. But no such "reversal" in the fossil record has ever been demonstrated. (Don't be fooled by long-demolished inventions, much beloved of creationist Web sites, about overlapping dinosaur and human fossil footprints—no such fossil traces have ever been found.)[3]

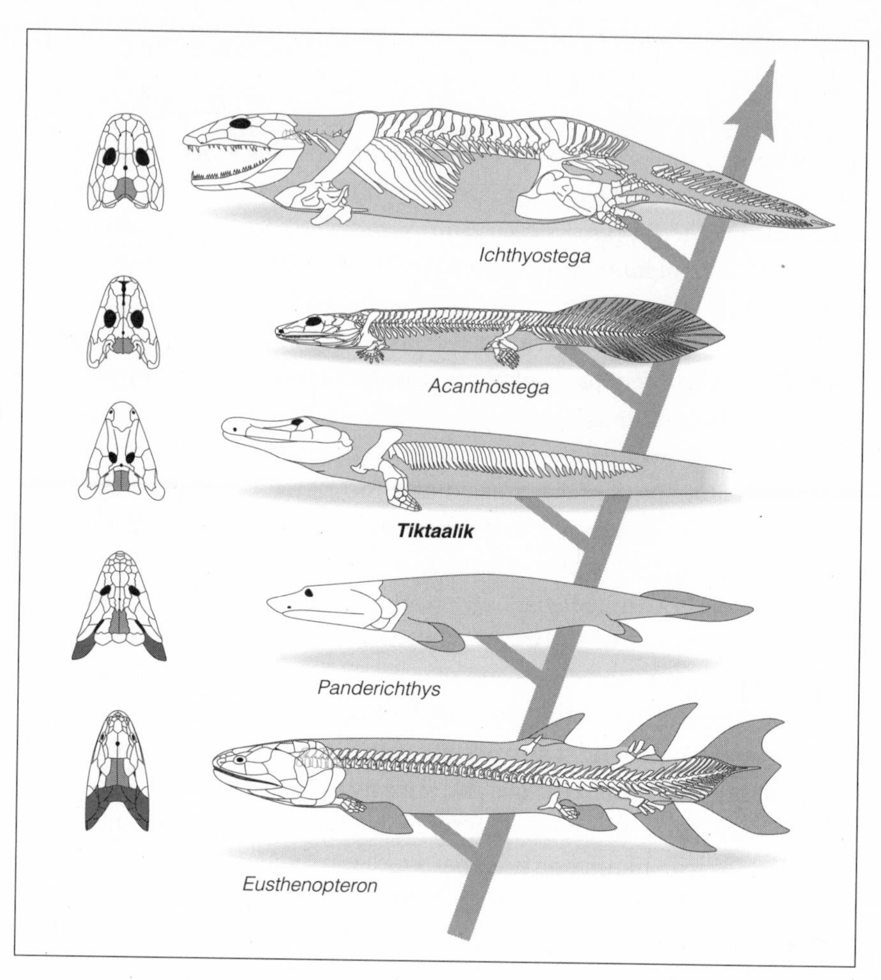

FIGURE 3-2. The newly discovered fossil *Tiktaalik* in evolutionary context with two lobe-finned fishes and two early amphibia. *Tiktaalik* shows signs of evolving into a tetrapod (a four-legged animal, in this case an amphibian), with front limbs that have leglike characteristics. Note the disappearance of the gill covers (shaded) at the back of the head and the gradual alteration of the bones of the skull. Redrawn from E. B. Daeschler, N. H. Shubin, and F. A. Jenkins Jr., A Devonian tetrapod-like fish and the evolution of the tetrapod body plan, *Nature* 440 (2006): 757–63.

The Human Line

Humanity's own history provides another excellent example of the discovery of many missing links. Darwin himself, in *The Descent of Man*, noted that the great resemblance between human beings and great apes meant "we can infer that some ancient member of the anthropomorphous subgroup [great apes] gave birth to man."[4]

Chimpanzees, the most humanlike apes, have brain volumes of about 400 cubic centimeters (cc). Modern people average about 1,350 cc. So where are the roughly 900-cc "missing links"? In 1856, just before Darwin published *On the Origin of Species*, a fossil of a different kind of human being was discovered in the Neander Valley of Germany. "Neanderthal man" was clearly different, but not that different. It had a brain size roughly like that of modern people (actually, when more specimens were discovered, it turned out the Neanderthals' average brain size was slightly *larger*). Interestingly, noting this large brain size in an ancient human being seems to have been the only written notice Darwin took of Neanderthals. They looked at best like a recent ancestor, and they were too much like us and not chimplike enough to seem like a real missing link. But Neanderthals were the first clear evidence that *Homo sapiens* wasn't the only "human being" ever to populate Earth.

Then the first and most famous human missing link was discovered in Asia, in 1891. That was *Pithecanthropus erectus* (now called *Homo erectus*), the "Java man." The classic Java people lived between roughly 1 million and perhaps 300,000 years ago. They were fully upright, as clearly testified by the structure of their fossilized legs, and they had brains that averaged 900–1,000 cc—pretty big, but not as big as ours. If she were properly dressed and cleaned up, you might not be startled when meeting *H. erectus* on the street, especially if a hairdo or hat disguised her quite sloping forehead.

The next, and perhaps most stunning, missing link was found in the 1920s. The first specimen was a youngster, found in a lime quarry in Taung, South Africa, and known as the "Taung child." The species was named *Australopithecus africanus* (southern African ape), and in the decades that followed, other species of *Australopithecus* were found in both southern and eastern Africa. As a group, they lived between about 4.1 million and 1 million years ago, and the fossil remains of more than a hundred individuals have now been found. Fully upright but small brained (400–580 cc) and apelike from the waist up, they resembled nothing so much as a bipedal chimpanzee. An adult seen on a street would stop you dead. They were clear links between ancestral apes and people, and they confirmed Darwin's early conjecture, based solely on our similarity to chimps and gorillas, that "it is somewhat more probable that our early progenitors lived on the African continent than elsewhere."[5] We personally will always have a soft spot for the Taung child, for years ago in South Africa we were privileged to hold her fossilized skull in our hands.

Half a century ago, the human family tree seemed pretty simple and linear—more a "family pole" than a tree: chimpanzee-like ape → *Australopithecus* → *Homo erectus* → *Homo neanderthalensis* → *Homo sapiens*. In the view of anti-evolutionists, of course, those discoveries simply created more spaces that needed missing links. Where, for example, they asked, is the intermediate form between *Australopithecus* and *Homo erectus*? The answer is that at least two links subsequently have been discovered in that gap: *H. habilis* and *H. ergaster*, with brain volumes in the 600- to 900-cc range. As you can see from figure 3-3, the human family pole—the hominins (the technical term for humans and their ancestors after they diverged from the chimpanzee line)[6]—has in the past fifty years become the "family bush." During much of the past, several human (hominin) species were living at the same time.

Among the additional australopithecine species discovered, the one represented by the famous "Lucy" (*Australopithecus afarensis*) is especially illuminating. Discovered in Ethiopia in 1973 and named after the Beatles' song "Lucy in the Sky with Diamonds," which was playing in the paleontologists' camp at the time of her discovery, Lucy pushed the time horizon of our earliest relatives back to 3.3 million years ago. Like her South African relative, Lucy was fully upright, although retaining some characteristics in her limbs that have led many scientists to think she was pretty good in the trees as well. She had a small brain and various characteristics of the skull such as forward projection of the face (prognathism) and teeth (relatively large eyeteeth) reminiscent of non-human apes. Lucy's kind might actually have been on the ancestral line leading to ourselves, but other australopithecines (shown in figure 3-3), in the genus *Paranthropus*, clearly were not. The characteristics of their skulls and jaws show that they fed on especially tough vegetable foods, and they survived until perhaps 1 million years ago, well after our own direct ancestors had emerged.

We now have even older fossils that help to fill the gap between the last common ancestor of chimps and hominins. At this point, the oldest widely accepted hominins are the chimplike *Ardipithecus kadabba*, which existed between 5.8 and 5.2 million years ago, and its presumed descendant, *A. ramidus*, which persisted until 4.4 million years ago and appears to have been a biped. Just why the chimp-human lines diverged, and

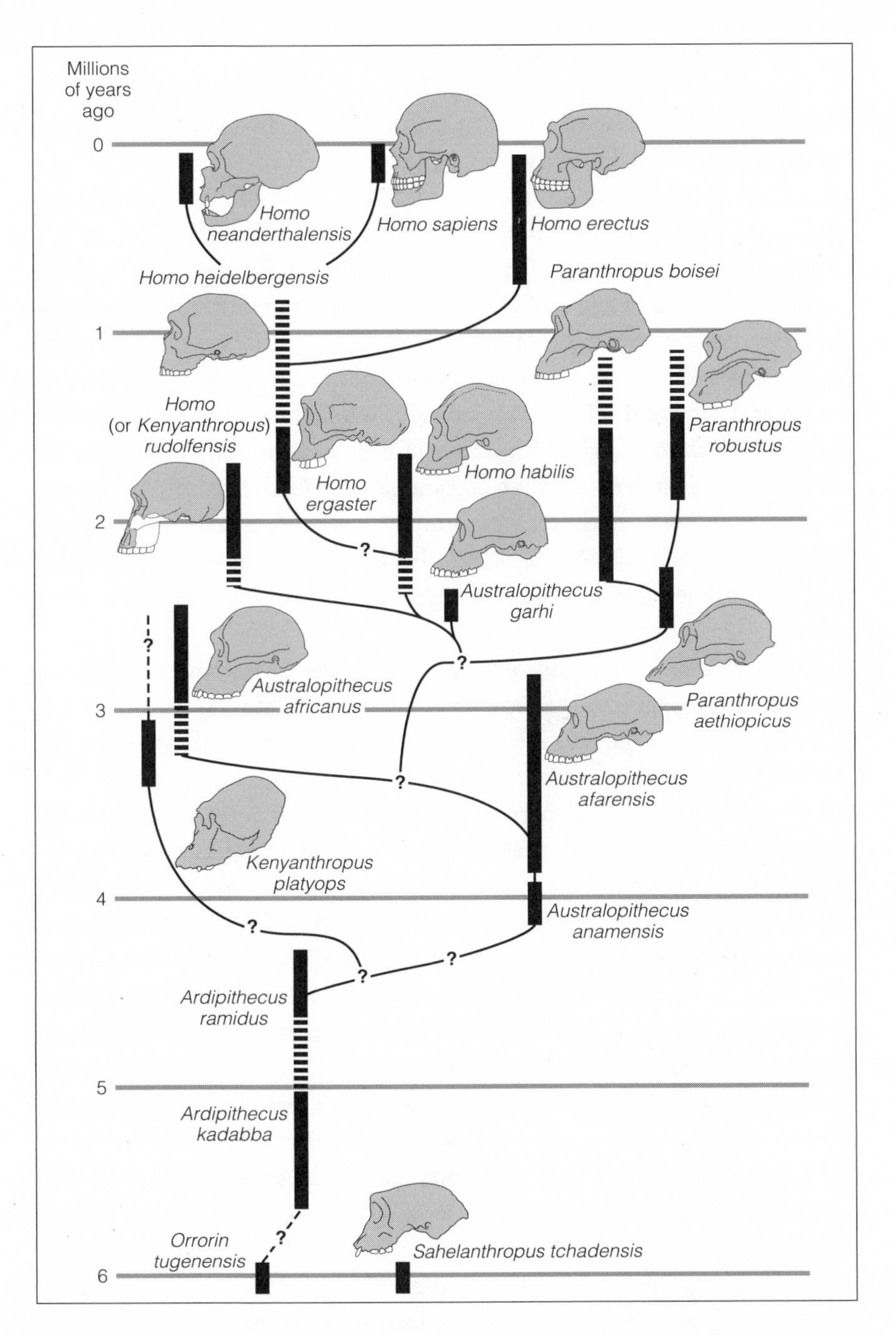

FIGURE 3-3. The human family "bush" as understood today, showing the hominin species now known that descended from the ancestors that split off from the chimps. An abundance of what once were considered "missing links" have now been found, and scientists now realize that several hominin species probably lived in Africa at the same time. Courtesy of Richard Klein.

especially what role was played by habitat alterations generated by climate changes, a leading environmental contender for the cause, is not settled. Scientists, using a variety of methods, including chemically analyzing the shells of single-celled organisms in ocean sediments (foraminifera), have concluded that about 10 million years ago the climate of Earth cooled, and forests in Africa thinned and shrank in area. Our ancestors may have increasingly taken advantage of enlarging savannas or mosaics of woodlands and open grasslands, and, in connection with that, may have been selected for upright bipedal posture for its putative advantages when spending time in the open—although recently some scientists have suggested that bipedalism evolved in the trees as an adaptation to walking on springy branches. In any case, recent evidence suggests that the upright *Ardipithecus* line gave rise to the other australopithecines, starting with *Australopithecus anamensis* and later *A. afarensis* (Lucy).

There also remains considerable debate about the exact positioning (in time and relative to our ancestral line) and relationships of named hominin forms to one another, especially those that lived between 7 and 3 million years ago. Problems are caused by the vagaries of fossilization as well as by difficulties in interpreting relationships derived from molecular genetic evidence. The scientific problems of interpreting human fossil remains, especially partial or distorted remains, with uncertain dating and generally small sample sizes, have been highlighted by the recent description of a "new species" of pygmy hominin from Flores Island in Indonesia. The find consists of a single, very small skull (400-cc capacity) and a femur indicating that the individual stood a little over three feet tall. *Homo floresiensis* putatively lived, along with other dwarfed animals, on the island as recently as 18,000 years ago. At the moment it seems possible that the principal specimen was drawn from a known modern human group with short stature, but that it represents a developmentally anomalous (microcephalic) individual.

The status of our first and most famous fossil relative has also changed in the past fifty years. Fossil and genetic evidence have demonstrated that, rather than being a recent ancestor of *Homo sapiens*, *H. neanderthalensis* split off from our ancestral line more than half a million years ago and evolved alongside it. Many questions remain about the Neanderthals, including how similar their speech was to that of modern *H. sapiens*, why they died out, and especially the degree to which our ancestors outcompeted or interbred with them. Recent evidence suggests that, at least in terms of quality of tools, the Neanderthals were in a league with our ancestors, though few paleoanthropologists think there was much, if any, interbreeding. Scientists have recently managed to extract ancient DNA fragments from Neanderthal bones, perhaps making it possible to sequence the entire genome of 35,000-year-old Neanderthals, and an attempt is already under way. That could help answer some of the questions.

The picture of human evolution today suggests that several of our most significant characteristics (and those of other apes and monkeys) trace all the way back to tarsier- or lemurlike lower primates that lived primarily in bushes. They subsisted mainly on insects, which they snatched in flight or on branches with grasping, humanlike hands. Thus binocular vision (giving fine depth perception) and hand-eye coordination may have originated selectively in response to a brushy environment rich with bugs. The ancestral line leading to the apes diverged from the line leading to African and Eurasian monkeys about 30 million years ago; the line leading to gorillas, chimpanzees, and people separated from the line leading to orangutans about 18 million years ago. That started a trend toward spending less time in the trees, a good thing from the human viewpoint, since a dominant animal would not have been likely to evolve in an arboreal habitat—too tough to develop agriculture, among other things.

The chimp line split from the human line roughly 6 to 7 million years ago. Some scientists have interpreted genetic evidence to suggest a fairly substantial hybridization between the chimp and human lines at first, with the chimps and bonobos (sometimes called "pygmy chimpanzees") diverging some 2.5 million years ago. Some of those evolutionary changes are thought to be related to the climate shifts that changed the distribution of habitats, with first gorilla ancestors and then chimp ancestors remaining behind in forests while hominin precursors became ever more erect as they moved into drier, more exposed savannas.

One of the persistent mysteries of human evolution is why, after splitting off from the chimp line, our early ancestors evolved an upright posture while retaining essentially chimplike brains. If they were invading open areas, what in a savanna or woodland mosaic environment created a selection pressure for bipedality? And what caused the later rapid (in terms of geologic time) expansion of human brains over the past 2.5

million years? Numerous explanations have been proposed for the evolution of upright posture, from allowing more efficient travel, freeing the hands to hold tools or weapons, and exposing less body area to the hot savanna sun to permitting upright displays, which would suppress competition in the new, relatively open and resource-poor habitats. Upright posture would also raise the eyes to a level at which they would be more able to spot approaching predators. Whether we'll ever know the definitive answer to the "why upright" question is doubtful.

The spurt of brain growth from about 400 to 1,350 cc probably was a response to selection pressures originating in the increasing complexity of human social behavior. Such changes in the *rate* of evolutionary change are common, so hominin brain size remaining more or less constant for a few million years and then tripling in a few million more is not an unusual sort of evolutionary phenomenon. In the absence of powerful selection pressures to change, a species' characteristics may remain constant for long periods of time but then, in the face of an environmental stressor, quite rapidly adjust genetically. This common pattern of variation in evolutionary rates has long been known. In the early 1970s a situation of relative stasis followed by rapid evolutionary change was given a special name, "punctuated equilibrium," which some people still misinterpret as some kind of novel evolutionary process.

The Fossil Record of Culture

Most paleontologists believe that simple chipped-stone pieces associated with the remains of *Homo habilis* indicate the dawn of toolmaking, about 2.5 million years ago. It was the first of our ancestral technologies to leave a record, and it was the start on the road to dominance that has produced technological "descendants" as varied as books, blenders, SUVs, antibiotics, and nuclear weapons. The first stone technology is called Oldowan, after the Olduvai Gorge in Tanzania, where anthropologists Mary and Louis Leakey first discovered these chipped-stone remains. The evidence indicates it took more than a million years after our ancestors started making tools until they began, with the aid of improving technologies, to expand significantly their own geographic distribution. It is generally assumed that there were two human "out of Africa" episodes—departures from the cradle of humanity to distant lands. The first exodus was more than a million years ago, almost certainly as much as 1.7 to 1.8 million years ago, and the emigrants were *Homo ergaster* or their immediate descendants, *H. erectus*.

The second departure occurred more than 5,000 generations after modern *Homo sapiens* evolved in Africa, about 200,000 years ago. Our own kind left their African birthplace, spreading first to the Near East (the Levant) perhaps 120,000 years ago, and then to much of the Old World some 65,000 to 40,000 years ago. They subsequently occupied the rest of Earth's land areas and eventually replaced all other hominins that had persisted until then. The role of climatic events in their dispersal and the part that invading modern *Homo sapiens* may have played in the extinctions of our hominin cousins are still debated. For instance, there have been extreme oscillations in Europe's climate over the past 100,000 years, and the severity of climate change may have reduced supplies of game and pushed less adaptable Neanderthals to extinction, with or without competition from modern *Homo sapiens* who lived in the same areas.

Interestingly, there is a connection between the ability to make stone tools and the ability to speak. Small-brained chimps fully understand the uses of tools and can smash rocks on a concrete floor and select sharp shards with which to cut a cord and gain access to food in a box. But they do not have the fine hand-eye coordination that would let them manufacture sharp tools by striking them from a rock core (a task difficult for even modern people to master). It turns out that the same kind of neuromuscular coordination required for stone tool manufacture is also essential to the ability of our tongues to undergo the incredible gymnastics that are required to produce speech. But when this ability arose in the past is unclear. Some scientists speculate that true language with syntax emerged suddenly with *Homo sapiens*, while others think that language started early with gestures and grunts, and that *H. habilis* would have been able to communicate verbally to a significant degree.

A puzzle related to these advances, as we'll see, is the reality and cause of what has become known as the "great leap forward," a cultural "revolution" that seems to have occurred about 50,000 years ago (or somewhat earlier) and greatly accelerated our rise to dominance.

CHAPTER 4 Of Genes and Culture

> "If you think we are hard wired [by our genes]—that is, everything is deterministic—there should be a lot more genes because we have a lot more traits. This makes me as a scientist both laugh and cry. I laugh at the absurdity of it and I want to cry because it is accepted by so much of our society."
>
> J. CRAIG VENTER (WHO LED THE SEQUENCING OF THE HUMAN GENOME), 2001[1]

UNTIL RELATIVELY recently, it was often said that what distinguished our species from all others was our ability to generate culture—to do such things as make and use tools. But it turns out that *Homo sapiens* is not alone in having the ability to pass along non-genetic information—culture—from generation to generation and among members of a generation. Recently, cultural transmission by imitation among chimpanzees was conclusively demonstrated in the laboratory by animal behaviorist Andrew Whiten and his colleagues. The dominant chimp in each of two groups was secretly trained in a different technique for removing a barrier that blocked a food-delivery pipe—by either poking or lifting a trigger. Then both were reintroduced into their respective groups, and fellow chimps were allowed to watch them unblock the pipe and obtain food. When the chimps in each group were allowed access to the food-delivery pipe, they copied the technique of the trained chimp, and by the end of two months two divergent food-getting cultures had developed. That such behavior could be passed from generation to generation was demonstrated by another experiment in which techniques were passed through a series of individual chimps, with the last chimp in the chain using the same technique as that learned by the first. Thus ways of dealing with their environments can be passed by chimps through time not only by transmission of changed genetic information (via natural selection) but also by changes in cultural information.

As do many other animals, chimpanzees and orangutans have cultures, stores of non-genetic information that are shared and exchanged among them and that may change over time (evolve). Just as human beings do, chimps learn from others, and teach others, how to deal with their biophysical and social environments. In some wild chimp societies the young find out in this manner how to open hard nuts; in others they learn techniques to "fish" for termites (which can be a very complex task) and even to make "spears" (possibly just probes) that they insert into tree holes to flush or kill the small nocturnal primates called bush babies that spend the day there. Some socially transmitted traditions are also found in other non-human animals. For instance, among birds called oystercatchers, the young of different local groups learn, by imitating their parents, quite different traditional ways of opening oysters.

The Evolution of Culture

Chimp culture (and that of oystercatchers) is very different from the complex cultures of modern human beings. The transmission and alteration of humanity's vast stores of cultural information have been the key to *Homo sapiens*' rise to dominance—the reason why the world isn't run by chimpanzees. First of all, non-genetic information that can be transmitted between generations of chimps takes the form of behaviors—not, so far as we know, ideas. And their behaviors often involve tools such as sticks and rocks, things that are common in the environment, rather than artifacts that they have created for special purposes. Should the cultural transmission (by imitation) be disrupted, one can easily imagine any given behavior, because of its relative simplicity, being reinvented by an individual chimp through trial and error and then imitated by others. But if even the human culture of Plato's day (let alone today's) were entirely lost, one would not expect an individual to be able to reestablish many of the behaviors by experimentation—be they building chariots, leading great armies in battle, or writing sophisticated dramas. It would instead take the contributions of large numbers of people and hundreds or thousands of generations—if it could be accomplished at all under whatever environmental conditions prevailed.

No other animal transmits culture on remotely the scale that human beings now do. It is incomparable not just in its scale but also in its spread to all corners of the globe, and in the devices that *Homo sapiens* has evolved to preserve it and give it continuity. And no other animal depends on culture so completely. As far as we can tell, if chimps had no cultural transmission of their technologies, they would still get along just fine. Yet paleoanthropologist Richard Klein believes, and we suspect he's right, that even groups of *Homo habilis*, if deprived of their simple stone tools, would have been in deep trouble, since they appear to have been heavily dependent upon those tools for processing carcasses for food.

Those big brains that our ancestors evolved over the past couple of million years have allowed human beings to enter an entirely new realm of evolution, one of large-scale cultural evolution: change in that unprecedentedly vast pool of non-genetic information stored in human brains and in the artifacts those brains have devised. Speech, which allows ideas to be transmitted, and writing, which furthered the conceptual and geographic range of that transmission, created an unbridgeable gap that separates *Homo sapiens* from all our living relatives. They allowed cultural evolution on a scale unimaginable in speechless, illiterate apes whose only tools are sticks and rocks and whose evolution has thus remained largely a matter of modification of the relatively small amount of genetic information that they (and we) possess.

Although the meaning of "information" itself can be hard to pin down, we will use the commonplace meaning of information as knowledge that is capable of being communicated or understood. It includes many forms of knowledge, from that stored in DNA sequences ("understood" by cellular apparatus), brains, and computer memories to that embodied in the structure of artifacts, ranging from hand axes to the Great Pyramids of Giza, from the remains of prehistoric irrigation systems on Polynesian islands to laptop computers and Boeing 747s. Information is facts, data, or patterns that a living being (in this case, a person), whose brain understands the conventions by which the information is coded, can interpret, understand, or be instructed by. Those who know the history of early Egypt are informed by what they see of pyramids; pilots can gain vast amounts of information from such things as the shape of the wings and the cockpit instrumentation in a 747.

While we have defined humanity's culture as its store of non-genetic information, that does not correspond precisely to various popular and anthropological definitions, as indicated by such terms as "Navajo culture" and "French culture." Anthropologists have struggled to define culture in that sense for more than a century. For example, in 1871 Edward B. Tylor, in his classic book *Primitive Culture*, defined culture as "that complex whole which includes knowledge, belief, art, morals, law, custom, and any other capabilities and habits acquired by man as a member of society."[2] In *The Tree of Culture*, Ralph Linton almost a century later defined it more restrictively as "an organized group of learned responses characteristic of a given society," and he defined a society simply as "an organized group of individuals."[3]

We'll basically go with Linton and define "a culture" (as opposed to culture itself) as a social group's collective behaviors that persist for generations and that are not, in essence, independent of social context (e.g., sneezing is a behavior not normally considered part of culture). Some people feel that the definition should also include a set of overarching ideas and beliefs that make the culture more than just a behavioral collection. It is the social group's shared bundle of non-genetic information that in large part determines the practices that characterize the society. Thus one can speak of having an affinity for baseball and being religious as features of American culture; drinking wine with lunch, wearing expensive designer clothing, and being proud of one's language as features of French culture; and making finely carved poles dedicated to ancestors and (for males) taking pride in head-hunting skill as features of Asmat culture on the southern coast of New Guinea. An overarching idea in the United States might be American exceptionalism and individuality; in France, pride in nation and language; and among the Asmat, the omnipresence of ancestral spirits and the headhunter's ability to absorb a victim's life force.

But identifying overarching ideas is problematic. As with "species" (and as we'll see later with terms like "ecosystem"), it is often difficult to define precisely the boundaries and features of a "culture," or for that matter a "religion," without robbing the term of its usefulness. Similarly, cultures clearly evolve—change through time—as one can see, for instance, in the gradual assimilation of wine drinking and soccer into American culture and the gradual disappearance of smoking from it. Our non-genetic information is in constant flux, and it changes on a much shorter time scale than does our genetic information, which you'll recall is essentially restricted to change over generations.

Early Human Cultures

The first records of human culture go back some 2.5 million years, very likely to *Homo habilis* and those simple Oldowan stone tools described in chapter 3. Interestingly, and probably not coincidentally, that was also roughly the start of the period of great brain expansion in our line.

One of the most interesting aspects of the human prehistoric cultural record, preserved primarily in stone tools until some 50,000 years ago, is how slow the rate of early human cultural evolution appears to have been (figure 4-1). In the Paleolithic period, or Old Stone Age (2.5 million to 12,000 years ago), the first cultural tool kit—flaked stones and sharp-edged flakes struck from them (Oldowan)—lasted 700,000 to 800,000 years. The second (Acheulean), which added more complex stone tools such as hand axes, cleavers, and picks, emerged with *Homo ergaster* and persisted for 1.5 million years. Culture as preserved in stone showed real refinement beginning only about 700,000 to 600,000 years ago, when Africans began to produce hand axes that were finely trimmed and often remarkably symmetrical when viewed from the side or the end. They also began to produce a wider range of flake tools, many of which were secondarily modified into scrapers, knives, and the like.

Not all of early human culture would have been preserved in stone, of course. For instance, chimpanzees use wooden tools and weapons, and it's thus very plausible that our earliest ancestors did too, even though their wooden tools and weapons weren't preserved. The earliest surviving wooden weapons are spears from Germany made some 400,000 years ago (whether the makers were *H. erectus* or *H. neanderthalensis* remains controversial). Sometime after our ancestors first started shaping rocks, they also began to make use of fire, although complete control of fire seems to have followed the invention of stone tools by a million years or more. The earliest compelling evidence for deliberate human use of fire dates from some 800,000 years ago.

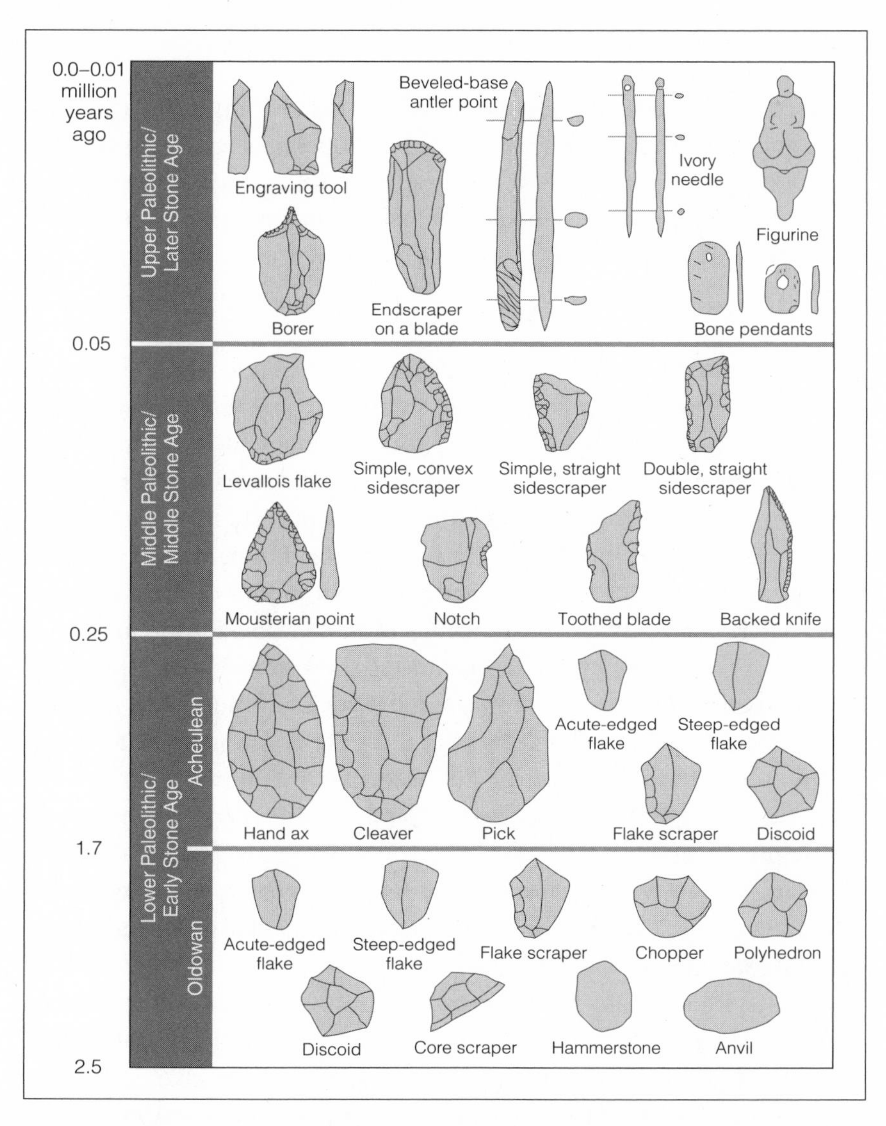

FIGURE 4-1. Cultural evolution as reflected in changes in stone tools. Note how slow the rate of early human cultural evolution appears to be; the first cultural tool kit, flaked pebbles and sharp-edged flakes struck from them, lasted 700,000 to 800,000 years. The second, which added more complex stone tools such as hand axes, cleavers, and picks, persisted for 1.5 million years. Courtesy of Richard Klein.

Some 250,000 years ago, the Acheulean culture (by that time produced by early *Homo sapiens*) entered a transition into what is called the Middle Stone Age in Africa and the Middle Paleolithic period in Europe. This technology displayed an emphasis on predetermining the size and shape of a sharp-edged flake *before* it was struck from the core. In some cases the flakes were then very skillfully retouched to make them even more useful. No one who succeeded the Middle Stone Age/Middle Paleolithic people has exceeded them in stone-flaking skill.

The Great Leap Forward

About 50,000 years ago, technology exploded in the African Later Stone Age and European Upper Paleolithic period, and new dimensions of human activity began to appear in the archaeological record. This transformation was one of the most dramatic and abrupt in the prehistoric record and may have been the most important in our history. Called the "cultural revolution" and popularized by Jared Diamond as the "great leap forward," it involved the development of new, sophisticated technologies with more diverse and standardized stone tools, and the first appearance of bone, ivory, and shell objects that were deliberately shaped into projectile points, needles, awls, and so forth. Added to that was the appearance of a diversity of very fine stone tools, a flowering of cave painting, sculpture, and body ornamentation, and signs of ritual (including burials), accompanying a spurt of population growth and, some scientists believe, a great advance in language skills. Perhaps more important, *Homo sapiens* after the leap rapidly replaced other hominin types in Eurasia, with little or no evidence of interbreeding or cultural exchange with them.

Richard Klein thinks that a rapid evolution of our ancestors' brains caused the leap forward. He believes the "wetware" of the brain (its neural structure, analogous to the hardware in a computer) changed. Some others think it was the same old brain developing new cultural "software." No one knows the answer for certain, but we tend to think Klein is right. A principal argument for the software view is that there have been periods of even greater technological change, such as from the horse-drawn chariots of Plato's day 2,500 years ago to today's jet aircraft and space stations, without evidence of significant corresponding changes in human brain structure.

But this doesn't preclude the possibility that such a change *did* take place some 90,000 to 50,000 years ago. As Klein points out, without such a wetware change roughly coincident with the period of the great leap, one must picture an entire world in which no one produced durable art, and then a sudden cultural change that produced a flowering of such art. Obviously a lot of wetware changes have occurred in the more than 2 million years of rapid human brain expansion from *Homo habilis* to ourselves, as we can tell from the size and shape of skull cavities. It seems quite reasonable that the great leap can be traced to a reorganization of the brain roughly along present-day lines that opened the door not just to art but to a cultural evolutionary sequence that started with replacing all other hominins and ended up producing books, computers, and (through sanitation and medical advances) population explosions. In short, the evidence doesn't demonstrate that human brain wetware underwent a rapid selectional shift before the great leap, but it certainly makes it seem likely. We also can't demonstrate that your brain isn't somehow more advanced structurally than Plato's—but we wouldn't advise you to make that claim in public.

Regardless of the cause, after the great leap forward cultures changed ever more rapidly, with our ancestors decorating cave walls and rock faces with their paintings and carving statuettes, improving their means of subsistence, and becoming terribly efficient hunters, to the point that they could exhaust local food supplies and alter continental faunas (and floras). They spread themselves and their more sophisticated technologies over most of the globe—finally occupying the Western Hemisphere.

Language and the Leap

Intimately connected with the questions surrounding the great leap forward are those of the development of language with complex syntax (that is, meaning embodied in relationships between the words in a sentence). Possession of such a language is the foremost characteristic distinguishing today's *Homo sapiens* from all other animals. It is also at the root of that other preeminent feature of humanity we've just been discussing: the possession and manipulation of vast amounts of nongenetic information—that is, a rich culture.

The physical infrastructure necessary for language, including a big, complex brain and structures in the mouth and throat (tongue, larynx, vocal cords), had to evolve genetically before modern verbal communication and the cultural evolution of language could occur. Especially important must have been the development of the neuromuscular machinery that enabled the tongue to perform the impressive movements required for speech. A disproportionately large part of the brain's motor control area is devoted to control of lips and tongue.

Beyond these physical prerequisites, there's little need for genetic evolution to explain the emergence of language. After all, many animals, from shrimps and cicadas to mockingbirds, monkeys, and chimps, communicate by making noises that carry information. It is not surprising then that, as brains grew in a social animal, selection favored improved neuromuscular coordination in the tongue. Indeed, we suspect that brain size and structure coevolved with language facility; the more culture human beings created, the greater the premium for being able to transmit and take advantage of it.

There may be an even closer connection between language and culture. Almost 100 years ago, Edward Sapir and Benjamin Lee Whorf proposed that language actually shapes a person's (and thus a culture's) worldview. An extreme version of this idea, that language *creates* the worldview, has been shown to be incorrect, but there is good evidence that in many cases language can influence how events or objects are perceived.

Linguists must depend on speculation about the evolutionary transitions from grunts, purrs, and barks that indicate mood or the presence of danger to the development of word symbols—purely arbitrary sounds to represent things in the real world or to express relations among things, or even among other words and concepts. The system of vervet monkey alarm calls shows that arbitrary symbolic verbal expression is not limited to people; to indicate the threat of a hawk, the monkey makes one noise, easily distinguished from another noise it makes to indicate a snake. While both noises are distinct, neither has any obvious connection with the predator (e.g., the hawk sound doesn't resemble a hawk's scream). The monkeys have developed a culture (and, by some definitions, elements of a primitive language) that associates these noises with information.

One can imagine that, for human beings, originally sounds and gestures (especially pointing) made the connection between speaker and listener. Once arbitrary sounds for things and actions came into use—when language was entering what linguist Guy Deutscher in his excellent book *The Unfolding of Language* calls the "me Tarzan" stage—the path to the complex languages now spoken by all human groups was open and, as he shows, their continuing evolution can now be explained.

Culture and the Brain's Evolution

Evolution of large and complex human brains made possible language with syntax and the storage of vast amounts of information, indeed made possible both "humanity" and "culture" as people commonly understand them. Modern human beings' big brains (compared with those of the great apes and our hominin ancestors) are our most important physical characteristic. Our brains evolved their impressive dimensions relatively rapidly over the past 2.5 million years or so, well after development of our other physical hallmark, upright posture, and over about the same period that our ancestors left stone clues to their culture.

Human brains, like those of every other animal, are made up of networks of nerve cells (neurons), each of which can potentially communicate with thousands of other neurons via chemically connecting interfaces known as synapses. Networks of neurons serve as control centers for the functioning and behavior of animals, from maintaining the chemical balance of the blood and controlling mobility to thinking. Electrochemical impulses are transmitted along neurons and across synapses. Synapses are actually very narrow spaces that can be bridged by perhaps 200 different chemicals called neurotransmitters, which neurons can secrete into the gap to pass on or inhibit the impulses.

The networks of nerve cells, synapses, and neurotransmitters bring information into the brain, process it, store it, and, in cooperation with hormones (chemical messengers that operate through the bloodstream, some originating in the brain), instruct the body how to behave. Human brains are the most complex and adaptable of those of any animals, but they mainly differ from those of, say, chimps, in their size and their numbers of neurons and synapses.

Big brains in general allow animals to adapt to environmental variability—they present their owners with an array of options in any circumstance. If a bear eating berries spots a fawn hiding in the grass, it can alter its diet instantly. If the fawn escapes and the berries run out, the bear can go down to the river and try its luck fishing. Tiny-brained insects probably can't make such choices, although they sometimes turn out to have more mental flexibility than is commonly thought.

We think the best current explanation of why human beings have evolved big brains is related to the growing ability of a highly social primate to discern, imagine, and project the thoughts residing in the brains of other members of their group. In other words, people have come to have a highly developed "theory of mind." Why did humanity's capacity for a theory of mind evolve? What was its selective advantage? The short answer is that it was needed to deal with the complexities of increasingly smart animals living in tight social groups. In the human evolutionary line, brains evolved not only to adjust, plot, plan, and maneuver their owners' lives but also to adjust their behaviors to those of other plotting, planning, maneuvering, and signaling individuals with whom they could empathize and cooperate (or compete) and whose minds they would continually try to read. And in evolving empathy, our ancestors' brains gave us the capacity to develop ethics.

A theory of mind appears to be present in a less developed state in our fellow apes. Some animals are known to have at least a primitive sense of self, which is the first step toward having a sense of others, since inferring the thoughts of others is clearly an extension of one's own mental experience. For instance, many visually oriented animals, when presented with a mirror, interpret their image as another individual and may attack it. Chimps, bonobos, and orangutans (but not monkeys and probably not gorillas), as well as elephants, however, recognize themselves. The apes will use a mirror to examine the backs of their heads or to direct their hands to examine a paint mark (placed there under anesthesia) when they cannot see it directly. In a different series of experiments, chimps were shown a human actor struggling with problems such as trying to gain access to bananas out of reach or trying to play an unplugged phonograph. When presented with paired photos, one of which showed the solution to the problem (using a stick to get the bananas, plugging in the phonograph), the chimps consistently selected the correct answer. That indicated that they recognized the people as actors with thinking ability considering a dilemma. This, and a growing body of other evidence, argues that apes have some grasp of the intentions of others, or at least behave as if they do. The issue turns out to be very complex.[4]

Something is now known of a possible physical basis for a theory of mind. *Homo sapiens* and other primates have elements in our brains (originally discovered in monkeys) called "mirror neurons" that respond both to our own actions and to those of another individual acting similarly. When we reach for food or when we see another person reaching for food, exactly the same neurons in our brain discharge. Indeed, some scientists have suggested that the mirror neuron system, which is involved in gesture imitation, may have been the evolutionary development tens of millions of years ago that eventually led to language. Hand and mouth gestures are related, and understanding the abstract gestures of their companions (e.g., shrugs) may have been the precursor to understanding abstract noises made by others. Indeed, some parents communicate with their children with symbolic gestures before the children have much grasp of spoken language. But so far no scientist has been able to connect mirror neurons directly with a theory of mind or with empathy—for which the theory clearly is a prerequisite.

Although scientists investigating phenomena such as mirror neurons are just beginning to understand the mechanisms by which brains store and process non-genetic information, what they and their predecessors have learned already has led to many interesting conclusions as well as much ingenious speculation. Several general points about the human brain with particular evolutionary relevance seem quite well established:

1. The brain is an organ that, like other organs, has evolved a structure that serves its various functions well, from regulating our digestion to solving problems and producing a sense of consciousness. The ancient duality of mind and body was a figment of the imagination of human brains. The brain is also very energy-expensive; that of an average adult constitutes about 2 percent of the total body weight but uses about 20 percent of the body's energy. Thinking doesn't come cheap.
2. The brain can compensate for partial damage and, often, keep thinking going. If an area is damaged where a specific function is localized, it is sometimes possible for that function to be taken over, at least in part, by another part of the brain.
3. The brain has many "programs"—connected sets of neurons—built in over hundreds of millions of years by natural selection and inputs from the social environment. For example, the frontal cortex of the brain regulates all sorts of behaviors; if it is severely damaged, inappropriate actions (using profanity in church, making unwanted sexual advances) dominate behavior. Interestingly, the frontal cortex is not entirely "wired up" until the mid-twenties—which partially explains the lack of inhibition (sometimes including criminal behavior) of adolescents and young adults.
4. Not only can the brain regulate the release of many hormones, but

hormones can also regulate brain function. They contribute to the feeling of emotions, subjective states of awareness usually producing a determination of the desirability (goodness or badness from the individual's perspective) of something. Emotions in turn play a large role in the way brain programs function. That has recently been popularized under the rubric "emotional intelligence"—the capacity to monitor your feelings and those of others and use that information to guide your thinking, decision making, and actions. In other words, this capacity allows you to make practical use of your theory of mind. Of course, emotions also make us human and give meaning to life. What kind of a social animal would we be if we couldn't feel joy, sorrow, compassion, anger, disgust, or love?

5. The brain remembers by undergoing chemical changes in neurons, by strengthening synapses, and by forming new synapses, and it almost certainly does so by constructing new patterns of neural networks.
6. More recently evolved programs in the brain enable us to deal with problems of relationships and causation that are difficult or impossible for other animals to solve, although there is more continuity between our capabilities and those of non-human primates than is often recognized. If intelligence is defined as the ability to solve certain kinds of problems, in some areas there is little sign of an "intelligence gap" between ourselves and other animals, even though in other mental areas the gap is immense.

   Fishes, pigeons, dogs, apes, and people alike, for instance, are able to solve a problem in which they are presented with a tray containing two food wells, one having a circular cover and the other a triangular one. When the food reward is always under the triangular cover at each trial, subjects learn to respond to that, regardless of the position of the well. In more complex tests, two food wells are covered by identical covers, and a third, under which the food is always placed, is concealed under a different cover. This problem can't be solved by pigeons, rats, or dogs, or even by some young children. Another level of complexity is added if the color of the tray signals whether the food reward is under the odd object or one of the matched objects. Non-primates can't handle this one, but some monkeys and apes do it easily and can solve even more complex oddity problems. These tests are of a sort originally designed to measure human mental abilities, and many human beings can't solve them.

Such results suggest there is not a difference in kind between other animals and people in brain programs that generate problem-solving abilities, although there is a difference in the complexity of the problems that can be solved. The gap in accomplishment traces more to our possession of neural systems that produce two characteristics unique among living animals: consciousness and a language with syntax. They allow us to produce a vast, complex culture and to generalize from our small intellectual triumphs to a much greater extent than any other living animal. Because of these characteristics we can do higher mathematics, organize political hierarchies, and design and build skyscrapers.

7. Natural selection has designed the brain to create programs that bias certain perceptions and behaviors, and to continuously develop beliefs on everything from military strategy to creation myths—and attempt to maintain them even in the face of inputs that reason suggests should cause them to be revised.
8. The genetic code cannot build specific instructions into the brain's structure for dealing with every conceivable behavioral situation, or even large numbers of them. Instead the brain's structure makes us smart and flexible, able to design strategies for dealing with problems when needed. Indeed, the long period of postnatal helplessness is associated with growth of the brain and its programming by the environment to do everything from learning to see correctly to learning to talk.

A closer look at memory and consciousness in particular can give us some background on how the brain has evolved to generate and deal with masses of culture-related information. How stronger or new synapses give us the impression of remembering—essential for cultural transmission (if you can't remember it, you can't pass it on)—remains mysterious. But some features of memory are reasonably well known, and understanding them can be of great practical importance in our lives. First, as you have doubtless noted, it usually takes some effort to encode things into long-term memory: repetition, association with a dramatic or painful event, and so on. This makes clear evolutionary sense; there is a limit to the capacity of our brain's neurons to form permanent memories. For this reason, the brain tends to store gist and to discard details. Furthermore, both gist and stored details tend to fade with time unless they are reinforced. Discarding of details may be well illustrated by an experiment you can do: Take a piece of paper and sketch the positions of the instruments on your car's dashboard and then compare the sketch with the real thing. Don't worry if you messed it up; what would be the point of remembering something you need only when it is in front of you?

Memories not only fade; they can also get confused so that what we think is an accurate recollection of the past may actually be dead wrong—you didn't see Sam at the New Year's Eve party, even though you're sure you did. He was in China, and it was George whom you saw and are now confusing with Sam. Suggestibility is another way that memory can be tripped up. Other people (or the media) can convince us that we remember things we don't. This is not surprising, since, just as our visual system routinely does, our brains sometimes fill in details of events that were not originally stored in order to make a composite whole. Scientists studying memory phenomena have concluded that fading memories and suggestibility make it highly desirable in a courtroom to supplement eyewitness testimony with other evidence.

Stereotyping is another important way in which our brains operate efficiently. Whenever possible, people file incoming information into already established categories. For instance, all societies divide the continuous spectrum of color into categories, but not all societies use the same categories. Experiments have shown it is easier to remember a color if it comes from the center of a culturally established category (say, fire-engine red) than if it is near a boundary (yellowish green). In another example, people connect the series of dots reaching their brains from the firing of cells in their retinas as they look at the corner of a room into the straight line their brain "knows" is a straight line. When the two of us, as former pilots and frequent air travelers, see airplanes, they are filed in our memories as such. Not so the New Guineans whom we saw in 1966 intensely examining the open rear cargo ramp of an Australian air force transport to see "if that was where its reproductive organs were." They put the plane in their mental category of "bird."

The built-in stereotyping and the urge to categorize make the mental life of our species simpler than it might otherwise be, but it also can have such negative effects. One of the problems we face as we try to understand the world is to avoid being confused by the "fallacy of misplaced concreteness": assuming that sometimes useful abstractions—such as genes, populations, species, cultures, mountains, religions, or races—are actual discrete entities. Thus genes were once stereotyped as being equivalent to beads on a string—a metaphor that concealed the diversity and structural complexity of the DNA arrays that carry genetic information. And today the vast diversity of people subsumed under "racial" terms such as "African American" are reified into units that some scientists claim can be useful in diagnosing diseases. In fact, examining the distribution of relevant genes and environmental conditions (which will vary in many cases along with socially constructed "races") in the world's population would be a more fruitful way of accounting for geographic variation in disease incidence.

The Mystery of Consciousness

Each of us (and, we suspect, you), when awake, is aware of ourself. We each have a perpetual narrative running through our head, interrelating this day's sensory inputs, thoughts, and events to ourself and to those in the past, and projecting possible story lines for the future. In other words, we are *conscious*. Anne's story differs from Paul's, and yet as far as we can tell, our two stories have many similarities, especially in the area of values. We share our views with each other and our daughter and thus have a family culture. We share our views with friends and colleagues as well and thus participate in ever more broadly defined subsets of American culture and, ultimately, human culture. Another way to view cultural evolution is as the changing pool of stories being narrated in the brains of human populations, just as their changing pool of genes constitutes genetic evolution.

The most mystifying thing about our brains is how the many trillions of connections among neurons bathed in hormones generate those narratives that we think of as "consciousness." British psychologist Nicholas Humphrey has developed an intriguing speculation about the evolution of consciousness. He, like many neurobiologists, ties consciousness to bodily sensations—the presence of mental representations rich with "feeling," of "something happening here and now to me." Sensations contrast with perceptions, which are bits of news without emotional content about "what's happening out there." The distinction is a fine one. The chemical exuded by a rose, for example, is perceived as having a sweet scent, but it also gives a person the sensation of being sweetly stimulated and may elicit further sensations of, say, being with a lover. Humphrey's hypothesis is that sensations occur at the boundary between an organism and its environment and in human beings are generally registered as sight, touch, sound, smell, and taste. When direct sensations are absent, mental representations are accompanied by reminders of sensation; for instance, some thoughts are "heard" as quiet voices within the head and would fade without that component of sensation.

To explain consciousness, Humphrey constructed a hypothetical physiological mechanism based on sensory feedback loops—nervous responses that are returned to the sites of nervous stimulation—a view supported by a variety of observations of nervous systems. But many of his claims can be tested only subjectively, and subjective testing may well not be a reliable guide to what actually is occurring. Do you hear some of your thoughts? We *think* we hear many of ours. We also seem to *see* many of them, visualizing them as in our dreams, although language is certainly lurking around the edges of even our most vivid pictorial representations. Do you find it hard to retain an imagined visual image of, say, a forest fire while looking at a cloudy sky? We do; the lack of sensation of heat and color makes it hard to retain consciousness of the fire.

Humphrey's short summary of consciousness is not the famous "Cogito, ergo sum" (I think, therefore I am) of René Descartes but rather "I feel, therefore I am." He reinforces that view with a quip from the novelist Milan Kundera: "'I think, therefore I am' is the statement of an intellectual who underrates toothaches." In Humphrey's view, there has been a gradual evolution of consciousness through a shortening of the sensory feedback loops. Originally, among our very distant ancestors—say, a simple worm—the input of sensation from the organism-environment interface triggered a nervous response to the stimulated part of the body surface. Then, gradually through evolutionary time, that circuit shortened, and the target of the responses moved inward to a surrogate area on the surface of a slowly evolving nerve center, which eventually became a brain. Nerve impulses arriving in the brain created conscious experiences. It seems a plausible evolutionary story, but it's at most a starting point for thinking about how consciousness evolved.

It seems reasonable to believe that conscious awareness of some sort must also be pretty widespread, at least in mammals and birds. Ravens, crows, and their relatives, for instance, show considerable ability to solve problems they face—such as finding a way to stopper a drain and restore a puddle in which to bathe. One imagines they can consciously picture the solution to the dilemma and then fix it and enjoy the result. And chimpanzees, with their sense of self and ability to learn from one another, surely have a form of consciousness—perhaps a narrative at the level of "see Fido, see Fido run, see Fido run after the ball." But we suspect that today only human beings have what Paul has described as "intense consciousness": "we analyze our physical and social surroundings, remember past events, and 'talk' to ourselves about those analyses and their meaning for our future. We have a continuous sense of 'self'—of a little individual sitting between our ears—and perhaps equally important, a sense of the threat of death, of the potential for that individual—our self—to cease to exist."[5]

That intense consciousness, which Humphrey wants to explain, is what scientists call an emergent property, and emergence is a phenomenon that is just beginning to be investigated. It is the appearance of unanticipated patterns, such as traffic jams with no apparent cause, when large numbers of relatively small units—sand grains, automobiles, or nerve cells—interact in a regular manner (according to the rules of the road, for example) under the influence of a flow of energy (often fluctuating regularly). An emergent property can't be predicted from the properties of the individual unit. Neurons can't think, and (as far as we know) they aren't conscious. But in a brain, billions of them constitute a complex self-organized system (one that isn't directed by an outside entity), and vast mobs of them working together can think and somehow generate what we call consciousness.

Arguably the most basic ethical issue considered by human beings relates to the emergent state of the human brain that we call consciousness (as opposed to the ideas it develops). It is the classic issue of free will, about which ideas have been evolving culturally for millennia and which can hardly be said to be fully understood by anyone. The question is: if the brain is an organic "machine," subject to the laws of nature and constructed by an interaction between genetic material and complex physical, chemical, and social environments, aren't all human behaviors simply programmed responses to those factors? Indeed, weren't they all preordained from the moment of the origin of the universe? Should anyone, then, ever be praised or blamed for her behavior?

We would not pretend to have a pat answer to this ancient question. Our view is relatively simple. Our minds are a function of our brains, which have been formed by an interaction of inherited genetic "instructions" with environmental information both in the womb and throughout postnatal life. That is an ongoing process, which in a social context means a constant dialogue with others and continual observation of events that lead to a constant updating of our views of the past, probabilities of future events, our opinions of others, and our ethical views. We learn the evolving norms of our social groups and adjust our views of what we will and (perhaps more important) what we will *not* do in various situations.

The world interacting with our brains is not totally determined (e.g., one cannot predict when any given radioactive atom will decay), and it seems reasonable to assume that our social behavior is not all preordained either. The question of how much of our activity is in some sense automatically programmed (evidence from new imaging techniques indicates that our brains can decide on many actions before we become aware of the decisions), or "constrained," and how much is "free" remains a philosophical one. But in many if not most cases we apparently can override automatic decisions, sometimes to act "responsibly." Neuroscientist Michael Gazzaniga said that "the brain is determined, but the person is free."[6] Intriguing, but, like many statements about free will, not totally helpful. Our brains are a product of genetic evolution, key to the evolution of our complex culture, and the organ responsible for our global dominance. But scientists, and perhaps philosophers, have a long way to go even to approach a full understanding of them.

Gene-Culture Coevolution

Do the great leap and the explosion of human culture mean that cultural evolution overrides genetic evolution in human beings? The issue is not dissimilar to that of whether the cause of an individual behavior is "genetic" or "environmental" (with much of the critical environment being cultural). The latter common question is a misleading one. Genes cannot function without an environment. Equally, there cannot be an environment without genes. Environments exist only relative to organisms, being all the external factors that interact with those organisms. Genes and environments work together to shape the phenotype, the actual observed structural and behavioral characteristics of an individual, and that individual in turn affects the surrounding environment. Genetic and non-genetic (environmental) information are as inseparable in producing cultures and cultural evolution as are the length and width of a rectangle in producing its area.

Inseparable though they are, a dramatic difference between genetic and cultural evolution is the amount of information each has to operate on. Human genetic evolution is a story of changes in a mere 20,000 to 25,000 genes, involving roughly 3 billion nucleotide base pairs (the "rungs" of the DNA "ladder" discussed in chapter 1), which are involved in encoding the structure of proteins. If we equate a word to a trio of base pairs, that would be about as much information as contained in some 5,000 books the size of this one. Stanford University's library holds 8 million or so volumes (most in English), or some 1,600 times as much information as in a human genome. If we consider the contents of books in other libraries the world over, warehoused documents, the data stored on computers, the flow of radio programs, and the experience and knowledge people share with one another, to say nothing of the information extractable from artifacts such as the cathedral of Notre Dame, the ruins of Persepolis, and potsherds in the soil of New Mexico, you can see that a university library contains only a fraction of the cultural information available.

The way cultural information dwarfs genetic information is inherent in the contrast between some 25,000 genes, each averaging 120,000 base pairs, and the information storage capacity of the human brain. The brain has hundreds of trillions of connections between nerve cells, so even if we very conservatively equate the information storage capacity of a connection to that of a trio of base pairs, the 1,000,000,000 base pair trios are up against hundreds of thousands of times as many brain connections, and each gene would need to program more than a billion synapses.

The sickle-cell situation described in chapter 1 is an example of how genetic and cultural evolution can interact—after all, the slave trade and malaria control are both cultural phenomena, and both have influenced the changing prevalence of the sickle-cell gene. Furthermore, the form of agriculture practiced by a society influences the risks of getting malaria and the selection imposed by it. For instance, flooded rice cultivation provides breeding places for malarial mosquitoes and should help determine where the sickle-cell gene is in high frequency. If a culture dispensed with domestic animals or eliminated wild animals that provide preferred blood sources for mosquitoes, that too should have an effect. So should cultural evolution that led people to build houses with window screens, learn the pertinent habits of the mosquitoes in order to improve control techniques, or provide chemical prophylaxis against the malarial parasites. All are factors in determining people's exposure to malaria and thus, over generations, could help determine the proportion of sickling and non-sickling hemoglobin genes they possess.

Many of human beings' most characteristic features, ones much more obvious than the shape of blood cells and necessary for survival as an organism, are undoubtedly programmed into our genes by natural selection and are not often significantly modified by the environment in normal individuals under normal circumstances. Failure of an organism to produce major phenotypic characteristics (e.g., head, legs, eyes, antennae, leaves, flowers) decreases the organism's reproductive potential in almost any environment. Selection pressure is strong against any genes that might produce wingless butterflies, plants without chlorophyll, or human beings who are brainless or whose blood will not clot. There is, however, much less reason to believe that many of the most interesting aspects of our natures—such as our differing sexual orientations, religions, languages, aggressiveness, intelligences, senses of humor, and so on—are primarily products of natural selection.

An entire field, called "evolutionary psychology," has sprung up based on the misconception that somehow genes, ostensibly through a long process of natural selection, are determining our everyday behavior and our personalities. Typical of popular presentations of this view is a statement by Nicholas Wade of the *New York Times*: "When . . . [the human genome] is fully translated, it will prove the ultimate thriller—the indisputable guide to the graces and horrors of human nature, the creations and cruelties of the human mind, the unbearable light and darkness of being."[7]

Many people have been persuaded of this view by reports that studies of identical twins raised separately "prove" that genes are responsible for at least 50 percent of the variability in our everyday behaviors; that is, that the heritability is at least 50 percent. This language of heritability, with expressions such as "genes are responsible for 50 percent of" or "genes contribute 50 percent of" a behavior, gives the erroneous impression that genetic and environmental contributions to human behaviors are actually separable. They are not.

The concept of heritability was originally introduced in the 1930s as an index of amenability to selective breeding in the context of agriculture, where environments can be controlled by the breeder. This index, now often termed "narrow-sense heritability," is the fraction of all variation in a trait that can be ascribed only to genes that act independently of one another and whose joint effect is the sum of their individual effects. One easy-to-understand way of measuring heritability is simply to do a one-generation selection experiment—breed from individuals at an extreme of a distribution (say, the 10 percent of a herd of hogs that are heaviest)—and then see how much the average weight of hogs in the next generation, raised in the same environment, increases. If it doesn't increase, the heritability is zero. A special statistical model is invoked to estimate, under these controlled environmental conditions, how effective selective breeding will be, and the "heritability" statistic is simply an index of that effectiveness.

In the 1960s the term "heritability" was adopted by some students of human behaviors whose interest focused on whether variation in these behaviors was primarily attributable to genetic differences or to environmental differences. Here, however, heritability was interpreted as an index of the extent of genetic determination. Because of the impossibility of controlling the environments of the subjects in human experimentation or observation, however, this new heritability *included* environmental effects; it did so in the form of variations due to interactions between genes and environments (e.g., the same gene acting differently according to the environment to which the individual is exposed). The new heritability's computation assumed no relationship between genetic transmission and environment, an assumption that is frequently false. For example, the IQ scores of parents may signify characteristics of those parents that could affect parts of the environment in which their children grow up (e.g., the number of books in the house or the content of dinner-table conversation). Any genetic influences on IQ, if they exist, would not be independent of that parental environment.

This lack of independence is called gene-environment correlation. Agricultural experiments to estimate narrow-sense heritability are designed to eliminate this correlation. But with human behaviors such designs are impossible—we can't ethically choose who gets to mate with whom and then raise their offspring in specified uniform conditions—and thus gene-environment correlations can't be separated out. Thus for human behaviors a less precise concept, "broad-sense" heritability, is used, and inclusion of gene-environment interaction and correlation effects into this broad-sense heritability inflates the fraction of variation that is interpreted as being determined by genes.

Many of the high estimates for heritability, and the resulting interpretation of strong genetic determination, of human behavioral traits have been derived from comparisons of twins that are either identical (from a single egg and thus identical in their hereditary endowment) or fraternal (from two eggs, sharing on average only 50 percent of their genes). Many assumptions can inflate twin-based estimates of broad-sense heritability. One is the clearly fallacious "equal environments" assumption, according to which variation in environments to which identical twin pairs are exposed should be the same as that to which fraternal pairs are exposed.

It might be thought that some of the problems with twin studies could be overcome if the twins under study were reared apart, that is, in different families. In a perfect experiment of this kind, all of the differences between each of a pair of identical twins reared apart should be environmental, and high levels of similarity of the pair should be due to their identical genes. It turns out not to be so simple. First, separated twin pairs are rare, and the reasons for the separation are not usually known. Second, the separation is frequently not carried out at birth, so some shared environmental effects would mistakenly be interpreted as genetic. Third, twins have often been placed in separate homes that are similar in respects that may be important for the traits under study—for example, in homes of relatives of their parents. The environments are therefore not a random sample of all possible environments. All of these effects add to that component of variation that is interpreted as genetic, with the result that estimates of genetic heritability based on identical twins raised separately are biased upward. These and other considerations mean that the logic of using heritability of some trait in a population to infer or predict something about a member of that population such as intelligence or sexual behavior is analogous to using population data to prescribe a medicine to treat an individual the doctor has never diagnosed.

Recent studies of intelligence in samples of twins of differing socioeconomic status strongly reinforce these restrictions on the generalization of heritability beyond its appropriate setting. For example, the estimated heritability of IQ in individuals from advantaged backgrounds is significantly higher than in those from disadvantaged backgrounds. That is because the better environments allow more variance in IQ to be expressed—potential geniuses have trouble developing into Einsteins in school-less slums. To see how this works, consider that the heritability of height in a normal human population is greater than that in a starved one, where everyone's growth is stunted and the variance in height thereby reduced.

The critical role of environment in determining behavioral characteristics is underlined by studies of people with known genetic anomalies. Individuals with Down syndrome, caused by an entire additional chromosome 21 (trisomy), develop as severely mentally handicapped if given no special treatment. But it turns out that the degree of handicap such people possess is extremely labile to the environment of rearing. Partition of variation in a trait such as IQ into genetic and non-genetic influences is therefore precluded.

It is especially inappropriate to talk about genetic "contributions" to such complex traits. It is no more appropriate to say, for example, that characteristic A is more influenced by nature than nurture than it is to say, as we indicated before, that the area of a rectangle is more influenced by its length than by its width. You should note, however, that a *change* in area can be assigned to a change or changes in the length of one or both sides. To put the trisomy case in these terms, one might think that the genetic anomaly of trisomy halves the length of the genetic "side" of an individual's capability rectangle. That halves the capability "area." Then the length of the environmental side could be doubled, restoring the original capability area. In neither case can we say that one side or the other contributes more to the area (or to capability), but we can say which contributes more to a *change* in area or capability.

None of this should be taken to mean that genes do not affect behavior; in fact in a sense they influence all behavior by, at a minimum, laying out the basic configuration of all individuals. For example, were it not for the genes that, in the course of development, interact with pre- and postnatal environments to produce the brain, the human behaviors that interest us would not occur at all. On the other hand, there is nothing to indicate (as some behavioral psychologists are wont to do) that genes that were favored by selection while our ancestors were hunter-gatherers are significant controllers of most individual behavioral characteristics and acts today.

For many behavioral traits, especially serious psychiatric disorders, the roles of genes in certain environments have been elucidated. Consider research by psychologist Avshalom Caspi and his colleagues on the effects of different forms of a gene involved in the transport of serotonin, a compound involved in transmitting signals along certain nerve pathways. Which form of the gene an individual possesses influences whether stressful events will produce depression. Having the "wrong" gene makes a difference only if an individual is exposed to a stressful environment—a beautiful example of gene-environment interaction.

Some specific cases illuminate the inability of genes to control behavior. One is that of the original Siamese twins, Chang and Eng, who were joined for life by a narrow band of tissue connecting their chests. Despite their identical genomes, they had very different personalities—Chang, for instance, was quick tempered and dominated Eng, who was submissive. Chang also was a drunk; Eng wasn't. Equally fascinating is the story of the Dionne quintuplets, five genetically identical little girls who were essentially raised in a laboratory in the 1930s under the supervision of a psychologist. When the girls were five years old, the psychologist wrote a book expressing his astonishment at how *different* the little girls were—something confirmed by their subsequent very different life trajectories. One had epilepsy, the others did not; some died young and others old; some married, others remained single, and so on.

Evidence for the lack of genetic determination of many common human behaviors is abundant. Perhaps most impressive are thousands of cultural cross-fostering "experiments" in which children from one culture were raised from an early age by adoptive parents from another. Invariably, the children matured with the language and attitudes of the adoptive culture. Similarly impressive is the ease with which culture overrides the only "tendency" we can be sure is contained in everyone's DNA. That tendency is to see to it that your kind of genes are better represented in the next generation, either by putting them there yourself or by helping your relatives (who tend to have the same genes) do it. Differential reproduction of genetically different individuals (not explicable by chance) *is* natural selection, the creative force in evolution. We wouldn't be here if our ancestors hadn't been effective reproducers of their genes: that is, if they hadn't had high "fitness."

But culture has led human beings to limit their reproduction as far back in history as we can trace, all the way to ancient Egyptians using crocodile dung suppositories (which might well have been very effective). Indeed, although some evolutionary psychologists like to imagine that rapists are programmed to assault women in order to reproduce themselves—that is, increase their fitness—there is little evidence of this. Most men culturally suppress that "genetic" urge, if there is one.

Perhaps the most powerful evidence that genes don't determine most common behavior has been implicit in our discussion already—the problem we call "gene shortage." Our roughly 25,000 genes can't possibly code all of our separate everyday behaviors into the human genome. After all, we have fewer than twice the number required to make a fruit fly, and just a few more than those that lay out the ground plan of a simple roundworm. Even if the human brain had evolved not for flexibility but to be programmed for stereotypic behavior, our genes couldn't store enough information to accomplish it.

Genes are not little beads with instructions like "grow up gay" engraved on them. They are instructions that, as we saw in chapter 1, can be encoded in different ways and that require a very complex mechanism to direct the assembly of a sequence of amino acid residues into a protein. It is virtually miraculous that these proteins, interacting with one another, operating in different physical, physiological, and social environments, and helping to control the production of other proteins, are able to produce a functional human body: not just muscle and blood and skin and bones but also an extraordinarily complex, fluid immune system that protects our individuality and the basic scaffolding for a brain with a trillion or so nerve cells communicating with and connected to one another by hundreds of trillions of intricate synapses. On average, each gene must influence very many characteristics. Obviously, there are enough genes, interacting with one another and with diverse environments at all scales, to guide the development of a brain that has the possibility of generating all observed human behaviors.

But this has confused some observers into thinking that, because, as we pointed out earlier, one gene often affects many functions, there is no gene shortage. That fact is actually the *basis* of gene shortage. It means that natural selection operating to encode one behavior would, in many if not most cases, change other things too—just as selection increasing, say, the speed of contraction of muscle fibers in the limbs would quite possibly change the speed of movement of the tongue. Because of the small number of genes in the human genome and the recent discovery that the code determining the structure of a protein may be dispersed over many areas of the genome, as well as the ubiquity of gene-gene interaction (genes controlling the expression of other genes) and of interactions among proteins and between proteins and environments, natural selection on one phenotypic characteristic will ordinarily entrain a multiplicity of changes in others. It must operate on a genome enormously "amplified" in development by multiple uses of the proteins produced by single genes, by alternative ways the proteins are assembled, by small RNA molecules that often control the expression of multiple genes, by variation in the number of copies of individual genes, and by epigenetic phenomena—changes in a gene's functioning, and thus in the phenotype—that persist over cell generations but are not caused by a change in the DNA sequence. The latter inheritance of chemical mechanisms is sometimes induced by environmental factors that regulate the functioning of genes and may have differing effects even on identical genotypes. We emphasize once again: natural selection has trouble doing only one thing—there may be no such thing as a genetic "free lunch."

This potential multiplicity of consequences from single genetic changes may be why it has been so difficult to demonstrate that natural selection has occurred on more than a tiny fraction of genes during the transition from chimpanzee to behaviorally modern human beings. It is also a major reason why most population geneticists reject evolutionary psychology as a theoretical paradigm: its predictions ignore how difficult gene-gene and gene-environment interactions make it for selection to operate on just one attribute of an organism. If we had trillions of largely independent genes, it might be possible for selection (given sufficient strength and very many generations) to program us to rape, be honest, easily detect cheaters, excel at calculus, or vote Republican. But, to repeat, *we have many fewer than 25,000 independent genes*. That small number indeed may be why natural selection operating on a very few genes was sufficient to produce the necessary gene-environment interactions to create the key differences that distinguish human beings from chimps. Changing just a few genes not only seems to mean changing a significant portion of the gene-poor genome but also involves altering interactions with the products of many other genes.

Perhaps the most interesting thing about all the attention paid to whether "nature or nurture" controls behaviors is not that individuals with identical genomes often behave very differently, but that those same individuals exposed to extremely similar environments also turn out to behave quite differently. This has been clearly demonstrated in mice, where genetically uniform strains exposed to laboratory environments that were made as identical as possible still behaved differently in different labs.[8] Human beings often show what appears to be a similar sensitivity to subtle environmental differences. Remember the Dionne quintuplets. Indeed, non-identical siblings, who share half of their genes, the same parents, and apparently very similar environments, often seem more different than unrelated people drawn from the same population. Think of all the "Isn't it weird that Johnny and Sammy Smith are so different?" comments—many more, it seems to us, than "Isn't it weird that Johnny and Sammy Smith are so similar?"

If genes don't "determine" our behavior, how can it be that obvious aspects of our (or a mouse's) environments don't either? We don't know for sure, but we can make some guesses. One is that researchers have not yet identified key environmental variables that are subtle to them but central to a behaving organism—be it a mouse with a genome that makes it love alcohol or Johnny trying to get along with Sammy.

Another is that prenatal influences may put individuals on quite different behavioral trajectories. There is a tendency to think that there's fertilization, and then some nine months later a baby pops out. But of course an incredibly complex series of events takes place during those nine months: cell-cell, tissue-tissue, and organ-organ interactions, pulses of hormones, responses to pleasant and unpleasant stimuli, the mother's and others' voices heard through the uterine wall, and in some cases interactions with another fetus or fetuses in the womb. Studies have already shown what dramatic effects prenatal environments can have. For instance, female fetuses whose mothers had minimal diets during the Dutch famine of World War II grew up into women who were more obese and had higher levels of "bad" cholesterol than did those whose mothers were well fed. Many of the differences we see between siblings could thus have prenatal origins.

Psychologists also have shown the dramatic effects that mothers' (or caregivers') treatment of young babies can have on the individuals' later behavior, and twins may seek to differentiate themselves from their siblings by seeking different life courses.

In summary, as we learn more about the human genome, the notion of "genes for behaviors" (which amounts to genes for cultural characteristics) must be discounted. For complex traits such as normal behaviors, few cases have been found of a specific gene or even many genes that greatly influence variation in the trait. It is becoming clear that when genes influence traits, and this applies especially to behaviors, they will do so in a way that is strongly mediated by the environment. Environmental circumstances during any phase of life may alter the way in which an individual's genes function at that time and later. This responsiveness to the environment is especially true for human behavior and cultural evolution because of the marvelous plasticity of the human brain—the center of cultural evolution and the source of our vast cultural proliferation and rise to dominance. In the next chapter, we'll look at some important issues of behavioral plasticity and some ideas about possible mechanisms of cultural evolution.

CHAPTER 5

Cultural Evolution: How We Relate to One Another

"Weizs hesitated, then said 'Forgive me Christa, but what you are doing, you and your friends, will it really change anything.'
"'Who can say? But what I do know is that if I don't do something, it will change me.'
"He started to counter, then saw it wouldn't matter, she would not be persuaded. The more danger threatened, he realized, the less she would run from it."

ALAN FURST, *The Foreign Correspondent*[1]

THE EPIGRAPH is just a bit of dialogue from a fine novel. But on inspection it fairly reeks of theory of mind, with one character thinking about what is going on between the ears of another. It simply accepts, as do we all, that one person knows that another is a thinking being—that all human beings have a theory of mind: a sense of self and the capacity to understand that other individuals have thoughts, goals, beliefs, and intentions, and that many of these differ from their own. That capacity lies behind almost all important aspects of human social behavior, from a child accusing a playmate of cheating in a game to a patriarch writing a will in an attempt to dictate the behavior of people considered to be kin. It is involved in simply coordinating with the activities of others, raising children, developing ethics, recognizing an insult, discerning the motives of a rival state's rulers, or buying a stock.

All people also have a related key mental characteristic: imagination, which is a major driver of cultural evolution. It would be hard to fall in love, for example, without imagination, and without believing that the object of one's affections has a mind that is aware of your mind, and could fall in love with you. Archaeologist Steven Mithen divides imagination into three types. One type, which doubtless long predated the split between our ancestors and those of chimpanzees, is simply the ability to mentally picture the results of future actions in order to make choices. Even rats exploring a garbage dump must in some way picture the consequences of various actions—whether to move toward a certain smell or not, or how to deal with a cat or a competitor.

A second type of imagination is probably restricted to human beings—the power to imagine a fantasy world in which the laws of nature no longer apply. It is the world of flying dragons and omnipotent gods. And finally, there is the "leap of imagination," which connects what we perceive of the world to new thoughts about it. Mithen gives as one example the nineteenth-century geologist Charles Lyell, who imagined a world extending back over enormous stretches of time—essentially introducing a new temporal dimension—on the basis of everyday observations of rocks, valleys, and erosion. His leap of imagination allowed Darwin to make his own leap and imagine that, if Lyell were right about time, natural selection could be the mechanism responsible for the great diversity of life.

Imagination and an active theory of mind are central to processes of cultural evolution as well as to engagement in most complex human enterprises, from falling in love to organizing states. People could hardly alter social organization, from hunter-gatherer band to modern state, unless both organizers and organized were aware of the thoughts and intentions of others. And virtually all major social and technological change has involved feats of imagination. In this chapter we'll examine how human imagination has been deployed over time. We'll do that by sampling the course of cultural evolutionary change at various points in history and discuss what the mechanisms of change may have been—starting with a form of aggressive behavior that extends far into the past and has been a feature of almost all human societies—warfare. This shifts our focus to human social organization, which, although it has some clear roots in the arrangements common to all apes, extends far beyond those of our relatives in its plasticity and its goal directedness (through planning, discussion, cooperation, and coercion). This has everything to do with how our ancestors were able to marshal resources and fuel our rise to dominance as a species.

The Roots of Warfare

Because of the human possession of a well-developed theory of mind, a capacity for plastic social organization, and, eventually, great technological advances, human aggression can take quite different forms from that in other animals. Passive aggression (making and not keeping commitments, doing a bad job at work to "get back" at a despised boss, etc.) is, for example, limited to human beings. It requires a lot of imagination. So does the ability to organize aggression beyond the local group and to develop specially trained warriors, as well as to direct aggression at distant, anonymous individuals (dropping bombs on suspected "terrorist" camps). Direct and active aggression—one individual threatening or attacking another of the same species, the equivalent of punching a rival in the nose—however, is far from restricted to *Homo sapiens*. Intragroup violence is common in non-human apes, especially in chimpanzee societies. It usually involves males fighting over dominance (and thus indirectly over females) or committing infanticide, and it may be manifested in the form of a coalition, in which, say, two male chimps team up to kill another male.

Chimpanzee violence happens not only within groups; it can occur between groups as well. An incident at Gombe Stream, Tanzania, where Jane Goodall famously studied chimpanzees in nature, illustrates this. In the early 1970s the Kasakela community of chimps split into northern (Kasakela) and southern (Kahama) communities. The Kasakela community had six males in their prime reproductive years and two old ones; the Kahama group consisted of seven males: four prime age, one past prime, one old, and one adolescent. Chimpanzees hold territories as groups, and they do so very aggressively. As is typical of territorial chimps, the Kasakela and Kahama groups patrolled their territories—compact groups of males, and often females in estrus, moved silently and purposefully through peripheral areas. During the early stages of fission in 1971, when Kahama males met Kasakela males, there were sometimes charging displays by the southerners, but generally interactions were peaceful. By 1974 encounters had started to become much more aggressive. Coalitions of males from the Kasakela community began a series of forays toward the south featuring protracted, brutal attacks on members of the Kahama community. By 1977 the Kasakela males had wiped out the Kahama community, with all the healthy prime-age males either killed or missing and presumed dead.

What about intent to kill in this case? Given the aggressors' gruesome behavior, Goodall thinks they probably were deliberately homicidal (or, more correctly, panicidal, after the scientific name for the common chimp, *Pan troglodytes*): "If they had had firearms and had been taught to use them, I suspect they would have used them to kill." It is not clear, however, how widespread such aggressive behavior really has been in chimp groups in the wild. Chimp cultures can differ from place to place, and the Gombe Stream chimps were living in an area of recently and dramatically restricted habitat. In 1960, when Goodall started her pioneering work at Gombe, the forest stretched unbroken some sixty miles back from the shores of Lake Tanganyika. A decade later, when the two of us arrived, it extended only about two miles to a ridgeline that more or less parallels the lakeshore, and most of the forest beyond the ridge had been cleared for cultivation. That drastic environmental change, conceivably leading to unusual crowding or resource shortages, may well have been a factor in generating the intergroup strife at Gombe.

It would also seem that the Kasakela-Kahama strife is not quite the equivalent of human warfare. Local feuds and vigilante attacks obviously pepper our species' history, but warfare as conventionally understood involves organized violence between groups, *authorized by their leaderships*, not against a malefactor or even his or her relatives, but directed against an entire society (or class of individuals within a society, such as adult males), whether related to one or more individuals who gave offense, or not. At its simplest level, human violence is capital punishment, a matter of direct one-on-one retribution in hunter-gatherer groups or ritualized execution in modern societies. In both feuds and warfare, a form of social substitutability is evident, in which injury intended for an individual or retribution are generalized from perceived malefactors to other members of a group or state.

That some violence is virtually universal within historical human societies has given rise to a school of thought that sees human beings as having an innate "drive" or a "military instinct" that makes them aggressive and is the root cause of warfare. This belief stems in part from observations of frequent *intra*community violence among people as well as the *inter*community violence or warfare that is nearly ubiquitous among human societies. In part, the belief that warfare is innate also traces to the observation that some other social animals besides the Gombe chimps engage in intergroup violence. For example, female-led spotted hyena clans in the Ngorongoro Crater of Tanzania defend well-marked territorial areas and kills of prey animals. This often leads to violent interclan battles in which individuals are badly mauled, sometimes killed, and on occasion also torn apart and eaten. However, many social mammals, such as peccaries, dolphins, and elephants, don't show intergroup violence, thus undercutting any argument that intergroup violence is somehow natural to the animal world in general.

Can we use our nearest living relatives as models to inform us about the likelihood that we have violent "natures"? Patterns of violence among primates are actually quite varied. Intergroup violence appears much less common among bonobos than among chimpanzees, for example; indeed, on occasion bonobo groups come together peacefully, with females initiating cross-community grooming and even copulation—without objection from males. Interestingly, bonobos are a physically similar but more slender species that split off from a common ancestor with the chimps some 2.5 million years ago. Being less combative, bonobo males don't generally dominate bonobo females, and tensions within and between bonobo groups are usually relaxed by nonviolent and pleasant means, such as mutual genital rubbing and sexual intercourse. Infanticide is unreported in bonobos, but it does occur in chimps.

Bonobo behavior, unfortunately, is much less well known than that of chimps, and further study might reveal more of a darker side to their nature. One element in the differences between bonobos and chimps may be that there is more nutritious vegetable food in the bonobos' environments, plants less laced with toxic tannins than those confronting chimps, who thus may face more difficulty obtaining food and more competition with each other for what is available. Bonobos, unlike chimps, do not overlap in distribution with gorillas and thus do not suffer competition with gorillas for food. Whether a change in the bonobos' environment might trigger more intergroup violence may never be known, as they are a highly threatened species, and the needed studies might never be done. But work with other primates suggests the environmental flexibility of aggressive behavior. Robert Sapolsky has shown that, in baboon troops, the degree of aggression is clearly influenced by local differences in baboon culture. When an epidemic killed off all the aggressive males that had been fighting over food at a tourist lodge dump, for example, remaining members of the baboon troop became much more peaceful. Similarly, males immigrating from troops with aggressive males adopted their new troop's more peaceful culture.

Could intergroup violence be a culturally evolved trait elicited by certain environmental conditions in chimps and people as well, and passed on generation to generation as long as the appropriate environmental conditions prevail? Remember, the Kasakela and Kahama chimps were living where the available habitat had very recently been greatly constricted. And recent observations have shown variability in the peacefulness of chimp communities elsewhere. West African chimpanzees are less prone to intergroup violence than are East African chimpanzees, possibly because of a more bonobo-like social structure that some think is related to the fact that these chimps, like bonobos, don't occur together with gorillas and thus don't suffer competition with them.

As with chimpanzees, intergroup violence among human beings often involves disputes over resources, including territory, or tribal conflict, as among males fighting over females. Anthropologist Lawrence Keeley has pretty much demolished the notion, held by some, that such violence is basically a disease only of highly organized societies such as chiefdoms and states. Indeed, careful examination of prehistoric human remains shows lots of depressed skull fractures, bones cleaned for cannibal feasts, vertebrae with embedded projectile points, and traces of fortifications that indicate that hunter-gatherers and early farmers were sometimes peace loving but often not. Recent work indicating that bloody warfare over water and land plagued prehistoric North America, especially in the Southwest, and that resource shortages and population pressures were involved in the generation of conflict, also supports the thesis that war was common among prestate groups.

Another anthropologist, Raymond Kelly (leaning heavily on the work of primatologist Richard Wrangham), speculated that intergroup violence went through several distinct stages of cultural evolution in human history. First, before people developed weapons that could kill at a distance, there was killing by coalitions of males similar to that seen in the Kasakela chimps. The death of adult males in a neighboring group provided a territorial advantage that translated into more food and perhaps more access to females.

When hunting parties started to be equipped with spears that could be thrown from ambush, perhaps among *Homo erectus* about a million years ago, the costs of intergroup aggression between hunting parties went up and the chance of lasting territorial gains declined. Kelly hypothesized that the advantages of intergroup cooperation in hunting big game (such as mammoths) that was now accessible combined with new respect for a potential adversary's defensive capabilities to produce an era of relative peace and friendship among hunter-gatherer groups. Then this cooperative spirit gave way to more frequent warfare when the agricultural revolution, about 10,000 years ago, permitted the development of military specialization, and groups of armed soldiers could attack villages where only a portion of the people present had and could use weapons. Here the aggression shifted from attacks on individual "enemies" caught near territorial boundaries to the consideration of entire groups as enemies. These ideas are interesting, but they rank as only hypotheses at the moment—more testing would be very desirable but also quite difficult to do.

Although war was certainly widespread in prehistory, anthropologist Gregory Leavitt's study of 132 cultures[2] suggests that societies have gone to war more often as they became more structured into classes, maintained cadres of professional soldiers, and developed more complex technologies. It could have been the other way around: more frequent battles might have stimulated stratification and technological innovation. In any case, bloody wars seem to have been common long before people built big cities.

Whether or not these speculations about the warlike behavior of our ancestors are correct, the recent work on primate conflict, in our view, makes the notion that human beings are genetically programmed for violence and warfare highly suspect, to say the least. While the capacity to be aggressive appears to be virtually universal in human individuals, so is the capacity to be friendly. And not all individuals (or societies) are aggressive. For instance, as author Barbara Ehrenreich points out, "throughout history, individual men have gone to near-suicidal lengths to avoid participating in wars. . . . Men have fled their homelands, served lengthy prison terms, hacked off limbs, shot off feet or index fingers, feigned illness or insanity, or, if they could afford to, paid surrogates to fight in their stead."[3]

In addition, massive data indicate that, at least in recent centuries in Western society, it has been surprisingly difficult to get many human beings to kill during wars. In the American Civil War and both world wars, huge numbers of infantry soldiers—possibly well over half of the men in combat—never fired their weapons or fired to miss.[4] During the slaughter of British troops in the ill-fated 1916 offensive on the Somme, German machine gunners ceased firing rather than kill wantonly, allowing lightly wounded British survivors to retreat to their own lines. Special training to kill was needed to substantially increase the rate at which American infantrymen in Vietnam used their weapons.

FIGURE 5-1. Wounded Canadian and German soldiers share a light in the mud of the Passchendaele battlefield, November 1917. In World War I the commanders on both sides had considerable difficulty getting men to fight each other, as "live and let live" arrangements had evolved. Only determined efforts by commanders kept the mayhem going. Photograph courtesy of Library and Archives Canada.

There is a paradox here—war is highly undesirable from the viewpoint of most sane people, yet warfare persists. In any case, it is uninformative to say either that people are genetically programmed to be violent or that they are "basically" peaceful. Cultural evolution is too complex for that.

Culture Since Farming

Following the evolution and diversification of language and the great leap forward some 50,000 years ago, by far the most important event in cultural evolution was the rapid change wrought by the agricultural revolution, which led to vastly greater stores of non-genetic information and transformed the social organization of *Homo sapiens*. About 12,000 to 7,000 years ago, in seven or eight different areas in the Near East, New Guinea, China, Central and South America, and eastern North America, something caused hunter-gatherer groups to begin settling down and take up agriculture. Perhaps the earliest cultivation was of figs at a prehistoric village near Jericho, about 11,400 years ago. Hypothesized reasons for the advent of farming include population growth, which made it increasingly difficult for groups to feed themselves as hunter-gatherers.

Whatever the reason, almost simultaneously different groups of human beings decided to stop depending on harvesting what gods and spirits provided in forests, woodlands, and meadows, and began intervening to shape nature so that it provided more of the foods they desired. They began to encourage the growth near their encampments of wild grains they traditionally had harvested, and they started to domesticate and herd animals. The process of spreading and intensifying agriculture continues to this day. And agriculture, by eventually freeing substantial numbers of people from the work of obtaining food and allowing them to specialize in other occupations (such as armed aggression), created societies in which huge amounts of non-genetic information could be stored and shared, one in which the writing of this book became possible.

That agricultural revolution, which began some ten millennia ago, represents much more than a new stage in human intergroup violence. It launched our species into an entirely new arena of cultural evolution, one that did away forever with the long-term human situation of everyone being the possessor of almost all of the non-genetic information of his or her society. Before people learned how to domesticate plants, they mostly had to move around to follow game or to find fruiting trees or other sources of vegetable food. Plant domestication allowed them to expand their tool kits beyond those items that were small enough to be portable. More important, farming for the first time made it possible for a group to produce more food than it required, opening the door to both specialization (not only soldiers but also carpenters, traders, priests, and teachers) and hierarchies (peasants, government officials, kings). As we saw, adoption of agriculture was at first very spotty geographically, and even today there are a few societies, such as the Inuit, that have not made the transition to agriculture.

If, over the past century and a half, enormous progress has been made in assembling a coherent picture of genetic evolution, changes in the non-genetic information possessed by humanity—cultural evolution—remain little understood. Despite some important first steps, no integrated picture of the process of cultural evolution has yet emerged that has the explanatory power of the theory of genetic evolution. We don't have, for example, a proposed overarching driver of creative cultural evolution to match Darwin's natural selection. Indeed, whether a unified theory of culture is achievable remains an open question.

Much of the effort in examining cultural evolution has focused on interactions between genetic and cultural processes—how, for instance, changing farming practices (culture) that affect mosquito abundance can influence genetic selection for the sickle-cell trait and its associated malaria protection. That focus, however, provides a sometimes misleading perspective. Most of our species' behavior of interest to the general public, policy makers, and scientists today is a product of the portion of cultural evolution that occurs so rapidly that interaction with human genetic changes is irrelevant. Fruit flies can undergo genetic evolution to become DDT-resistant in ten generations (a few months). Ten generations for people would be a couple hundred years, and we doubt that the required 95 percent mortality / infertility of susceptible people resulting from DDT poisoning would make this genetic evolution seem worth it. For practical purposes, most significant genetic evolution in human beings would require thousands of years.

The rise of *Homo sapiens* to dominance and our species' profound power to affect all of life on Earth make human cultural evolution almost always now more significant than our genetic evolution. In a world of environmental deterioration, overpopulation, intensive competition for natural resources, changing climate, and the threat posed by thousands of nuclear weapons, it is crucial to understand the process of human cultural evolution per se. Understanding how cultural changes interact with individual actions seems to us central to informing democratically and humanely guided efforts to influence civilization's future path.

Human culture has been evolving for some 6 or 7 million years, since the line leading to *Homo sapiens* split off from that leading to modern chimpanzees and bonobos. That evolution accelerated as our forebears' brains expanded and they gradually began to master the manufacture of tools, the control of fire, and language with syntax. They spent more than 99 percent of that time in small hunting and gathering groups of perhaps 25 or 30 who lived together and had a more extended social group of 90 to 220, perhaps averaging around 150, within which individuals might circulate. That's a somewhat controversial number, estimated on the basis of conjectures by anthropologist Robin Dunbar about the relationship between the size of the brain's perceiving-thinking structure (the neocortex) and the factors that bind primate societies together. Curiously, though, in giant modern societies, the number of one's close acquaintances also hovers in the vicinity of the magic 150.

Within that group size, every individual in a hunter-gatherer society knew every other one, and most of them were kin (exceptions may have been, for instance, females brought into the group in intertribal exchanges or through successful raids). Kin recognition has been demonstrated in animals as unlike us as tunicates (simple marine relatives of animals with backbones, called "sea squirts"), tiger salamanders, and ground squirrels, often using chemical clues. Chimpanzees have some ability not only to determine visually their own kin but also to match fathers with sons simply by viewing pictures of unknown individuals. Human beings seem to judge kinship primarily by sight, although odor also supplies clues.

However kin recognition is accomplished, living in groups of kin increases the likelihood of genetic kin selection operating, that is, selection favoring relatives of individuals who behave in certain ways because they carry many of the same genes. For instance, kin selection is one possible explanation for the evolution of some kinds of altruistic behavior in people; by helping other members of the group, an individual could promote the passing on of some copies of her own genes (the proportion of genes that are the same, depending on how close the kinship is). That would be an example of what is known as inclusive fitness—fitness that involves not just the genes an individual passes on but also the copies of those genes that the individual can help transmit to the next generation by encouraging the reproduction of her relatives. But living in closely related groups also maximizes opportunities for cultural transmission through social learning (learning by observing others rather than by trial and error) and through deliberate instruction and direct exhortation. Groups of kin (family) can thus be viewed as important units in both genetic and cultural evolution. And while the agricultural revolution opened the door to humanity's forming much larger groups, those large groups have remained highly subdivided into families, clans, classes, and so forth.

Family Definition and Structure

So what, exactly, are human families, and are there universal or near-universal principles upon which they are organized? Dictionaries tend to define families as parents and their biological or adopted children: Mom and Pop and the kids—that is, nuclear families. The definition can be expanded to various relatives living in the same household (Grandma in the upstairs room) or elsewhere (Uncle Leo in Miami), or all identified kin and their spouses (the invitation list for a family reunion—sometimes referred to as an extended family). This expanding definition reflects the great variation that exists among human societies, virtually all of which have genetically and maritally related individuals cooperating more with one another in their productive and reproductive lives than they do with other individuals.

But the flexibility of human behavior is underlined by the one well-studied exception to this rule: kinship and social structure in the Na (pronounced "Nay"), a minority people of southern China. The Na culture has neither husbands nor fathers. The majority of the society lives in matrilines (groups in which descent is traced through females), which are the units of production. Reproduction, however, is mostly accomplished by a system of promiscuous male covert "visiting." The overt goal is sexual pleasure, but males are also viewed as giving gifts to the matrilines by allowing women to get pregnant. The necessity for intercourse to produce children is recognized, but a genetic contribution from the male is not. Children are not associated with their fathers, and in many cases they are unaware of their father's identity (not surprisingly, since women frequently have dozens of lovers). Girls are associated with their mothers, boys with their maternal uncles. There is a powerful anti-incest taboo within each matriline, with sanctions that include death.

Over the centuries, pressure from the dominant Han culture of China has attempted to alter the Na system, but with minimal results. In 1956 the Communist Party government made a major attempt to institute monogamous marriage as the standard structuring feature of the family for all of China. The enthusiasm of officials pushing for "communist morality" was based on an evolutionary theory of marriage patterns in which the Han system (monogamy) was seen as the highest stage. The certitude of the Chinese communist cadres was very much like the certitude of religious fundamentalists in the United States determined to deny marriage to gay couples and to condemn extramarital coupling.

Among early hunter-gatherer bands, there probably was little distinction between the group and the extended family. Everyone was related to everyone else, and if historical hunter-gatherer societies give us a clue, children were often raised cooperatively either by an assortment of close relatives or by the entire group, and emotional closeness often was supplied by other group members as well as by parents. Furthermore, it is likely that close kin, particularly grandmothers, were important to children's survival by helping to obtain food for the family. (This may be the reason for the evolution of menopause—grandmothers could pass on more genes identical to theirs by helping daughters than by attempting late-in-life reproduction.) Perhaps more important, the group was coherent enough for everyone to recognize and know everyone else.

In the simplest, "unsegmented" hunter-gatherer societies, there was no further structure beyond cooperating families living together. But many bands became segmented by, among other things, dividing the array of families into related sets that anthropologists call clans—groups that share descent from a real or imagined ancestor. The core clan members usually are fathers, sons, and brothers—ordinarily it is females who move from clan to clan—and kinship categories in tribal groups map quite well onto actual genetic relationships. Kinship, real or imagined, remains important in all societies today, as demonstrated by events such as the recent genocidal violence between Hutu and Tutsi kin groups (clans with *imagined* common ancestors—largely assembled by Belgian colonialists) in Rwanda and between Shia and Sunni kin groups in Iraq.

The surpluses created by the development of agriculture permitted further segmentation (specialization) into occupations, and it also led to the evolution of an enormous variety of family structures, each to some degree patterned on the tasks that families had to accomplish. Trying to discover the environmental correlates of family arrangements has become a cottage industry in anthropology. Why, for example, does a wife in some societies move into her new husband's home, and in others the husband move into the wife's home? Why do some families trace descent through the male line and others through the female line?

It has been hypothesized that matriliny, descent traced through females, developed with the shift from hunting and gathering to agriculture. Women, who in hunter-gatherer bands primarily foraged, became attached to the land as they took over the care of crops and domestic animals, and the less they moved around, the more productive the family could be. Thus the woman staying in her birthplace at marriage became advantageous.

From Families to States

Why weren't all early farmers, after they settled down to cultivate crops, satisfied with relatively relaxed, low-conflict subsistence farming lives? For example, how did the subsistence farmers of the Nile Valley start some 5,000 years ago to evolve into a state that could build the pyramids? How did Genghis Khan, around AD 1200, weld groups of nomadic horsemen not just into a state but into one of the greatest empires in world history? Most scientists believe that population growth, environmental deterioration, or rising social pressures (including domination by leaders or neighbors), or all three, made early subsistence farmers' lives increasingly uncomfortable in many areas—especially as other groups began raiding them and expropriating their surpluses and women or trying to push them off their land. Population growth, with birthrates climbing in response to the sedentary lifestyle that both freed women from the need to carry children on treks and supplied grains that could be made into mush for weaning, caused the environment to "fill up" with people and agricultural activities. That meant that forest agriculturalists had to return too frequently to lands left fallow for regeneration of the nutrients necessary for productive cultivation. It also meant conflicts over valued farm sites. Nutrient depletion, plant diseases, pest problems, or combinations of such environmental changes reduced yields, increasing the need for land. Pressures from neighbors for food in exchange for tools or brides may have raised demand for food, or incursions by marauding pastoralists may have required aggregation for defense. In each case, a new level of intensification of food production was required to solve the problem, and many groups of early agriculturalists were able to respond to the challenge.

Some common themes run through discussions of the cultural evolutionary path from coalitions of farm families to states. States, as anthropologists Allen Johnson and Timothy Earle defined them, are societies that cover entire regions and include hundreds of thousands or millions of people, often drawn from different ethnic groups and engaged in diverse economic activities. States first evolved in Mesopotamia and Egypt around 4000–3000 BC, and the first genuine empire (a state that conquered and occupied other states, making them vassals) was that of the Assyrians (744–612 BC), in what today is Iraq. States have centralized governments that are based not on kinship (although they may have hereditary monarchs) but on a professional ruling class, and that have a near monopoly on the use of force.

States and empires represent the culmination of humanity's rapid post–agricultural revolution trend away from being a small-group animal whose social organization dealt exclusively with kin. The human experience as a small-group animal is over today for almost everyone. Instead of bands of a hundred or so people, contacts now are often measured in many thousands, and beyond the immediate family household even close kin may be seen rarely, if ever. Modern states are also characterized by well-defined upper classes that profit from the domination of lower classes, as well as by attempts to substitute pseudokin (people not genetically related but dealt with or discussed as if they were) for the genuine article in societies too large for kinship to be an organizing principle. This tendency for extended kinship was first reflected in legends of common descent of tribal groups. Its persistence is patent in the rhetoric of states about "fatherland" and "motherland" and about leaders as "little fathers," "Uncle Joe" Stalin, "Uncle Sam," the "Dear Leader," and so on. The language of pseudokinship is widespread within states as well, used to bind together unrelated people in social or religious groups: for example, fraternity "brothers," sorority "sisters," the Holy "Father," "Mother" Teresa, Jewish "sisterhoods," and the use of the term "family" in describing everything from Japanese corporations to the Mafia.

Perhaps the most interesting theme in discussions of how states developed is what has become known as circumscription. This idea can be traced at least back to nineteenth-century English social philosopher Herbert Spencer, but its modern version owes its formulation to a classic paper published in 1970 by anthropologist Robert Carneiro of the American Museum of Natural History. Carneiro's basic thesis of circumscription was simple: something had to prevent future subjects from fleeing their would-be rulers before a state system could evolve (the dictionary meaning of the word "circumscription" is confining within bounds). When proposed, Carneiro's theory stood in contrast to the widely held view that the emergence of states was somehow "automatic." In that view, states were seen as a stage in the evolution of society, made possible simply by the production of surplus food by agriculture, which permitted specialization and greater division of labor within societies. More specialization in turn permitted further agricultural intensification. For instance, the building and maintenance of large irrigation systems allowed the support of even larger populations and required more extensive managerial functions to maintain the state (whose power and control the rulers arrogated to themselves). In this view, whether or not the society was circumscribed was irrelevant.

Carneiro postulated three kinds of circumscription—geographic, resource, and social. Geographic (or environmental) circumscription occurs when barriers such as oceans, deserts, or mountain ranges restrict the movements of farming peoples. Island societies are obvious examples, as are dwellers in steep mountain valleys. The Hawaiian Islands have a limited area, and they were populated by the Polynesians quite early—for instance, before New Zealand was reached. As a result of the geographic constriction, thoroughly stratified incipient states developed in Hawaii—the most complex social organization in Polynesia.

Resource circumscription occurs when migration is limited by a sharp gradient in resource or environmental quality. For example, people farming rich river-bottom soil cannot spread to poorer soils on higher ground. Resource circumscription also occurred in Hawaii because of great variation in the suitability of areas for agriculture. New Zealand,

FIGURE 5-2. By the time of European contact in 1778, early Hawaiian political organization had evolved virtually to the state level, producing leaders such as King Kaumuali'i (1778–1824). He was the last independent king of the people of Kaua'i and Ni'ihau. In 1810 he added his kingdom to that of Kamehameha, becoming his vassal. © 2008 Herb Kawainui Kane

with thirty times the land area of Hawaii and colonized some 500 years later, also presented its inhabitants with problems of resource circumscription. Polynesian colonists found suitable farming areas only in the northern part of North Island, where marine resources also were easily obtained. The rest of New Zealand was too cold for most of their traditional crops. High population density on North Island produced a more complex social structure than did the lower density on South Island.

Social circumscription occurs when a society is surrounded by other peoples, thus limiting its expansion. Hostile neighbors of some groups on the North Island of New Zealand produced social circumscription, preventing expansion into resource-rich areas. In ancient Palestine, social circumscription caused by the Philistines of the southern coastal lowlands may explain the rise of David's monarchical Israelite state. The Philistines prevented the Israelites from fleeing the imposition of a ruler who could control their lives.

According to Carneiro's theory, where there is no circumscription, people can simply move into unoccupied lands. To a degree, North America was populated by people fleeing Europe to avoid governments they detested; the Native Americans tried and failed to apply social circumscription. Improvements in ship design had made circumscription by oceans less effective. But state organization followed the European emigrants across the ocean, although changing in form. When a circumscribed society grows to the point that no unused terrain remains, squabbles over land lead to conflict, and losers often are subordinated to winners, since the former can no longer simply move away. That, Carneiro speculated, is the force driving social stratification and political evolution, moving societies from the family-oriented band organization through tribes and chiefdoms to states. Interestingly, circumscription also depends on population growth to bring its force to bear.

Evolution toward statehood is correlated with more frequent warfare, as we've seen, possibly as a result of quarrels over increasingly scarce resources, and intensive warfare seems to have first been associated with chiefdoms that were precursors of states. War captives, which at first were made brides or sacrificed, or sometimes eaten, eventually were enslaved and formed the basis of a lower class. Successful warriors and other faithful followers of the chief were awarded slaves for their service, sowing the seeds of further stratification. Eventually society evolved from a status pyramid to clearly delineated, more or less fixed classes or castes. Political evolution also led to taxation, conscription, bureaucracies, and the other state features that subordinated people object to. For all this to happen, the labor of the lower classes had to produce goods beyond those needed for their own subsistence. Control of such surpluses by a small upper class, which was one of political economist Karl Marx's central foci, was what allowed class divisions (stratification) to emerge. Whether states today dividing up a "supercircumscribed" planet (from which no one can realistically flee) will evolve into some sort of planetary regime of governance remains to be seen.

Norms and the Mechanisms of Cultural Evolution

The changing face of warfare, taking up of agriculture, diversification of family structures, and development of states are all examples of cultural evolution. They are also examples of modifications of norms (which we define broadly to include conventions or customs)—the typical patterns and rules of behavior in a human group.

How do norms evolve? The question is at the very center of the quest to understand cultural evolution and resolve today's human predicament, the threat posed by the weight of great human numbers coupled with our unprecedented technological capacity. Norms such as treating the planet's resources as if they were limitless and assuming that the economy can grow forever, or treating human beings in other groups as if they were less important than members of one's own group, must be modified if a sustainable global society is ever to be achieved.

Norms provide a cultural "stickiness" or viscosity that can help sustain adaptive behavior and retard detrimental changes in society. Standard ways of dividing up game in a hunter-gatherer society would be an example, preventing free-for-alls over the spoils of the hunt. Equally, though, stickiness can inhibit the introduction and spread of beneficial behaviors—military organizations' treatment of nuclear weapons as if they were just bigger versions of conventional explosives comes to mind. Uncovering general rules of how norms in particular, and cultures in general, change is itself a daunting problem. Some scientists who made early attempts to understand cultural evolution sought parallels to the population genetic models used to analyze genetic evolution. A tempting and facile analogy has been that there are cultural analogues of genes—termed "memes" by evolutionist Richard Dawkins. Memes are supposed to function as cultural units that can be replicated, just as genes are genetic units that can be replicated. Memes can be ideas, behaviors, patterns, or units of information that can be passed from mind to mind. They are thought by some scientists to be, like genes, subject to natural selection and thus to account for patterns of cultural evolution. But the differences between genes and memes make the analogy inappropriate, and "memetics" has not led to any real understanding of cultural evolution. Genes are relatively stable, mutating rarely, and those changes that do occur usually result in non-functional products. In contrast, memes are extremely mutable, often transforming considerably with each transmission.

Among human beings, genes can pass only unidirectionally from one generation to the next via reproduction (vertically). But ideas, postulated to be "memes," now regularly pass between individuals distant from one another in space and time, between generations from parents to offspring, between non-parents to succeeding generations (obliquely), and within generations (horizontally), or even backward through generations (when children give ideas to parents or grandparents). Through mass media or the Internet, a single individual can influence millions of others simultaneously in a very short period. In addition, individuals have no choice in which genes are incorporated into their store of genetic information, and the storage is permanent. In contrast, people are constantly filtering what will be added to their stored cultural information, and the filters even operate differently according to the way the same idea is presented. Cultural information can be transmitted, but it also must be absorbed, as any parent or teacher can testify. Moreover, individuals often deliberately reduce the store (when computer disks are erased, old books and reprints are discarded, etc.) or do so involuntarily, as when unreinforced names or telephone numbers are dropped from memory.

Such qualitative differences ensure that simple models of cultural evolution based on the analogy to genetic ("Darwinian") evolution will fail to capture much of the relevant dynamics and that a model framework addressed to the specific challenges of cultural evolution is needed. But models for norms don't need to be based on genetic analogies. One quite informative set of such models is based on a "game" called the "prisoner's dilemma," described in Robert Axelrod's classic book *The Evolution of Cooperation*. The details need not concern us, but those models show that when people are "playing" against one another attempting to maximize their own self-interest, that can counterintuitively lead cooperative behavioral norms to evolve. In many other models now being developed, the most basic assumption is that the spread (or non-spread) of norms shares important characteristics with epidemic diseases. In particular, like diseases, norms spread vertically, horizontally, and obliquely through infectious transfer mediated by webs of contact and influence. Like the rise and fall of infectious diseases, norms may wax and wane as people's ideas and beliefs change. And just as exposure doesn't necessarily mean contraction of a disease, transmission of an idea does not necessarily entail reception or adoption of behaviors based on that idea.

There are clearly some major challenges to a full understanding of the evolution of norms. They include, at the most elementary level, defining exactly what is changing in cultural evolution—which Paul and Simon Levin called the "meme dilemma" in honor of Dawkins' pioneering but regrettably problematic notion. A second major challenge is discovering the mechanism or mechanisms by which novel ideas and behaviors are generated and spread. A third is discovering the most effective ways to change norms.

One way to address those challenges is by formulating hypotheses about the evolution of norms that can be tested with historical data, modeling, or even (in some cases) experiments. For example, it seems that the evolution of technological norms will generally be more rapid than that of ethical norms. Technological changes are usually tested promptly against environmental conditions—a round wheel wins against a hexagonal one every time, and the advantages of adopting it are clear to all. But "winning" technologies may be conserved by selection. Circular wheels have been in use for a long time. Ethical systems, on the other hand, normally cannot often be tested against one another.

Paul and his colleague Deborah Rogers recently decided to test a weaker but related hypothesis: that, because of the environmental tests to which they are put, technological norms would evolve at a different rate from norms not so tested. They were able to examine this issue in Polynesian canoes, which have both structural features (presumably tested against the environment) and decorative features (much less so, if at all). And indeed it turned out that the decorative features of the canoes evolved much faster than the structural ones. Natural selection favoring conservation of cultural features that helped avoid disaster slowed the differentiation of structural characteristics; decorative designs were not under such selective constraints.[5]

Those results seem to us to fit well with other patterns in cultural evolution that have not yet been tested rigorously. For much of hominin evolutionary history we have been social, small-group animals, with a recognition of "self" and a theory of mind to match it. When biophysical environmental pressures are not constraining the direction of evolution, human groups seem to have a tendency to differentiate themselves from other groups—as seems to be indicated in the results with the decorative properties of canoes. Such signs of group efforts to individuate are obvious in many aspects of culture, as in the different stories and ceremonial procedures of religions—even those that recognize the same god or gods. Religious groups show both an astounding persistence (consider the durability of Christianity, under attack for three-quarters of a century in the Soviet Union) and a tendency toward fragmentation into subgroups treasuring what, to outsiders, are trivial differences in myth or ritual. Within groups, individuals often strive to differentiate themselves from other group members—somewhat, but not too far, since they are constrained by the social environment, the culture, of the group. Patterns of this sort seem evident in Western society in historical times, in cultural areas such as the evolution of clothing styles, coiffures, music, and art.

Understanding cultural evolution (and how our brains evolved to make it possible) is both fascinating and critical to an understanding of how *Homo sapiens* achieved its global dominance. While satisfactory syntheses in these areas have yet to be achieved, building blocks now exist to help us gain insight into some critical aspects of brain and cultural evolution. We'll next turn to something at the center of cultural evolution: how human brains perceive their environments.

CHAPTER 6

Perception, Evolution, and Beliefs

"Knowledge derived from physics informs us that the world from which we obtain sensory information is very different from the world as we experience it."

IRVIN ROCK, 1984[1]

"All societies are sick, but some are sicker than others. . . . the existence of traditional beliefs and practices that threaten human health and happiness [is found] more in some societies than others."

ROBERT B. EDGERTON, 1992[2]

EACH INDIVIDUAL is evolving culturally in response to changing environments. That sounds simple, but it isn't. Individuals differ in the information they receive from their physical, biological, and cultural environments, and they filter that information in various ways. Consider this as famous neurologist Oliver Sacks has: some people who were blind from birth or soon after birth were able to regain their sight in adulthood, but they generally didn't turn out to be pleased. In fact, they had great difficulties and in some cases even were relieved when they lost their sight again. At first this might seem strange, but on second thought it makes sense. For one thing, visual input in infancy seems to be required to program the visual cortex of the brain to interpret scenes properly. Initiating input from the retina to the brain later apparently does not provide visual experiences similar to those possessed by people with normal vision from birth.

Then, too, imagine you suddenly had your perception of frequencies of electromagnetic radiation expanded beyond the small range we call visual. Not a "visual" extension so that you could see infrared and ultraviolet—though even that would be incredibly confusing. No, suppose a new organ appeared that filled your brain with all of the radio and TV messages in which our bodies are bathed now, unnoticed. Suddenly you would directly receive all the radio and TV channels in all languages that radio and TV sets in your area can receive over the airwaves. That would probably drive you nuts immediately—you would beg to have the organ removed. If, on the other hand, you had been born with that input and gradually learned to interpret it, disregard most parts, and focus on a few, you might feel deprived if that input from your environment disappeared.

So it isn't the environment as a whole that interacts with our genomes to influence our behavior and cultural evolution; we respond only to the subset of environmental information that is passed on to the brain directly or through its influence on our hormonal systems. The human nervous system has evolved to screen possible perceptions in certain ways so that individuals are cognizant of and influenced by only a small part of the potential stimuli that are "out there." Our visual perceptions have been strongly shaped in turn by the evolutionary course taken by our ancestors, especially during that long bush-clambering, insect-snatching period. That lifestyle gave our ancestors stereoscopic color vision connected with highly controllable mobile fingers with very sensitive touch pads. We don't orient ourselves or put together a picture of the world primarily with smells as dogs do, by echoes as do many bats, or by distortions of electric fields as some fish do, even though smells, echoes, and electromagnetic radiation are abundant in the human environment. Suppose we were actually "smell animals" rather than "sight animals" and could easily detect pesticides in our food, hormone-mimicking chemicals in our drinking water, and carcinogens in the air we breathe. In that case, the thousands of potentially toxic chemicals humanity has released into the environment would very likely be much higher than they now are on the list of perceived threats to human well-being. Or, even more likely, most of them would never have been released in the first place.

Clearly, there isn't a one-to-one correspondence between what's "out there" and what we perceive. Because our perceptions are an interaction between the external world and the evolved characteristics of the human nervous system, we miss a great deal that is detectable by other organisms. Of course, dogs, electric fishes, and bats miss a great deal of what we perceive. But we can partially compensate for the limitations to our senses by the use of clever devices our species has invented, such as microscopes (which now let us perceive and attempt to dominate the microbial world), telescopes, radio and TV receivers, PET scans, magnetometers, devices for chemical analysis, and the like. But the evolution of our large, pattern-seeking, planning brains has allowed us to expand both our physical and our perceptual abilities far beyond those of other animals, allowing us to become the dominant animal and (we hope) to remain so. And knowing how we perceive things, and the limits to our perceptions, may be central to the latter task.

A simple experiment can show how the brains we evolved help to determine what we see. Move your head from side to side or hold your head still and sweep your eyes over the page. Note that you can still read the text—the words stay still. Now, shut one eye, and with your finger tap the corner of the other eye. See the words jump? Why the difference? Because your brain receives information from special sensory receptors that inform it about the position and movement of body parts. Your brain knows that your head is moving, and it is programmed to compensate for that movement and keep your mental representation of the book's page stationary. Because the movement in your eye when you tapped it was not caused by your neck or eye muscles, the receptors didn't send a message to your brain that would allow it to compensate.

Our brains are also programmed to let us see what we expect to see. Figure 6-1 is a photograph of two people in an "Ames room"—they look like a giant and a midget sharing the same space. But the individuals are both of similar height; the room is abnormally constructed to create the illusion of size difference when the scene is viewed with one eye from a particular point, without moving the head (something a standard camera on a tripod simulates perfectly). We *expect* the room to be normally constructed, with ninety-degree angles at the corners and with rectangular windows, and it is those assumptions that shape the perceptions of anyone viewing an Ames room photograph. Put another way, from the visual (and other) sensory stimuli that the brain registers, it forms a

FIGURE 6-1. A specially constructed "Ames room" distorts the apparent size of people. The illusion works when the room is viewed through a peephole because people from "carpentered" societies assume that the walls and windows are rectangular. Photograph courtesy of Susan Schwartzenberg, © Exploratorium, http://www.exploratorium.edu.

hypothesis about the state of the real world. The hypothesis that is inferred from the data is what we call a perception—the perception, for instance, that the Ames room is constructed normally.

Our reactions to ambiguous images such as the vase-faces drawing in figure 6-2 attest that a given set of inputs can lead to different hypotheses and that those hypotheses can easily change. In such figures, the visual inputs—that is, the flows of incoming data—remain the same, but our minds construct two hypotheses that we "flip" between as our interpretation of the input changes. In contrast, we can also form a single, solid hypothesis on the basis of a combination of different inputs (data streams) and experience. Thus we interpret a letter as "A" regardless of font and whether it is rendered in italics, roman, or script, capital or lowercase.

In a number of ways, called "Gestalt laws," our brains tend to organize incoming visual data. Several examples are shown in figure 6-3; one, for instance, shows how your brain groups together dark and light dots so that you see the figure organized into columns. These laws probably

FIGURE 6-2. Is it a vase or two faces in profile? Such ambiguous images can elicit different hypotheses, between which one's perceptual system may alternate.

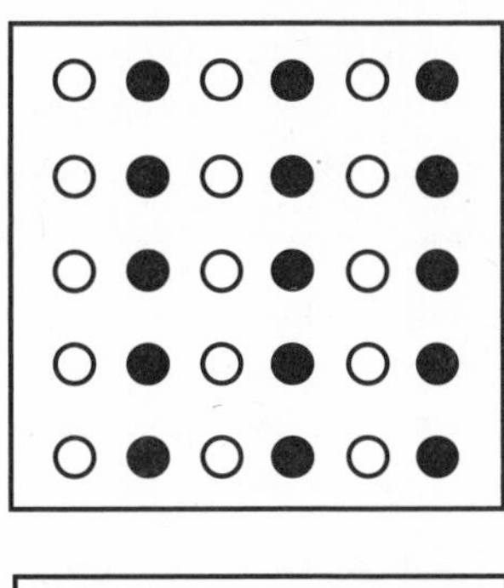

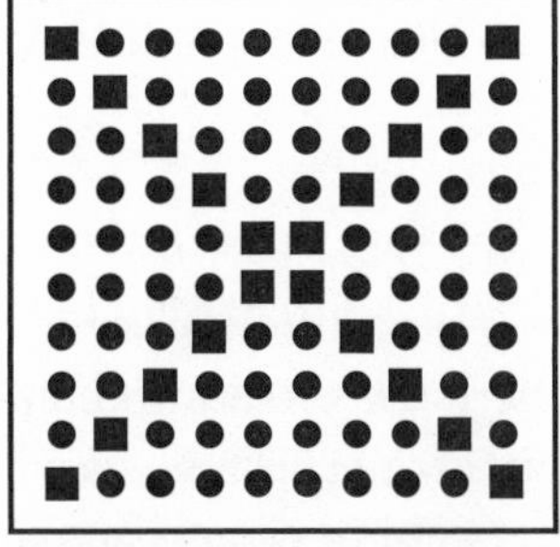

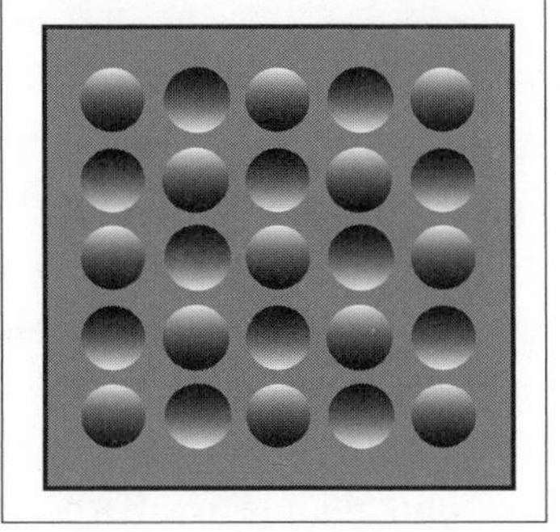

FIGURE 6-3. *Top and middle*: We mentally group together figures that are similar in size and shape. In the middle figure we see columns even though the dots in the rows are closer together. These figures illustrate Gestalt laws. *Bottom*: We also generally interpret figures as if they were illuminated from above. Turn the page upside down and see the bumps change into pits and vice versa.

are partly genetically determined elements of our hypothesis-generating machinery, though ones that are activated by early visual experience. Without such programming, based on the evolutionary experience of our non-human and human ancestors over hundreds of millions of years, it is difficult to imagine how the arriving millions of electrical impulses that physically constitute our visual input could begin to be assembled into a useful subjective view of the world.

A special perceptual capacity that seems tightly tied to our social evolution is that of face recognition. A great deal of research has been done on such perception, which turns out to be controlled at least in part by different areas of the brain from those that deal with the recognition of inanimate objects. It appears that we are born with a great talent for perceiving differences among faces so that we recognize many other individuals of our species quickly and automatically and can read their emotions remarkably well from visual clues—critically important skills for highly social hominins.[3] The evidence for a special brain area for face recognition is quite strong, given that brain damage can cause a total loss of ability to recognize faces (for example, to name previously known individuals from their photographs), while sparing the ability to recognize common inanimate objects (e.g., a coat or a pair of scissors). Damage to a different area of the brain can cause exactly the opposite result—retention of face recognition but loss of object recognition.

Cultural Differences in Perception

Although certain aspects of everyone's perception arose from humanity's common biological evolutionary heritage—making abilities such as skilled facial recognition nearly universal in our species—varying cultural environments play a large role in determining what is actually perceived and, through that, influence the course of cultural evolution. At one level, cultural modification of perception blends with learning—as when scientifically trained individuals instantly "perceive" the significance of a line on a graph or experienced naturalists readily identify a species of bird or butterfly.

But a particularly interesting and more pervasive kind of modification produces culture-specific differences in perception. For example, some cultures do not commonly use pictures that are two-dimensional representations of three-dimensional scenes, and people from these

FIGURE 6-4. People from non-picture-using cultures might think that the bison in this picture is closer to the man than it really is, and that the arrow will be shot at it. That's because they don't share our ideas of how to interpret two-dimensional representations of three-dimensional scenes.

cultures do not employ cues of object size, superimposition (as when an individual partially blocking another in a picture is seen as being closer), and perspective in interpreting pictures, as people from picture-using cultures do as a matter of course. For example, when viewing figure 6-4, a person from a non-picture-using culture might say that the bison is closer to the man—that is, she would give a response that is correct if the picture is interpreted two-dimensionally. Interestingly, two-dimensional representations of three-dimensional objects are a cultural invention that appeared fairly late in humanity's history: art historians point out that the emergence of architectural perspective in the western European tradition did not occur until the fifteenth century.

Intriguing research on intercultural differences in visual perception was carried out in the 1960s by psychologists Marshall Segall, Donald Campbell, and Melville Herskovits. Subjects from fifteen different societies were shown a series of geometric figures that those in Western cultures see as containing optical illusions. The susceptibility of people in non-Western cultures to the illusions depended heavily on whether the built environments of those cultures were carpentered (i.e., had dwellings with rectangular rooms, walls that meet in right angles, etc.) and

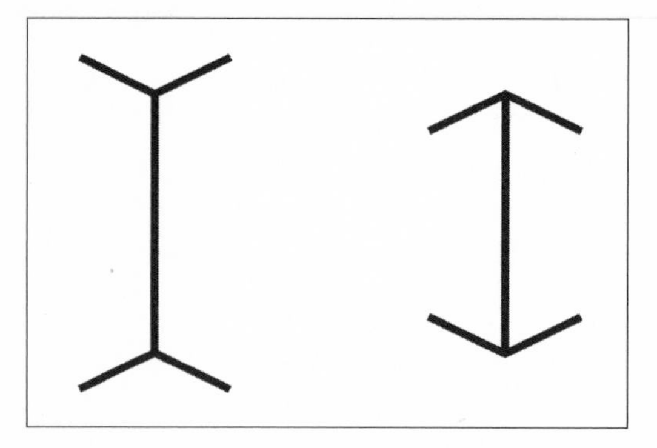

FIGURE 6-5. People who are used to right-angle joinings, when looking at this Müller-Lyer illusion, see the left vertical as longer than the right—apparently interpreting it as a receding corner of a room farther away, while the right is interpreted as the closer corner of a building. Thus each is interpreted according to its apparent distance.

whether the people lived in areas with wide vistas (plains) or restricted vistas (rain forests). In the Müller-Lyer illusion (figure 6-5), people from carpentered environments usually see the left vertical as longer, presumably because it resembles the corner of a room receding from the observer, while the right-hand one is interpreted as the corner of a building coming toward the observer, even though the lengths of both vertical lines are exactly the same. Once again, we see that perceptions are combinations of the way our neural apparatus has evolved and the environments in which that apparatus develops from infancy and operates subsequently.

Perception and Environmental Problems

Several aspects of our perceptual system provide some insight into humanity's general failure to come to grips with environmental problems. One aspect, as noted in the earlier head-movement example, is that perception tends to hold the environmental backdrop constant. Related to this is the phenomenon, widespread in animals, of habituation. That aspect refers to the removal of a constant stimulus from consciousness—one may hear an air conditioner start up, but its continuing hum is soon "tuned out." Habituation can apply to other kinds of stimuli as well. For instance, when smog first became a feature of the Los Angeles sky after World War II, it was a topic of much discussion among people and in the press. Now city dwellers don't pay much attention to usual levels of air pollution (which has actually improved in Los Angeles from its nadir), and it has been relegated to the weather report section of the media. Our fear of atomic bombs was great when they were first used, but the slow accumulation of more than enough explosive power to end civilization has largely been ignored by the general populace. Our nervous systems develop "filters" and "feature detectors" that ensure that all the possible stimuli in the environment don't reach our consciousness indiscriminately. Thus the noise of a running brook is filtered out, but a soft cry from a baby at midnight penetrates the mother's sleep.

Keeping the environmental background constant through habituation makes it easier to perceive new threats or opportunities as the ecological play proceeds. Our ancestors lived in situations in which that was of paramount importance, and we're still quite good at dealing with sudden changes in our environments, be it a car swerving at us from another lane, a baby's cry of distress, or an unexpected bold proposal by a potential lover. In contrast, for our ancestors the ability to detect background alterations occurring over years or decades would have had little or no adaptive value. But that situation has changed, and holding the background too constant over long periods appears to be an unfortunate genetic evolutionary hangover.

The most serious threats now faced by humanity are slow, deleterious changes in the environmental background itself, changes our perceptual systems have evolved to encourage us to ignore. These are changes that take place over decades—population growth, gradual alteration of the climate through global warming, loss of biodiversity, land degradation, accumulation of hormone-mimicking chemicals, and the like. It is very difficult for people to react to global warming, for instance, and not just because our perceptual systems can't detect the rise in greenhouse gas concentrations in the atmosphere. Even if the gases were visible, we wouldn't be likely to detect the change because the increase has been too gradual. The only way to recognize the change is by interpreting graphs made by scientific instruments that were designed to extend our perceptual systems and track the changes. Those graphs condense representations of changes made over many decades into visual differences occurring over the space of a few inches—changes we *can* perceive. A remarkable aspect of human cultural evolution, however, is that such evolved mechanisms can be brought explicitly to people's attention. Then humanity can corporately take them into account as it struggles deliberately and openly to alter the course of its own cultural evolution—a process that psychologist Robert Ornstein and Paul once termed "conscious evolution."[4]

Perception and Belief Systems

However we perceive the world, all human beings establish a body of ideas they accept emotionally as true—a belief system, a set of convictions about how the world works. We all share with one another, and with many animals, a most important conviction, that is, that effects have causes. Indeed, perhaps the most interesting perceptual discovery is that a concept of cause and effect seems to be programmed into our brains early in the course of development, quite possibly as part of the genetic-evolutionary patterning of major mental features. Children as young as two can make the association. If we see a rock tumble down a slope and hit another rock, which in turn starts off in the same direction, our brains automatically assume cause and effect.

In the 1940s, psychologist A. E. Michotte became one of the first investigators of human perception of cause-and-effect relationships. He asked people to watch filmstrips in which a black square was shown moving toward a stationary red one. When the black square reached the red one, it stopped and the red one began moving on the same trajectory on which the black one had been, with the red square moving either more slowly or at the same speed. Even people who knew that there was no physical connection between the squares, that the black square was not bumping into the red one and "launching" it, tended to perceive such a collision and launch. On the other hand, the causal illusion disappeared for virtually all subjects if, after the black square made contact with the red square, the red moved off at right angles to the trajectory of the black one. If the red square moved off more rapidly than the black one had been moving, the illusion was of "triggering," and if the two squares moved off together and eventually stopped together, it was of "entrainment," with black pushing red. That such experiments were necessary to show that cause and effect are "built into" our nervous systems may seem strange, however, to generations that have now been conditioned to see cause and effect in video games.

Although there has been some question about universal susceptibility to the Michotte illusion, recent work makes it seem highly likely that notions related to causation are part of the repertoire of the human brain acquired during our long evolutionary past (and are doubtless present to a degree in many animals, including pigeons). Anyone who has a dog or cat can see that, in certain situations, it learns about causes and effects, for example. Chimpanzees clearly understand that a blow from an appropriate rock against a palm nut placed on a rock anvil is a cause that will have the effect of opening the nut and making its contents accessible.

Our early ancestors also must have spent much time associating, or trying to associate, effects with causes, even when those causes could not have been transparent. Why does the sun travel across the sky? What causes rain to fall? Why do conical mountains sometimes blow off their tops? Why did a member of the clan suddenly become so aggressive? Why did another stop breathing? Small wonder that, with their powerful brains, our ancestors invented belief systems that featured a pantheon of supernatural agents to play the part of "causes." Those agents could be mischievous, evil, weird, or benign spirits with limited power dwelling in rocks, trees, or other individuals, or a diversity of more powerful gods with various domains. While spirits and gods are attributed different powers and characteristics, many of them are clearly modeled on people themselves and loaded with human motives—Earth goddesses who are "mothers," chief gods who are "fathers," jealous gods, forgiving gods, vengeful gods, horny gods, virgin gods, fishing gods, and on and on.

For many, if not most, *Homo sapiens*, the existence of death was (and is) puzzling and frightening—and there can be little doubt that, in addition to a hardwired penchant for seeking causes for phenomena, puzzlement and anxiety about death spurred early people to build religious constructs. It seems reasonable to assume that, as human thinking powers increased in the course of evolution, people would begin to invent causes of other observed, immediately inexplicable effects, if for no other reason than to quell the anxiety that the mysterious and the fearsome often elicit.

There are other widespread perceptions that would have led our early ancestors to seek causes. They doubtless experienced dreams, trances, and hallucinations and saw shadows and reflections of themselves in still waters. Hallucinations, such as hearing voices that are not there, occur when substances similar to the neurotransmitter serotonin, such as the drugs mescaline and psilocybin, are ingested. They travel from our guts through the bloodstream into our synapses and excite downstream neurons. Such substances are found in nature in some mushrooms, and eating those mushrooms and plants containing natural psychoactive chemicals doubtless added to the "supernatural" experiences that are at the root of religion. Interestingly, no one knows why we perceive voices in hallucinations rather than just random sounds! Myths and some legends often have a dreamlike quality, doubtless because they actually originated as dreams, and were recounted as such.

Neuroscientists have shown that a brain disorder called temporal lobe epilepsy leads to extreme religiosity, hearing of voices, and experiencing of mystical visions. Our neuroscientist colleague Robert Sapolsky has speculated that shamans display schizotypal behavior—milder forms of the disordered thoughts and hallucinations that afflict those with full-blown schizophrenia—at a much higher rate than does the general population. And, until quite recently, it was common even for scientists to believe in a dual nature of human beings—that a mind could exist independently of the body that gave rise to it. That made the idea of a mind persisting after the body has ceased to function seem reasonable to many people. All of these factors could suggest a separate existence, a subconscious independent part of the individual that gives it motion but can depart from it—a soul. The permanent departure of that animating element would leave the body lifeless. Death might well have inspired the early cultural evolution of the concept of a soul.

And if people had souls animating them, why couldn't trees, rocks, and winds? If one perceives the existence of a variety of sentient forces of natural or supernatural beings, it would make sense to seek ways of manipulating them to benefit oneself or one's family or group. An organism that had developed a Machiavellian intelligence to deal with other members of its social group could quite easily try to negotiate with supernatural beings in various ways, through prayer, magic, rituals, offerings, sacrifices, or other kinds of behavior, in order to achieve desired results.

Sadly, there is little direct evidence about the origins, original functions, and early evolution of religion, but we can build a partial picture of them based on inference and common sense. The Neanderthals apparently buried at least some of their dead—one carefully disposed grave has been excavated by scientists in Kebara, Israel—and they may have been the first hominins to do so. Less dependable evidence comes from three graves in southwestern France in which the intact, articulated condition of the skeletons implies a deliberate effort to protect them from natural decay processes and predators.

There is, however, no dependable record at Neanderthal sites of the decoration of corpses or the inclusion of goods in the graves, as have often characterized the burials of *Homo sapiens* over the past 50,000 years. Nonetheless, it seems reasonable to conclude that the Neanderthals shared with us a strong awareness of death and a belief that there was something significant about the remains of group members that elicited a need to protect them from scavengers or the elements. Neanderthals might also have had the idea of a soul or a spirit that should be somehow honored or mollified, an idea that almost certainly has been present in our species at least since the great leap and very likely before. Burials of anatomically modern *Homo sapiens* that include animal remains (offerings?) arguably go back almost 100,000 years. Burials with certain kinds of art objects, almost all more recent than 50,000 years ago, surely signal the presence of religious belief among early human beings, since they suggest some sort of continuation of the mind after the body has disintegrated.

Religious belief systems, those based on supernatural actors, must have been especially diverse among early modern *Homo sapiens*, judging from the beliefs of historical hunter-gatherer cultures and the low population densities and geographic dispersion of our prehistoric ancestors, which would have made widespread unifying beliefs difficult to propagate and maintain. There are many different ideas about souls in hunter-gatherer cultures. For example, people need not have only one soul, souls can wander (or not) during dreams, they are located in different organs, and so forth. Some Inuit (Eskimo) groups thought that animal souls resided in the bladder, while human souls occupied the entire body of a living person, yet were immortal. The diversity of notions about souls and other beliefs about sacredness that characterized hunter-gatherer religions has been perpetuated to some extent through the state religions of antiquity to religions of today.

Whenever religions arose, despite their likely early appearance and diversity, it seems reasonable to assume that they had the same two roles they serve today. One is explanatory and manipulative: designating forces as the source of events perceived in the world that seem mysterious, and trying to influence them. The other is integrative and controlling: organizing groups to deal with those forces, dictating appropriate behavior, and justifying power gained by some individuals over others within those groups.

People often have taken sensory inputs that are beyond their limits of rational explanation and woven them into stories that they believe are accurate and explain their origins, aspects of their existence, and fundamental truths about their natures. Those stories, which we call myths, help to shape people's view of themselves and their environment and justify their norms and behavior. They become part of their belief system.

Many of the myths of Western culture, including its creation myth and the story of Noah, have clear predecessors in Egyptian and Mesopotamian myths. That is beautifully exemplified by the story of a humanity-destroying flood in the *Epic of Gilgamesh*, most completely recorded on a series of eleven clay tablets in the library of an Assyrian king (ca. 650 BC). In *Gilgamesh*, the flood is survived by a Noah figure, Utnapishtim, his wife, and the animals he took on board his ark. The account on the *Gilgamesh* flood tablet (and in the biblical story) may trace to a great flood that formed the Black Sea, which for most of the past 2 million years was a freshwater lake. That lake was huge when it received vast amounts of meltwater from retreating glaciers, but it is thought that, as continents warmed up, the lake contracted while sea levels continued to rise. Some evidence indicates that about 6400 BC there was a breakthrough of the Mediterranean into the Black Sea basin, leading to a catastrophic flood.[5]

Why human beings originally created gods in their own image is obvious. Less obvious is why so many human beings have clung to religion as a way of orienting to the world even as the causes of most perceived effects have become clear. Part of the answer doubtless has to do with persisting imponderables (is there a real world "out there"?) and a frequent need for solace. It certainly is not for lack of detailed intellectual analysis of religious issues—just consider the enormous literature on theodicies (explanations of the existence of evil). It's not a simple task to follow the logic of the army of theologians who have proffered different explanations for why millions of innocents are allowed to suffer and why only some of the sick are healed by miraculous cures.

One speculation is that the contradictory nature of a loving god who tortures many of his creations for eternity—the counterintuitive quality of this and many other religious ideas—in itself has some sort of built-in appeal. Science could have an appeal with similar roots. What could be more counterintuitive, less in tune with what our sensory perceptions tell us, than flight in heavier-than-air machines, that solid rocks are actually largely space, or that there is no absolute time and space? As scientists try to unravel the mechanisms of cultural evolution, they would do well to examine the role of the counterintuitive. And that could have profound consequences for the way people treat their environments. As a single example, perhaps the idea that it is possible to have perpetual growth on a finite planet is acceptable to some people *because* it is counterintuitive.

Another theory is that at least some religions, especially orthodox ones, provide structure and an array of rules, prohibitions, rituals, and demands that many people find reassuring. Although religion has long been considered irrational by some, a convincing case can be made that in fact the consumers of religious products (e.g., practices and doctrines) feel they get their money's worth. The behavior of people in the arena of religion seems as rational as human behavior in other areas.

Perhaps another reason for the persistence of religion is the inability of science to provide a clear basis for ethics, as well as its failure (perhaps permanent) to answer those important questions related to solipsism, the view that the only thing we can know is ourselves: How do we know that the others we perceive—indeed, as mentioned earlier, the world we perceive—are actually out there? What makes us think there is a past or a future? Would there be a world if I didn't exist? Scientists, like religious, and possibly all, human beings, simplify the dizzying confusion of their sensory inputs by accepting some things on faith: there is something "out there"; natural effects have natural causes; the physical laws that operated a million years ago still operate and will operate into the future indefinitely.

It is amusing that some scientists, embedded in their own belief system, love to attack the unsupported assumptions of the religious, while calmly accepting their own. Good scientists, though, try to keep those assumptions in mind and to find ways to test them. But perhaps there have always been people who have solved the problems of daily life with an Enlightenment-style "question-and-test" approach to the world layered onto their own faith base—as scientists, many religious people, and many atheists do today.

Like various other cultural traditions, religions often serve as support for those wishing to maintain a social order. Consider the notion of a great chain of being, described in the prologue. When the proper order was preserved, all was tranquil; kings and fathers loved it because it preserved their authority. The frequent use of religion today in an attempt to preserve a social order can be seen, for example, in the invention of "intelligent design" by a right-wing think tank to give political advantage to conservatives. Similar behavior can be traced far back in human history—and is clearly present in the relatively new religion of "scientism," belief that the methods of natural science can resolve all human issues and problems. But, of course, in some circumstances religions can become engines of change—as in the great period of "liberation theology" in Latin America in which Jesus Christ was promoted as a liberator of the poor and oppressed.

Anyone interested in how human societies function and how culture evolves must always pay attention to the myths, faiths, and religions that to one degree or another motivate us all. Similarly, if humanity is to solve its environmental dilemmas, it must pay careful attention to the difficulties of perceiving deleterious gradual change and the ways in which various belief systems can influence those perceptions, as well as to how cultures will evolve to meet environmental challenges. One of the main consequences of our perceptual systems being focused on sight is quite familiar to you already—since the colors of human beings' skin play such a major role in what they believe about, and how they treat, one another.

Race and Cultural Evolution

One of the delights of life is that all human societies are not the same, that different environments and divergent cultural evolution have produced groups of people with very diverse attitudes, behaviors, and sometimes (if you share the sentiment of anthropologist Robert Edgerton in the epigraph) social "sicknesses." In fact, thanks to the influences of environments, cultures, and genes, every individual is also unique, endowed with both pluses and minuses from the viewpoint of other members of her culture. With our categorizing minds, however, we like to put people into pigeonholes and to categorize ourselves and others by race, religion, sexual preference, nationality, political party, economic status, and so on—and usually make value judgments about the qualities of members of those groups.

In closing this chapter we look at a prime example of those evolutionary divergences, in aspects of the visible phenotype, that has become a major element in lessening humanity's chances of remaining the dominant animal. Racial prejudice is one of the true social sicknesses of today, along with gender and religious prejudices. We first ask some questions about how genetic evolution created those phenotypic divergences and how cultural evolution has altered the way we describe and deal with them. And we'll point out that the way they are handled today can affect future cultural evolution and the chances of resolving the human predicament—the complex of problems that threaten the ability of civilization to persist.

One of the pervasive myths of our species is that humanity can be divided into relatively discrete biological units called races. The myth traces in part to people being sight animals, placing great emphasis on visually obvious characteristics such as skin color and hair form. Hominin skin colors very likely became prominent more than a million and a half years ago in populations of our ancestor *Homo ergaster*. Members of that species were upright tropical savanna dwellers and probably faced the problem of keeping cool when they were active out in the sun. Fur coats don't help much in that situation, so it is thought that around that time our chimplike covering may have largely finished fading away. Our forebears were evolving a finely tuned temperature-regulating system, suited for cooling an upright animal that was striding over open areas and that had to maintain an even temperature in its arterial blood. Those characteristics are critical to developing a big brain that would be sensitive to overheating. Skin got naked, but it was probably at first a light skin, which, under their fur, is characteristic of chimpanzees (but not bonobos) today.

That light skin color exposed individuals to ultraviolet (UV) radiation from the sun, which might have induced cancers and damaged sweat glands (UV radiation damages or kills cells in the glands), which are essential for cooling. The cancers would not have been terribly potent as selective agents, since they mostly kill people in their post-reproductive years, but the loss of sweat glands would have been. Even more important might have been UV-caused disruption of the body's synthesis of folate, a B vitamin necessary for sperm production and for proper embryonic development (its lack in pregnant women, for instance, can cause spina bifida in their children, a serious and incurable condition of the spinal cord). So selection led to darkening of human skin with deposits of melanin pigment. We almost certainly had become a dark-skinned species when the first groups of *Homo sapiens* burst out of Africa about 50,000 years ago.

But when groups of our species moved to less sunny climes, more of a certain type of UV light (the shorter-wavelength kind, UVB) needed to be absorbed from less radiation to carry out sufficient synthesis of vitamin D. That important vitamin aids absorption of calcium by the intestines and thus contributes to the building of a strong skeleton, and it is critical for the health of the immune system. The need for vitamin D presumably created a selection pressure back toward lighter skin. Darker-skinned people who have recently migrated to northern latitudes (e.g., Pakistanis in Scotland) often suffer from vitamin D deficiency. Arctic-dwelling Inuit have darker skin than their high-latitude lives would suggest, but they have been in place only for some 5,000 years (250 generations), and their diet, heavy in fishes and marine mammals, is rich in vitamin D. So selection against dark-skinned individuals in that culture has probably been minimal. In contrast, light-skinned Europeans living in tropical northern Australia are plagued by skin cancers and quite possibly folate deficiency. Indeed, in Australia the lifetime chance of an individual getting skin cancer is close to 50 percent, several times that in the United States.

Women are usually lighter skinned than men. Some anthropologists have explained this as a sexual preference of men, but a more reasonable explanation is women's greater need for calcium, particularly when they are pregnant or nursing. Selection would not be kind to the offspring of women who could not produce infants with strong bones.

Today, not only is human skin color geographically variable, but it also has doubtless been changing in many directions under the influence of selection (and both mating choices and interbreeding) as our ancestors left Africa and moved through diverse environments over the entire planet. The palest racist in Europe or the United States is descended from dark-skinned Africans. The African American in Cambridge, Massachusetts, using cream to lighten her skin is unknowingly anticipating a likely selective trend if her descendants remain in such a relatively gloomy climate, although the cultural tendency of visually oriented human beings to mate with people of relatively similar skin pigmentation may damp the results of selection.

Complex as the geographic distribution of skin color is, the patterns of distribution are not matched by geographic variation in hair structure (e.g., curly vs. straight), height, or head shape (long and thin vs. short and broad). This discordant variation is the basic reason why the races that early scientists and most laypeople today imagine are biological units, or even incipient new species, simply aren't. Which "races" one gets when attempting to categorize people depends almost entirely on the characteristic chosen as the first criterion. If you start with skin color, you get one array of different units (which tend to blend into one another). If hair type is selected, an entirely different array of "races" appears. If blood types are studied, yet another way of dividing up *Homo sapiens* is the result.

Similar variability can be found within other "races" based on skin color divisions, which can lump together people as different as Indonesian pygmies, Genghis Khan's horsemen, Hunkpapa Sioux warriors, and Greenland Inuit. Yes, humanity varies greatly geographically; but no, it isn't divided into discrete races. Recent analysis of the distribution of many genes shows a similar pattern of gradients in gene frequencies crisscrossing the planet, not a picture of the genes varying geographically in lockstep. In summary, although "races" based on arbitrarily selected, mostly visual, characteristics have long been defined and treated as social or political units, there are no such units biologically.

Socially, of course, race can be a very real thing, the source of a plethora of myths, because of pervasive racism. Biological evolution has produced obvious visible differences among human populations, related to such things as sunlight exposure, temperature regulation, and disease resistance (remember genes for sickle-cell anemia and the presence of malaria). Cultural evolution, built on a brain structure that uses categorization for efficiency, has used obvious attributes such as skin color and language to divide up the huge numbers of people whom most individuals don't know personally but have observed or learned about secondhand.

Ours is a much more complex world than that of hunter-gatherers, where only a few distinctions (e.g., member of group or not) had to be made among at most hundreds of individuals. The geographic expansion of *Homo sapiens* and the explosion in our numbers have vastly altered not just the physical but also the social environments in which human culture evolves. Today groups are no longer geographically segregated into discrete packets. Vast movements by human beings over centuries have swamped many of the selection pressures that produced the original geographic differentiation of cultures, and in some cases have produced convergent cultural evolution. A person of Sioux descent doesn't need to go to the Old World tropics to meet someone with very dark skin; someone of Zulu descent needn't travel to Norway to find someone with a pale skin; and chances are substantial that they can all communicate in English and listen to music on an iPod. Interbreeding and, in the social context, what is technically known as positive assortative mating (people tending to mate with others of similar skin hue or other characteristics) have further complicated a picture already confused by the mixing of people of different cultural origins. We suspect racism was part of a small-group animal's attempts to retain small groups to belong to—that is, to define many other people as out of the group, perhaps to justify exploiting them. While skin color is a handy tool for doing that, religion also comes into play, as, of course, does gender, or even imagined differences in descent or class membership.

The results of extreme discrimination, from eighteenth-century slavery to the Holocaust, the extermination of the Kulaks in the Soviet Union, the genocide in Rwanda, and the continuing abuse and degrading of women almost universally but especially in a variety of religious sects, are an all too familiar part of human behavior. A key factor in producing this behavior in an empathic species seems to be defining some people as belonging to an undesirable out-group—on the basis of differences in physical appearance, nationality, ethnicity, religion, genitalia, or what have you—and usually making up stories of innate differences to justify that stance (e.g., that blacks and Irishmen are stupid, Jews are venal, women can't do math and are subject to "hormonal storms," Muslims are terrorists, Japanese are treacherous, Scots are cheap, etc.).

But cultural evolution continues. In the United States in the 1930s and 1940s, lynchings of African Americans were common, and it was acceptable to mail horrifying photos of the events on picture postcards to friends. Professions for women were pretty much restricted to school teaching and nursing. The military was segregated, and during World War II black American soldiers could not eat in Southern restaurants that fed groups of white Nazi prisoners of war. When African American Jackie Robinson was hired to play major-league baseball in 1947, there was much discussion of whether blacks could compete in the sport on an equal footing with people with pale skins. Blacks were essentially barred from performing in movies except as maids and janitors.

In part because World War II opportunities demonstrated that blacks could fight well and do industrial jobs well, and in part because President

Harry Truman courageously desegregated the military, cultural evolution in the area of American race relations sped up. The change was accelerated by protests against segregation by blacks who no longer would tolerate treatment as third-class citizens, combined with important court decisions, especially *Brown v. Board of Education*, which mandated desegregation of schools in 1954. Although accompanied by some terrible violence, the evolutionary trajectory in American culture was set, at least temporarily, and society changed—not far enough to eradicate racism, but much further than could have been imagined in the 1940s. Unfortunately, some reactions to people of Middle Eastern dress or appearance in the United States and parts of Europe since the terrorist attacks and scares in the 1990s and the early twenty-first century, as well as de facto resegregation of many schools in the United States, show that positive trends can sometimes be reversed. Indeed, the recent focus in American politics on whether gay people should get married indicates that the cultural tendency to define out-groups is alive and well.

All this points to a very fundamental ethical issue that is sure to be central to much of future cultural evolution in an increasingly large, globalized, and threatened society. To what degree is it appropriate to impose one's perceptions and ethical views on others? Should ethnic background or skin color be included in official government or health records? Should genital mutilation of women (or men) be looked on benignly because it is a cultural tradition in some groups? Should we respect an indigenous group's religiously based desire to continue hunting an endangered whale species? We have no pat answers to these dilemmas, but we think participation of well-informed citizens is critical to discussion of and decisions about them. A key characteristic of our species is the ability, thanks to the evolution of language with syntax, to make collective decisions about the world of the future. And we believe that human beings, as the dominant animal, have an ethical imperative especially to carefully consider the environmental dimensions of our dilemma, become informed about them, and start deciding. But unless humanity can overcome some of these debilitating prejudices and practices and start working together on its critical environmental and social problems, we are pessimistic about *Homo sapiens*' tenure of dominance and the quality of life of future generations.

CHAPTER 7

The Ups and Downs of Populations

"Of all things people are the most precious."
MAO TSE-TUNG

IF OUR ANCESTORS' numbers hadn't proliferated substantially from a relatively few thousand individuals in East Africa, *Homo sapiens*, no matter how smart, could never have occupied the entire planet, formed cities and states, and become the dominant animal on Earth.

Human population growth has been so prodigious in recent centuries that it has also become a major driver of environmental deterioration, in the extent of pollution, consumption of nature's resources, and destruction of habitats needed by other species. On top of other effects, population size is also the "elephant in the living room" on the issue of global heating, for if the size of the human population were only half what it is today, the chances of avoiding a climate catastrophe would be much better; many fewer people would be emitting greenhouse gases into the atmosphere; many fewer people would be crowding onto vulnerable coastlines. These and many other issues vital to the future of civilization are inextricably connected to trends in human population size.

Population genetics is concerned with changes through time in the composition of gene pools (all the genetic information possessed by populations), which we discussed in chapter 1; population dynamics (or demographics, in the case of human populations) is about changes in numbers of individuals and in the forces that produce those changes in a given population. The essential factors in the dynamics of any population, whether human, trout, or butterfly, are the same. A population's

FIGURE 7-1. A mob scene like this is the way most people envision overpopulation. But the most important standard is how the numbers of people are related to the basic resources and ecosystem services required to support them. Crowding can be a local problem even in a region that is not overpopulated. Photograph courtesy of iStockphoto.

size is basically the result of an input-output system. The inputs are births and immigrants; outputs are deaths and emigrants. The difference between inputs and outputs produces population growth or shrinkage (the latter is sometimes called negative growth).

Understanding how populations change in size through time can be critical for successfully carrying out many human activities. We need to know something of the population dynamics of particular species to find ways to harvest fish stocks sustainably, to control pest populations, to preserve endangered species, to predict the occurrence and spread of epidemics, and so on. Moreover, shifting population sizes have important consequences for the evolution of those populations.

Butterfly Dynamics

Like the human populations of many nations, populations of other organisms may be relatively isolated from one another, making it easier to detect environmental factors (as opposed to factors connected simply with migration) responsible for population size changes. For example, three populations of checkerspot butterflies (*Euphydryas editha*) that our group studied at Stanford University's Jasper Ridge Biological Preserve were separated from one another by only a hundred yards or so. When we first started to study them, the population located between the other two was shrinking in size, while one of the flanking populations was expanding and the other was fluctuating in size. This suggested that the populations were not in close contact with one another since they were changing in size independently. We were able to confirm this by observing the rarity of movements from one population to another of the thousands of individuals we captured, marked, and released. Even more rare than movement between these three adjacent populations was migration from more distant ones, especially as one after another nearby population disappeared as a result of suburban development and highway construction.

The lessons we learned in almost a half-century of study also focused our attention on the sensitivity of both the butterflies and the plants their caterpillars feed upon to the vagaries of climate. In fact, changes in the climate in the San Francisco Bay Area interacted with the varied topography of Jasper Ridge to alter the delicate relationship between the timing of the butterflies' life cycle and that of their food plants, leading in some years to much caterpillar starvation. Where there were hill slopes of different orientations (e.g., north or south facing), one slope or another might have a microclimate that allowed the butterflies and plants to be synchronized, and so populations persisted longer in areas of varied hills and valleys than they did in those that were flat or on a slope with a uniform orientation. But, like small, isolated populations of any organism, the checkerspot butterflies were highly vulnerable to extinction, and by 1997 underlying climate change gradually moved all three Jasper Ridge populations to extinction.

Human Population Dynamics

The dynamics of human populations follow the same basic principles as those of checkerspots, although people have developed a vastly greater ability to alter their environments or to migrate than have other animals. Those abilities have helped vault *Homo sapiens* into a position where it not only can greatly diminish or wipe out populations of other organisms, but also can threaten its own future. Thus our focus in this chapter is mostly on the dynamics of human populations—their role in our past, their critical contribution to today's global problems, and their importance in shaping the future of civilization.

In prehistoric times, the human population consisted of scattered groups of hunter-gatherers, and those groups only slowly expanded and spread around the world until the agricultural revolution and its attendant domestication of plants and animals began some 10,000 years ago. At that time, the *global* population of *Homo sapiens* was only 5 to 10 million. When people settled down to grow crops, the population started to grow faster, and probably not because health conditions were better. As we pointed out earlier, the more likely primary reasons were that people then had processed grains and other suitable foods to use in weaning children, and women did not need to carry young children around, as they had in mobile hunter-gatherer groups. With a settled lifestyle, women could bear children more frequently and thus have more in total. Even if there was no change in the population's mortality rate, the increased fertility led to somewhat faster growth of the human population.

As a result of increased growth, the population reached perhaps 250 to 350 million by the time of Christ, and after that, population growth gradually accelerated further. Human numbers worldwide reached half a billion about AD 1650, by which time relative peace and improved agricultural practices (tied to the then recent European economic expansion known as the commercial revolution) helped to reduce death rates. By 1850 the population had doubled, to reach 1 billion. In the middle of the nineteenth century, health actually improved in Europe and North America, despite horrific conditions in the factories and mines of the industrial revolution. The reason was largely the establishment of better water and sewage systems. The number of people grew rapidly in the late nineteenth and early twentieth centuries as the industrial revolution proceeded, accompanied by further improvement in sanitation and health conditions, causing death rates to fall.

In the 1800s, after death rates began declining in industrializing nations, birthrates started to follow them, initiating the oft-cited "demographic transition." That was a transition from a regime of high birth- and death rates to one of low birth- and death rates, typically with a lag between the improvement in infant and child mortality rates and the reduction in birthrates. The causes of the change in birthrates are not fully understood, but family finances seem to have been important.

Children gradually became less important economically as more and more people moved into cities and as agricultural machinery replaced farm labor, making children more expensive to support and educate. Lower infant and child mortalities and the increased costs of raising children reduced incentives for large families, and the smaller families in cities apparently often were emulated in the countryside. The average age at marriage rose as more people, especially women, were educated through high school and beyond; and as women increasingly participated in the labor force, they opted to have fewer children.

This modernization process, which led to a decline in population growth rates (and later in many cases to an end to population growth itself, leaving aside migration), had been largely completed in Europe and North America by the middle of the twentieth century. But it had not yet begun in the rest of the world; worldwide, population growth continued to accelerate. The global population reached 2 billion around 1930, 2.5 billion in 1950, and 3 billion in 1960.

The Family Planning Movement

Early in the twentieth century, as people in industrializing countries increasingly sought to limit their production of offspring, the family planning movement emerged and pressed for the right of couples to determine the timing and number of children they would bear. The economic and health benefits to families of being able to control their reproduction were and are very clear, although, especially in the late nineteenth century and the first half of the twentieth century, the motives of some of those who advocated limiting reproduction were less than admirable by today's standards. They held strong sentiments about too many children among those of the "wrong" skin color or class, or from poor, recently decolonized nations who might emigrate to the rich ones.

Nonetheless, gaining the right for people, especially women, to control their reproduction was the principal motive of the family planning movement. In the early twentieth century, contraception and abortion, and the dissemination of information about them, were illegal in most of the world. During those decades, family planning leaders fought a long series of battles for acceptance and legalization of contraception and, later, abortion—battles that are not yet entirely won. The Roman Catholic Church and a number of other religious groups are still opposed to the use of "artificial" contraception, and abortion to them is anathema, though very low birthrates in overwhelmingly Catholic nations such as Italy, France, and Spain testify to the relative ineffectiveness of their positions.

While contraception is accepted and widely used in the great majority of countries now, abortion is still far from universally accepted or legal. In an extreme case, El Salvador recently outlawed abortion under any circumstances, including for victims of rape, and began prosecuting not only providers (medically qualified or not) but also the women who obtain abortions.

Much of the struggle for the right to family limitation has been carried out in the United States, where the last state law against provision of birth control information and materials was struck down by a court decision in 1962—two years after introduction of the birth control pill. A state-by-state movement toward legalizing abortion for unwanted pregnancies beginning in the late 1960s culminated in the famous *Roe v. Wade* Supreme Court decision in 1973, which overnight legalized abortion nationally.

Since then, conservative anti-abortion groups have attempted, with some success, to establish state laws that hinder access to abortion, while carrying out intensive propaganda and lobbying campaigns as well as demonstrations and picketing of family planning facilities that provide abortions. The campaigns have led in a few cases to bombings of clinics and injury and deaths of clinic personnel. The same "right-to-life" groups have also fought the release of RU-486, a chemical means of terminating an unwanted pregnancy, and "Plan B," known as "morning-after contraception."

While the political battle over abortion rages on, an interesting social change may well be linked to the legalization of abortion in the United States in the early 1970s: a sudden, substantial decline in the nation's crime rate. The decline appeared around 1990, seventeen years after the *Roe v. Wade* decision. That unwanted children often grow up to be socially handicapped adults has been known for a long time. Some social scientists, statistically analyzing a great deal of data, have concluded that the availability of abortion was an important factor in the subsequent drop in the crime rate. Nonetheless, we should always remember that correlation does not necessarily indicate causation.

Worldwide, an estimated 42 million abortions are performed each year, of which nearly half are performed where they are illegal. A recent study concluded that abortion rates are similar among countries regardless of abortion's legality. The principal difference between nations in which abortion is legally available and those where it is severely restricted is whether the procedure is conducted under safe medical circumstances —and it usually is not where it is outlawed. Tens of thousands of women die each year from complications of abortions performed unsafely. High abortion rates, moreover, are correlated with a lack of access to effective birth control methods. As one journal article put it, "the fastest way to reduce the number of abortions is to provide access to reliable contraception."[1]

Curbing the Population Explosion

Following World War II, modern medical technology (especially antibiotics and pesticides to use against malarial mosquitoes) was exported to non-industrialized regions in Asia, Africa, and Latin America. As a result, human death rates there plummeted, although birthrates remained high. For example, birthrates in Mexico in the mid-1960s were in the vicinity of 45 per 1,000 and death rates were about 10 per 1,000, giving an annual growth rate of some 3.5 percent. Similar high growth rates prevailed in many less developed countries during the 1960s and 1970s.

Globally, the population reached a peak growth rate of over 2 percent per year in the 1960s, although differences in growth rates continued to exist among nations and over time. While 2 percent per year doesn't sound like much, population grows like compound interest in a bank account. Just as the interest in such an account earns interest on itself, so children themselves eventually reproduce. This means that a population is set to double every 35 years when the annual growth rate is 2 percent (citing doubling times is often a more effective way of showing the magnitude of population increases, since annual growth rates seem misleadingly small). Even more dramatically, if Mexico's growth rate had remained at its 1960s level, its population would have doubled to 82 million in only 20 years, and by 2005 it would have reached 164 million. But both birth- and death rates by then had fallen by half, while perhaps 15 million people emigrated, leaving the Mexican population at only about 106 million in 2007.

By the 1960s, concern about rapid population growth had begun to focus seriously on the problems it was creating for modernizing societies: environmental deterioration, limited arable land and concern over capacity to feed an exploding population, and increasing inability to provide social services for expanding families. In response to all this, family planning programs were established in many developing countries, often first by private groups and later under government sponsorship and with financial support from industrialized nations. By the mid-1980s nearly every developing nation had a family planning program, which provided contraceptives and advice to people who wanted them. The numbers of couples who adopted them increased in many countries, but not everywhere. In many places children were still economic assets, and family planning facilities often were confined to cities or offered too few options. So the overall decline in birthrates seemed discouragingly slow compared with the continuing rapid growth of populations.

In the 1990s the focus of efforts to reduce population growth in developing regions was broadened to include assistance to improve family health, encourage and support education for girls as well as boys, and provide economic opportunities for women—all factors that have been found to influence people to have smaller families. Indeed, the importance of educating and providing job opportunities for women can hardly be overstated; education of girls is the strongest single factor found to be associated with falling birthrates. A little thought explains the significance of educating girls. Men normally apply their educations to breadwinning, but women especially will use their knowledge to enhance the health and well-being of their families (which also helps to reduce infant and child death rates further). Nonetheless, school enrollment of girls still lags significantly behind that of boys in much of the developing world.

Women's participation in society outside the family, moreover, besides being a factor in reducing birthrates, seems to be essential for the successful economic development of societies. It also provides women with broadened horizons and occupations and thus puts them in a position to make choices such as "Will my family's quality of life be more enhanced by another child or by investing in better housing?" Another important factor that encourages people to have fewer children is societal provision for support in old age, which traditionally has been the responsibility of one's children, so having more children was akin to having an old-age insurance policy.

In the 1990s the efforts of the family planning movement (which included encouraging the improvement of women's status) finally appeared to be paying off as birthrates fell significantly in most regions of the world, including many industrialized nations that already had relatively low birthrates. Nevertheless, worldwide population growth was far from halted; the total world population had reached 6.7 billion by 2008 and was still increasing at about 1.2 percent per year. At that rate, some 80 million people are added each year—the rough equivalent of a new United States every four years.

After the turn of the twenty-first century, when the George W. Bush administration took office, it, like its Republican predecessors since 1980, withdrew most United States government support for family planning assistance to developing nations and subjected the rest to stringent restrictions mainly focused on preventing any association with abortions, legal or not. Even programs intended to combat HIV/AIDS were hindered from making use of contraceptives as a preventive measure. During the 1990s and especially after 2000, population programs in developing countries, along with activist non-governmental population groups, increasingly downplayed the importance of contraception while emphasizing women's education, health, and economic empowerment. The resultant loss of momentum in fertility control may well have caused a significant setback in ending global population growth.

The explosive growth of the human population, especially in the past century and a half, from 1 billion in 1850 to 6.7 billion in 2008, has had profound impacts upon Earth's natural systems, as we will see in later chapters. Human beings have already now taken over and transformed most of the planet's land surface to support themselves and increasingly have exploited the oceans as well, all at the expense of the rest of nature. Although the rate of population growth has slowed markedly since the 1960s and has stopped and even been reversed in some countries, the global population is projected, by demographers of the United Nations Population Division, to continue expanding in the twenty-first century by another 2 to 3 billion, although at a slackening rate, reaching a mid-range estimate of some 9.2 billion by 2050 and continuing ever more slowly for some decades beyond. Both birth- and death rates in the future could rise or fall more than expected; thus the global population in 2050 could be as large as 10 billion or more or as (relatively) small as 8 billion or less.

Nearly all the prospective population growth in the next several decades is expected to occur in developing nations, mainly in Africa, Latin America, and parts of Asia. Populations in the developing world will expand by more than 50 percent by 2050 on average, and those of some countries will double and even triple their numbers. The latter populations are unlikely (barring some disaster) to stop growing by 2050. Even though birthrates have fallen in almost every country in the past few decades, they have not fallen equally fast or far everywhere, and this puts an enormous burden on many societies that can ill afford it.

The Arithmetic of Population Growth

As indicated earlier, in demographic studies growth rates are customarily expressed as a percentage of the population, whereas birth- and death rates are expressed as a number *per thousand* people in the population. Thus in 2007, when the United States' population had passed 300 million, the birthrate was 14 per 1,000, the death rate was 8 per 1,000, and the rate of natural increase (which excludes migration) was 0.6 percent (6 per 1,000). But the United States also had a net immigration (immigrants minus emigrants) of roughly 4 people per 1,000—which meant that 40 percent of the nation's annual growth was due to immigration. Immigration (legal and illegal), in other words, boosted the U.S. annual growth rate to 1 percent.

Migration has been a factor in the dynamics of many if not most historical human populations and undoubtedly was for many of our prehistoric ancestors as well. For many national populations today, the extent of migration can have a significant effect on overall rates of population growth and many related trends.

Nearly all industrialized nations—those of Europe and Japan—have significantly lower birth-, death, and growth rates than the United States. Although most also have moderate rates of immigration (usually between 1 and 5 per 1,000 annually), growth rates have neared zero or even been negative. In an extreme case, Russia's 2007 population was about 142 million; its birthrate in 2007 was 10 per 1,000, its death rate was 15, and the growth rate was –0.5 percent—a shrinking population. At the other extreme, a developing country, Nigeria, with 144 million people in 2007, had a birthrate of 43, a death rate of 18, and a growth rate of 2.5 percent.

Birth-, death, and growth rates can provide a "snapshot" of a population's trajectory at a moment in time, but alone they won't tell you much about past or future trends. To understand those, it is important to know something about the population's age structure (how many people are in different age classes), sex ratio, and total fertility rate (the average number of offspring produced by each woman in her lifetime at current reproductive rates, roughly equivalent to average family size).

In rapidly growing populations, a high proportion of people are young. Often as much as 40 to 50 percent are under 15 years old and the majority are under 30, while the fraction of people over 65 is typically no more than 5 percent. By contrast, in nations with slow or no growth, the proportion of people under age 15 has become smaller (usually making up about 15 to 18 percent of the population), and the over-65 age group has gradually increased to about the same proportion. Those between 15 and 65 are considered to be members of the "productive" ages, capable of holding jobs and supporting children and the elderly.

Consider the population profiles of three countries, beginning with a developing country with a rapidly growing population, Nigeria (figure 7-2). In Nigeria, young people under age 15 make up 45 percent of the population, and nearly three-quarters are less than 30 years old. Barely 8 percent of women use modern contraceptives, and one in five children die before age 5. Because of high early death rates, in 2007 life expectancy at birth was only 47 years. Even so, with a total fertility rate (TFR) of 5.9, Nigeria's population is projected to double by 2050.

The second profile (figure 7-3) is that of the United States, which in 2007 had a TFR of 2.1 and a relatively low death rate of 8, with a life expectancy of 78 years. Recent net immigration has had a small effect on U.S. demographic rates and structure: the median age is slightly lower, fertility is a little higher, and the proportion of people of working age is a little higher than it would have been without immigration. The combination of relatively high fertility (for an industrialized nation) and a continued annual net immigration of about a million people is projected to

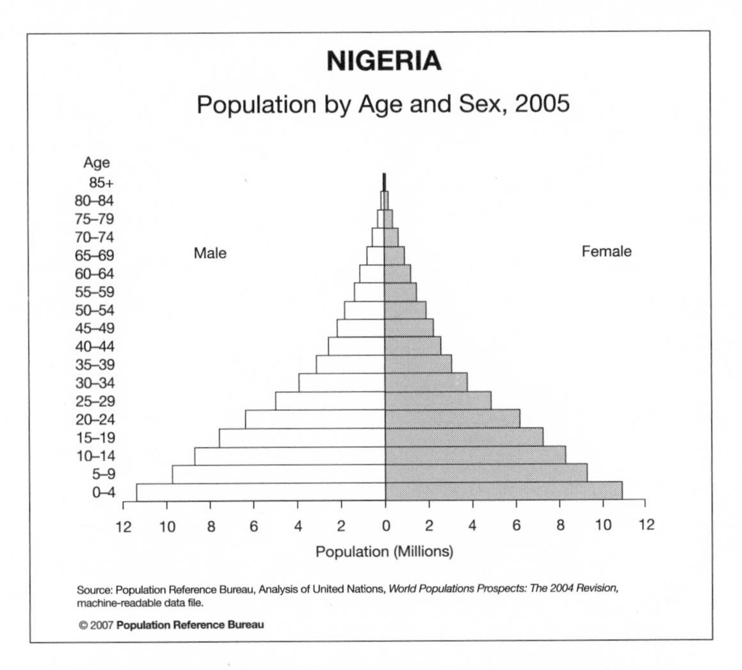

FIGURE 7-2. Population profile of Nigeria in 2005, showing its age structure. Note that a very large proportion of people is under age 20.

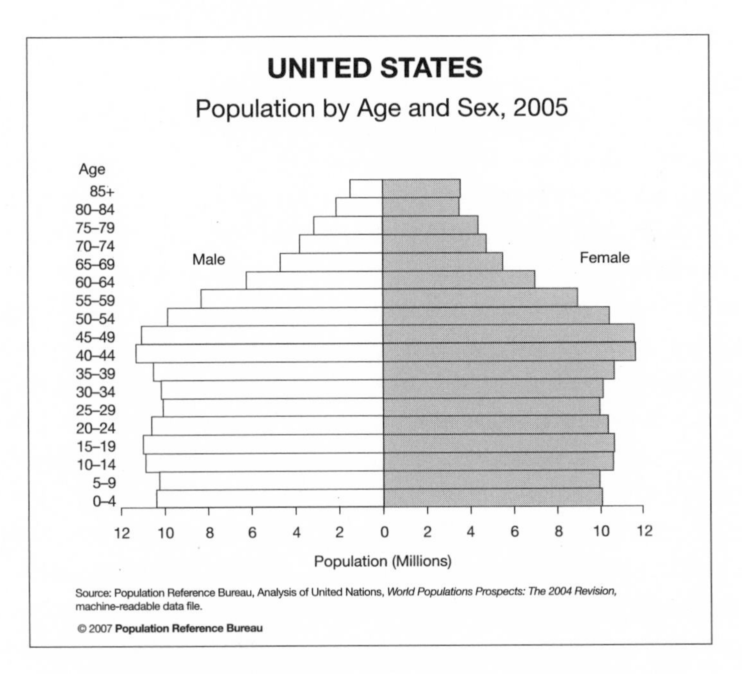

FIGURE 7-3. Population profile of the United States, which is still growing slowly, partly from immigration and partly from a moderate birthrate.

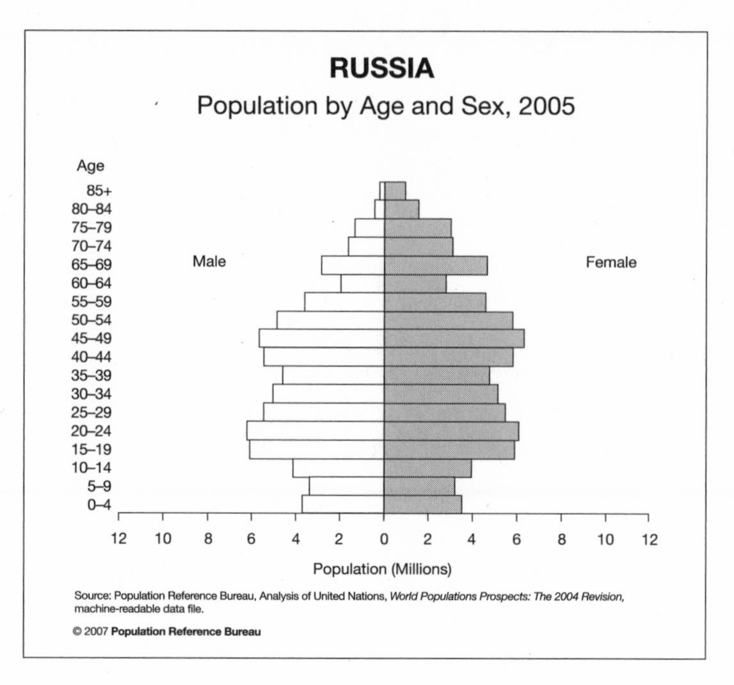

FIGURE 7-4. Population profile of Russia, in which deaths each year outnumber births.

produce a population of nearly 420 million in 2050—about 50 percent larger than it was in 2000.

The third profile is that of Russia (figure 7-4), with its shrinking population. There, people under age 15 represent only 15 percent of the population, and a nearly equal proportion are over 65. Russia's TFR, 1.3, has been below replacement level for a generation or more. In addition, Russia experienced a significant level of net emigration, primarily of young adults, after the collapse of communism, although that seems to have reversed recently. By 2050 the population is projected to be less than 110 million, compared with the 2007 population of about 142 million. Life expectancy for Russian men is only 59, whereas women's is 72. The high mortality rate for men is attributed to high rates of alcoholism, smoking, and suicide, among other social problems. Greater exposure to pollution and toxic substances may also be a factor. Age structure in Russia, as in the United States and Nigeria, will very likely have profound effects on its economy, resource use, and future development unless vital rates change rather dramatically.

Population Momentum

Even though a substantial and growing portion of the human population now has below-replacement fertility, the expansion of human numbers will continue for many decades because of the momentum of population growth. For a population to be "stationary" (neither growing nor shrinking—"zero population growth") the TFR must be at replacement rate, slightly above 2 children per family (in countries with good health care, about 2.1). Replacement rate is just enough offspring on average to replace the parents and compensate for children who die before they can become parents themselves.

Just reaching replacement rate is not sufficient to end growth; the momentum generated by previous population growth must be dissipated. Much of humanity still has a TFR well above replacement reproduction (in 2007 the global average TFR was 2.7). In rapidly growing nations, such as those in sub-Saharan Africa and several Middle Eastern nations, the proportion of young people under the age of 15 can be as high as 45 to 50 percent, as we've seen, corresponding to TFRs of 5 to 8. Those youngsters are the gunpowder of the population explosion. They will be the parents of the next generation and most likely will produce an even larger cohort of children, who all too soon will produce grandchildren. Today's young people will then live alongside both their children and grandchildren before reaching the older ages subject to high mortality rates. That high proportion of young people is what supplies the momentum of population growth, meaning that a population will continue growing (usually for sixty years or more) even after the TFR has been reduced to replacement rate. In societies with reproduction significantly *below* replacement, growth will end sooner, but there will still be a substantial lag.

China, for example, has had below-replacement fertility since about 1990, and its TFR in 2007 was 1.6. Yet it may add another 150 million people—roughly equivalent to the present populations of Germany, France, and Denmark combined—before its growth ends and is reversed around 2025. China initiated a mandatory no-more-than-one-child-per-family program in 1980 after more than a decade of experimentation with a two-child family policy. The one-child program was a triumph in terms of its effect on population size. The present Chinese population of about 1.32 billion would have been an estimated 350 million more today,

FIGURE 7-5. Three generations backpacking in the Sierra Nevada. The idea that people are generally unable to be economically productive over the age of 65 is very out of date. The grandfather in this picture was 74 at the time, and continues to be active as a hematologist at the age of 79. Photo by David Schrier.

and would have added several hundred million more in decades ahead, had the program not been put in place. That reduction is credited by Chinese commentators as having helped to spur the country's recent economic success, although that has come at a price.

In China, as in many other Asian countries, sons are much preferred over daughters. Among other things, sons have traditionally been responsible for the care of their parents in old age. So many Chinese couples, limited to a single child, did whatever they could to ensure that their child was male. Ultrasound machines deployed to check whether women were retaining their assigned intrauterine contraceptive devices (IUDs) or to monitor pregnancies were soon used illegally to determine the sex of fetuses. Many female fetuses were then aborted. China was forced to relax the one-child requirement for couples who bore daughters. Nonetheless, the ultrasound machines are ubiquitous and cheap, and illegal screening persists. The upshot is that, in some areas, in a generation now reaching marriage age, there are as many as 150 young men for every 100 young women.

That portends some difficult times because of the large numbers of single, frustrated men in the age classes that spawn most criminals and terrorists. Indeed, the majority of terrorists behind the 9/11 attacks on New York and Washington were young adult men, the demographic age group responsible for most crime globally. Terrorists, especially suicide bombers, in other events have been of similar or younger ages, usually between 16 and 26.

Aging Populations

In most industrialized nations and some developing nations, however, momentum has largely or entirely been wrung out of the populations' dynamics as low fertility has prevailed for many decades. Most European countries, Canada, and Japan are approaching zero population growth, and several populations, including those of Germany, Russia, and several eastern European nations, have already begun to shrink. Some developing countries, including China, South Korea, Thailand, and several Caribbean islands, are not far behind. United Nations demographers project that, as birthrates continue to fall, the numbers of people over age 60 will more than triple, from about 600 million to nearly 1.9 billion in 2050, accounting for more than 20 percent of the global population at that time and as much as 30 percent or more in some countries with dwindling populations.

This momentous change has aroused alarm in some circles. A few demographers and many politicians and commentators have expressed grave concern about the future of social security programs to support the elderly, predicting dire problems for people in the comparatively small productive age groups, who will be burdened with caring for their aged parents. They want to keep populations growing, to avoid the shift toward an older age composition. Concern about aging is sometimes accompanied by near panic over the prospect of a shrinking population. Businessmen and politicians worry about too few customers for stores, too few workers to keep wages low, too few soldiers for their armies, a loss of national prestige, and so on. A number of countries, including Germany, Russia, France, and even Australia, have considered or established financial incentives for childbearing, although such measures seem to have little effect. But this worry neglects the trade-off represented by having proportionately fewer children to educate and support, and smaller cohorts of young people seeking jobs or engaging in criminal activities.

In developing nations that still have high birthrates, the youthful age composition of the population, in a context of poverty, high unemployment, poor health care, limited education, gross inequity, repressive government, and other factors, creates fertile conditions for a desire to challenge the power of the affluent. High population growth rates are expected to persist in many developing nations, with a projected annual growth rate of nearly 3 percent for people aged 20 to 34. In the face of such growth, job opportunities in developing nations, already often scarce, may fall even further below the numbers of applicants. This situation provides the impetus for large numbers of job seekers to migrate from poor countries to rich ones, legally or not.

A decrease in the percentage of people in younger cohorts and an increase in older ones is an inevitable consequence of significantly slowing and then ending population growth. Except to those foolish enough to believe that the population can grow forever, it is obvious that sooner or later the problems of a changing age structure must be faced. There is no compelling reason to postpone the inevitable and every reason to welcome it. After all, most people over 65 are not dependent in the sense that children are; most of them can take care of themselves and even contribute significantly to society, either in paid work or as volunteers. In today's industrialized nations, older people also are significantly healthier and stronger than were those of previous generations. Rather than attempting to turn back the clock and revive population growth as some observers propose, societies with aging populations could consider revising their retirement and social security arrangements.

In any case, it seems highly unlikely that either exhortations about "the crisis of the birthrate" or government bribes will lead to further population growth in most rich countries. Germany has hardly undergone an economic or social breakdown because of a dwindling population. Economic woes in Russia and eastern Europe can be ascribed to factors other than a shrinking population, and they may be more a cause of the region's very low fertility than an effect. In western Europe, limited immigration from developing countries has been tolerated as a way to augment working-age groups. Moreover, scarcity of labor could also reasonably be viewed as an incentive to increase efficiency and productivity.

Population shrinkage in Europe, Japan, and other wealthy nations is an incredibly positive trend in our view. It is, after all, the high-consuming rich who place disproportionate demands on humanity's life-support systems, to say nothing of the costs of maintaining the economic power to try to keep those demands satisfied, without regard for the costs to the world's poor people and to future generations. Changing age structures and labor pools, however, will present genuine problems of equity, with consequences for patterns of consumption, migration, and the like—all tied to the ancient Socratic question of how we should live our lives. Is it fair to expect twenty-five-year-olds to pay very high taxes to support perfectly healthy seventy-five-year-olds in retirement? Is it wise to import large numbers of young, low-wage laborers from poor nations to readjust national age structures? These and other issues are serious, as are the economic dislocations that rapidly changing age structures can cause. They demand open social discourse in all nations. Yet these are trivial difficulties compared with the problems that continued rapid global population growth is now bringing to a planet that, in the view of the vast majority of knowledgeable scientists, is already overpopulated.

CHAPTER 8

History as Cultural Evolution

"It would indeed be the ultimate tragedy if the history of the human race proved to be nothing more noble than the story of an ape playing with a box of matches on a petrol dump."

WILLIAM DAVID ORMSBY GORE, 1960[1]

ONLY HUMAN beings have evolved a sense of history, and a sense that history matters. Much of human culture evolves under the influence of religious myths generated thousands of years ago, or of historical events that happened long before any *Homo sapiens* living today was born (just think of the Battle of Kosovo of 1389, still a touchstone of Balkan politics in the 1990s, or the arguments over competing homeland claims in the Middle East). We have the capacity to carry culture across generations to an unprecedented degree, and the forces that shape us are thus, in part, of our own making. That in many ways is relevant to the way we became the dominant animal, and to how we treat one another and the environment that supports us.

History, the recording and analysis of past events, is the principal record of human cultural evolution and of how our species became a force of nature. Yes, we can make some statements about the increasing power of humanity on the basis of artifacts such as stone tools, boneyards of prehistoric human prey, ancient ruins, and myths about prehistoric times, but the written historical record yields the greater understanding of past behavior and thus can help both to inform the present and to construct a better future. Imagine, for example, trying to understand the way people treat one another in the Middle East if we had no records of the history of the region. George Santayana's statement "Those who cannot learn from history are doomed to repeat it" is, in our view, one of the most accurate of all clichés.

The factors that cause groups, institutions, and societies to take similar or divergent cultural paths can be divided into two classes: cultural macroevolutionary and cultural microevolutionary. These factors are not necessarily cultural in themselves, any more than factors that affect genetic evolution—like pollution in English woods—are genetic.

Macroevolutionary factors are constraints on, or enablers of, patterns of behavior arising from resource, geographic, or other features or circumstances external to a human society. Landlocked countries are unlikely to become sea powers, even if their educational systems give people the skills that make great admirals. Such factors were first taken seriously by Baron de La Brède et de Montesquieu (1689–1755), who could be considered the father of cultural macroevolution. He described climate, natural barriers, and soil fertility as factors that shape societies and politics.

Interestingly, Montesquieu also can be thought of as the father of political science, and he is credited with many of the ideas of governance, especially the notion of the separation of powers, that informed numerous writers of constitutions, including the founding fathers of the United States. "In republican governments, men are all equal," he once commented; "equal they are also in despotic governments: in the former, because they are everything; in the latter, because they are nothing."[2]

Microevolutionary factors are those that depend directly on the behavior of human actors, on their motives, abilities, and actions. A talented programmer will build a better climate model for an appropriate computer than will a bad programmer; a charismatic individual such as Winston Churchill can wield tremendous influence over a powerful, industrialized nation; a brilliant Montesquieu could enunciate in the eighteenth century a principle of balance in republican government that would be in the headlines in 2007 as the executive branch of the U.S. government attempted to assert control over the legislative branch.

Individuals can change the course of cultural evolution and sometimes do. Micro factors can work on the scale of a small group, as when a family plans a vacation or military leaders plan a campaign. And they can operate on a much larger scale, as when a cruise ship's itinerary is set or when Churchill in 1941 decided to protect Greece from the Nazis, and, as we will see, changed the course of the war in North Africa.

A wonderful example of the interaction of micro and macro factors can be seen in the introduction of smallpox vaccination into Japan in the mid-nineteenth century. Japanese physicians were aware of the procedure (infecting people with cowpox, caused by a nearly harmless relative of the smallpox virus), which had been invented by the English physician Edward Jenner a half-century earlier. But the physical (macro) and political (micro) isolation of Japan delayed the importation of samples of the delicate cowpox virus. Once they arrived, vaccination against smallpox spread rapidly, illustrating many key features of cultural microevolution, such as the importance of backing by powerful politicians, the existence of networks of interested individuals or small groups, and the ability to publish treatises on the procedure quickly, all of which helped overwhelm the natural isolation (macro) of the parts of an archipelago nation. And it shows the influence of a macro factor (the virus) on the (micro) transition of Japan from political isolation to the role of a world power.

Cultural microevolutionary factors can in some cases *create* macroevolutionary forces. For instance, the cultural preferences of influential leaders (e.g., wanting to enrich themselves and their friends) sometimes put powerful macroevolutionary constraints on a society. One example is the change in transportation and settlement patterns that occurred in the United States in the decades before and after World War II. The most dramatic instance was the removal of the interurban rail network in Los Angeles County and its later replacement with freeways, a move promoted by automobile manufacturers and oil and rubber companies. This change was soon emulated in cities around the country, reinforced by housing developers who built suburban subdivisions connected to city centers by freeways. This process has produced the leapfrog developments we see today around most large American cities. Suburban sprawl, with its wastefulness of both energy and productive land, virtually forces people to own cars and spend hours weekly driving to and from workplaces and to shop for food and other necessities, whatever their individual proclivities. The suburban lifestyle is a microevolutionary cultural influence, originating in decisions by a relatively small number of politicians and businessmen, that became macroevolutionary because the cultural trend it unleashed greatly altered the physical (external) options of millions of Americans as well as of future generations. The spread of suburbia, for example, limited people's movement patterns and opportunities for social contact and helped, through increasing greenhouse gas emissions from automobile use, to alter the climate, which puts external constraints on many human activities.

Among the main determinants of the broad patterns of history are the biological and physical environments in which people find themselves. They are the factors that Jared Diamond brought to wide public attention in his classic book *Guns, Germs, and Steel*. The basic idea is that, after modern human beings left Africa some 60,000 to 100,000 years ago, the environmental circumstances in which dispersing groups found themselves were critical in determining their future cultural evolution. Their fates were governed in part by the ease with which local plants and animals could be domesticated, or the diseases they faced. Such features clearly seem to explain some major differences in the patterns of development of the resulting societies.

History contains many unsolved puzzles of cultural evolution beyond those Diamond laid out. For instance, Argentines and United States citizens have had similar biophysical environments, including access to similar domesticates and technologies, for the past 250 years. Nonetheless, their societies have very different cultural histories, which are difficult to account for in terms of geography. To explain the U.S.–Argentine divergence, many historians would focus upon the different political, economic, and administrative styles of the occupying colonial powers before independence. Questions relating to why some countries today are rich and others are mired in poverty, and still others, such as Argentina, are in between, are complex and still hotly debated.

This lack of agreement is found in many other issues in cultural evolution, such as the interminable debate on whether or not slavery led to the American Civil War or whether economic disparities were more important. But here we think it can be sorted out, for as is so often the case (remember autos and sprawl), micro factors seem to have morphed into macro factors. The claim that slavery was not at the base of the split is contradicted by statements of Southerners before the war started. In addition, slavery (a micro institution based on interpersonal relationships and laws) permitted development of a plantation agricultural system that could not have existed without slaves to pick cotton (a social production system that was a macro factor, creating biophysical constraints on what kind of agriculture could be profitable). Understanding the processes of cultural change is more important than finely categorizing macro versus micro, but it is always important to keep in mind that cultural evolutionary courses followed by individuals, groups, and governments are partly determined by macro factors that are not easily or rapidly changeable.

The Beginnings of History

Herodotus (ca. 485–425 BC) is often credited with inventing history, that is, with beginning to describe and analyze, in writing, human cultural evolution. He was an inveterate traveler whose *Histories* recorded events in a war between Greek city-states and the Persian empire under Xerxes, who wished to conquer those states and enslave their people, some 2,500 years ago. It is from Herodotus' work that we can learn the stories of such famous events as the battles of Thermopylae and Marathon and about Egypt in the fifth century BC. But his work is also replete with gods and myths, explanatory stories of origins and cultural events that often contain some impossible elements—stories rather than analyses. Thus he might be viewed as transitional from preliterate societies, in which oral histories were passed down through generations and more often morphed into myths than did later histories based on written documentation.

Thucydides (ca. 460–395 BC), known from his single classic history of the Peloponnesian War, could be considered the first "modern" historian. He attempted to analyze events objectively, depending largely on contemporary sources, using interviews, checking carefully, and abjuring fantasy. He searched for, analyzed, and discussed causes for the war and recommended methods historians should use. He may have been the first scholar to recognize that causation was a factor in historical patterns of human affairs, that they could be *explained*. He was the first known cultural evolutionist.

Since then, of course, many schools of history have themselves evolved, varying in the patterns they did or didn't recognize or emphasize in history's course. Georg Wilhelm Friedrich Hegel (1770–1831) and Karl Marx (1818–83), two of the most profound thinkers on the subject, saw history as developing through a series of conflicts—in Hegel's case, between ideas as a great plan unfolds. In the Marxist case (much shaped by Hegelian thought), the conflict was between previous ruling classes and those who developed new means of production and overthrew the previous rulers. Plutarch (ca. AD 45–120), the famous Greek historian who lived just after Christ, focused on the lives of great men and was more interested in their morality and its effects on history than in discovering patterns in historical events themselves. It seems fair to say that much of history taught in schools today has followed Plutarch, the archetypal cultural microevolutionist, in the sense that it focuses more on successions and decisions of kings and other leaders than on history's broad forces.

Historians today who are interested in the "big picture" are essentially students of cultural evolution. They seek to discover major patterns of change in human history and to uncover mechanisms that explain them. They are, in a sense, analogues of scientists who seek broad patterns in genetic evolution. Those scientists are necessarily not distracted by the continual environmental variation and resultant genetic change that roils the gene pools of all sexually reproducing populations. Instead, they select model systems to understand change at a detailed level, and build on those to elucidate broad patterns.

But the analogy shouldn't be taken further. As we suggested in the discussion of memes in chapter 5, the ways in which Darwinian evolution works can be a poor model for understanding processes of cultural evolution. There is every reason to believe that change in humanity's store of non-genetic information involves a monstrously more complex process than changes in its store of genetic information. It is unclear whether broad causal mechanisms can ever be elucidated at the relatively simple level at which we understand genetic evolution. The difficulties are so severe that one outstanding evolutionist, Richard Lewontin, and a distinguished historian, Joseph Fracchia, have claimed that history should not be considered a record of cultural evolution. On this point we disagree. We think, for instance, that Robert Carneiro's testable (and partially tested) theory of the origin of states, discussed in chapter 5, is a convincing indication that culture evolves in a meaningful sense.

Standards of History

Jared Diamond's *Guns, Germs, and Steel* approach to history shares many features with an earlier reaction to the "sequence of kings and battles" approach to history. It crystallized some decades ago in the Annales school (a group of historians associated with the French journal

Annales d'histoire économique et sociale, founded in 1929), which focused attention on the context of everyday life in societies within which the rulers operated. The most famous Annales historian, Fernand Braudel, discussed features that tend to persist with the passage of time: geographic relationships, economic arrangements, cropping systems, population growth and density, and the like, which he termed "structures." He saw history as operating on three basic time scales: the short term of battles, elections, and fads; the mid term of cyclical processes that stretch over decades (such as the alternating dominance of the U.S. Congress by the Republican and Democratic political parties); and the long term, which he made famous as the *longue durée*. The *longue durée* could stretch over centuries and, in the spirit of Montesquieu's ideas, involved the geology, geography, flora, and fauna of an area as well as the people's diets and systems of trade. From the viewpoint of the *longue durée*, the most consequential cultural macroevolutionary event in human history was the invention of agriculture, which changed everything from diets to birthrates to social structures to human effects on the environment.

Another historian who has attempted to study history on a large scale is George Basalla, who has analyzed technological evolution. Changes in the devices we create is one of the most studied aspects of cultural evolution, and perhaps the most thoroughly documented conclusion about it is Basalla's, that technological innovations virtually never occur as a response to fundamental human needs. Just as the vast majority of other animals get along without any technology, so could have human beings, although the habitats in which we could thrive would have been much more limited, and human biological evolution might have gone in a different direction. Like chimpanzees and foxes, we also could get along in much of the world without that supreme invention, control of fire. And not only could our ancestors have survived without the wheel, which many claim to be the second most important invention, many did, and a very few possibly still do. The great Inca, Maya, and Aztec civilizations did—they knew the principles of the wheel but never put them to practical use. Presumably this was because of the lack of suitable mammals in the Americas to domesticate as draft animals. The perceived advantages of the wheel appear to have been most appreciated by the culture of the West during the past couple of centuries as societies wholeheartedly embraced railroads and automobiles.

There is a lot of anecdotal support for Basalla's contention that the fulfillment of basic needs doesn't explain the drive toward technological innovation. Just consider carefully what basic "needs" are met by cell phones (to drive those around you crazy with your babbling?), ever more advanced automobiles (in which to sit out traffic jams?), computers on which to have virtual sex (to replace the real thing?), and myriad other gadgets now deemed necessary. If fulfilling basic needs doesn't explain them, what does? The answer appears to be that *desire* (human aspirations, imagination, creativity, fantasies, wants) rather than necessity is the principal source of technological change. Individuals, or in some cases groups, familiar with certain processes imagine ways to make a modified artifact that performs the task of an existing artifact "better" by their standards or that will perform a brand-new task. For instance, Eli Whitney's famous "cotton gin," invented in 1794, for the first time made possible the separation by machine of the lint and seeds of short-staple cotton, whose fibers stuck tightly to the seeds. Whitney wanted to solve the problem of ginning that kind of cotton, and his invention was a brilliant solution that transformed cotton production in the United States—but in itself it didn't satisfy any basic human need not otherwise met. To the contrary, it gave new life to slavery by making cotton so profitable that it increased the demand for slave labor to plant and harvest it. Whitney's device led to the invention of a large number of improved gins. It was not created out of nothing, however, but was based on a variety of other gins that did the same job for long-staple cotton.

Similar stories can be told about other famous inventions. When James Watt developed his steam engine in the 1760s and 1770s, many steam engines of different designs were already in use in Great Britain. The transistor, invented just after World War II at Bell Laboratories by John Bardeen, Walter Brattain, and William Shockley, traces to the semiconductor devices (ones that conducted currents better than insulators but not as easily as good conductors) used in crystal radio sets that Paul built as a child.

Such continuity of innovation makes the cultural evolution of artifacts appear similar to genetic evolution. The resemblance is furthered because the environment in both cases "selects" the variants that will persist and lead to new kinds of technologies or organisms. Social and economic conditions, just like climate and the availability of animals to domesticate, play major roles in determining the technologies that survive and "reproduce." For instance, after making a series of important technological inventions (printing, the magnetic compass, and gunpowder) that would transform medieval Europe, the Chinese failed to exploit any of them to the same level themselves. Many factors have been put forth to explain why the Chinese lagged behind the West technologically in the eighteenth and nineteenth centuries. One of the most interesting is that the national fragmentation of Europe provided a series of small, flexible, competitive economies, which was more conducive to the rapid evolution of technologies. China, in contrast, was a single, bureaucracy-bound entity in which education focused heavily on classics.

But the way in which new ideas are formed in the brain, *ideation*, is still pretty much a mystery. We can see how steam engines already in use influenced Watt, but we don't know how or why his brain developed the idea of his new device. As Basalla put it, technological evolution has its Darwins but lacks its Mendels. Darwin was able to produce a mechanistic explanation of the origin of organic diversity without having uncovered the genetic mechanism itself. That had to await the monk Gregor Mendel's experiments with peas that showed that inheritance was particulate (via genes) rather than blending.

It is important to note that caution must be employed in dealing with beliefs about technological "progress," just as with genetic evolution and non-technological aspects of cultural evolution. We think the idea of technological progress can be used only in a limited sense. It is not in the sense of general improvement of the human condition, since, for example, it is debatable whether a hectic but technologically and culturally rich life under the threat of large-scale nuclear destruction is generally better than the relatively leisurely life of some hunter-gatherer groups. But in a narrower sense, a 9-megaton hydrogen bomb (with the explosive force of 9 million tons of TNT) is certainly an example of "progress" as a weapon of mass destruction relative to a stone axe or even to the atomic bomb that destroyed Hiroshima, which was barely a thousandth as powerful. Similarly, supersonic transport (SST) aircraft certainly represent an improvement over Boeing 747s in terms of speed—often cited as a measure of technological progress. But in other dimensions, such as noise creation and environmental destruction, the fleet of SSTs proposed in the 1970s was judged *not* to represent progress by people of the United States, and public opposition killed the project.

Of course, no shortage of historians in the past volunteered to place value judgments on cultures that in their opinion were better or worse or more or less "advanced." And not just historians. A judgment persists among many groups today that some peoples are intrinsically superior to others and thus are set on different historical (cultural evolutionary) trajectories. But because we know that nearly all genetic variation in *Homo sapiens* today is within rather than between populations, and since no one has ever found a genetic difference between peoples that was somehow tied to differences in intelligence, creativity, or political dominance, we feel safe in assuming that observed differences in historical trajectories are rooted in culture—be it in such things as shared attitudes of societies or the availability of animals to domesticate.

And, since many people in our own society promote its superiority (along with various versions of racial superiority), it behooves us to look carefully at the yardsticks of superiority employed. Is a society that has achieved the capability to destroy itself and others with nuclear, chemical, and biological weapons "more advanced" than one that depends on spears and clubs to settle its political differences? Should westerners believe themselves superior for their scientific advances if those advances most likely would not have occurred except for the invention by the Hindus of Arabic numerals (so called because the Arabs brought them to the West)?

It is, of course, virtually impossible to compare the results of historical sequences. How should we decide which society is the "most advanced" today? Scale of military capabilities? The United States wins hands down. Life expectancy? Japan's, of eighty-two years, leaves the United States well behind with its life expectancy of seventy-eight. Highest level of literacy? Several eastern European countries claim 100 percent adult literacy, which may not be far off the mark. Most equitable income distribution? Denmark, Japan, and Azerbaijan are in the top few of the most equitable nations, depending on the year and the metric. Below these countries, the United States is around number seventy on the list. Longest period without participating in a war? Switzerland. Income per capita? The United States is at or close to the top, depending on how per capita income is measured. Most generous nation with foreign aid? Among industrial countries, the United States is one of the stingiest in official government aid, but private aid in part makes up for it. Happiest? Surveys suggest that Mexico and Nigeria are near the top, the United States is well down the list, and Russia is near the bottom. Happiness is a tough thing to measure, though. Maybe Bhutan (not surveyed), with its government program of "gross national happiness," is the real winner. It seems there is as little basis for claiming overall "progress" in cultural evolution as there is in genetic evolution. Progress in both cases is clearly in the eye of the beholder.

The Paradox of Culture

As we said, human beings have evolved the capacity to carry culture across generations to an unprecedented degree, and the forces that shape us are thus, in part, of our own making. The paradox is that cultural stickiness, that "carrying forward" by individuals and groups, seems to be highly beneficial, at least over certain periods (think of public health technology or agriculture itself), whereas at other times cultural continuity can lead to disaster (overuse of antibiotics threatening to make them useless or, as we'll relate, farming and grazing leading to a potentially catastrophic loss of essential biodiversity). One can see the paradox clearly on short time scales in times of cultural stress. Let's consider, as an example, some aspects of World War II in the North African desert in light of the personalities and cultures involved. At the start of the war, Italy was a country with relatively little industry—it had perhaps 15 percent of the industrial capacity of Britain or France. Italy's leader, Benito Mussolini, was an opportunist, lacking in self-confidence and courage; as historian Douglas Porch put it,[3] he was an egomaniac with an inferiority complex. His personality, and its contrast with the more determined and vicious attributes of Hitler's (those reflecting, among other things, the cultural stickiness of anti-Semitism), strongly affected the course of the war in the Mediterranean theater, and perhaps even the entire war.

Despite Mussolini's "Axis of Blood and Steel" pact with Hitler, Italy remained neutral after the Nazi invasion of Poland and declared war on Britain and France only in June 1940, when General Erwin Rommel's army in France reached the English Channel. While Hitler was threatening England, Mussolini decided to grab some British possessions and ordered Marshal Rodolfo Graziani to attack Egypt from Libya. The Egyptian invasion turned out to be a disaster for the Italians, who were going up against a man some consider to have been one of the great generals of British history—Archibald Wavell, commander in chief of the British army in the Middle East. The Italians might have captured Egypt and the Suez Canal and threatened all-important Middle East oil but for some British leadership skill and Graziani's personal shortcomings, lack of supplies, and the generally poor quality of the Italian armed forces under his command.

Graziani moved into Egypt on September 13, 1940. He had perhaps 250,000 men, and the British could muster only about 30,000 to defend Egypt. Not knowing how weak the British were, Graziani penetrated to Sidi Barrâni, sixty miles inside Egypt, with a grotesque parade-ground sort of movement, and dug in.

General Wavell had a big job. The distinguished British historian Correlli Barnett outlined the cultural macroevolutionary elements involved—in this case the geophysical factors:

> The desert became a battlefield in the Second World War because it was the western flank of the British defence of the Middle East and the Axis attack from Italian Libya had to pass over it. For the British the Middle East was only just less important to the waging of the war than their own homeland; for it contained around Mosul, in Iraq, and at the head of the Persian Gulf the oilfields without which the Royal Air Force, the Army and the Royal Navy would be paralyzed. . . . [T]he long campaign of 1940–43 was not fought for the Suez Canal . . . but for oil.[4]

On December 9, 1940, the British launched an attack in Egypt called Operation Compass, originally planned as an extended raid. The British then had only some 35,000 troops to face what were by then 500,000 Italians. Despite this, the British succeeded in cutting off the Italian forces. Bold, brilliant, and unorthodox General Richard O'Connor commanded Wavell's Western Desert Force. O'Connor, with the aid of excellent signals intelligence (code breaking), managed to advance some 500 miles, taking all of far eastern Libya and reaching El Agheila on February 9, 1941. Tens of thousands of Italian prisoners were taken. But the opportunity to chase the Axis out of Libya was lost when Churchill insisted on diverting troops to defend Greece.

A few weeks later, the Afrika Korps arrived in North Africa; General Rommel was sent by Hitler to reinforce Mussolini's forces. Under Rommel's direction, the Germans and Italians pushed the British back to El Alamein. By pure chance, on April 6 a critical cultural microevolutionary incident occurred, a disaster for the British. O'Connor had been sent back to the front in Libya by Wavell, but he lost his way in the dark, ran into a German unit behind his own lines, and was captured. The British lost their most astute commander. After O'Connor's brilliant campaign and accidental loss, the British army began to display its cultural rigidity —a severe failure to adapt as the environment changed. The British army had a long and largely successful tradition of upper-class officers leading lower-class soldiers in tradition-bound regiments. As Barnett said, it was "spiritually a peasant levy led by the gentry and the aristocracy."[5] The leaders had to be stolid, brave, and dedicated to their regiments, but not necessarily bright or quick-witted.

Moreover, the revered and once effective regimental structure was ill suited to the combat environmental conditions on the desert "oceans" of North Africa. What was *not* needed was the modern equivalent of cavalry regiments trained to do reconnaissance, shock attacks, and exploitation of breakthroughs by infantry and artillery. What *was* needed were coordinated armored divisions that relied heavily on artillery for both attack and defense. The British focused on anti-tank guns in defense and failed to use their artillery offensively to destroy Axis anti-tank guns and thus pave the way for the advance of their tanks. The Germans used their fabled 88-mm anti-aircraft gun with enormous effectiveness as an anti-tank weapon. But even though the British had superior anti-aircraft guns, the 3.7 inch, in considerable quantity, they never used them against tanks, where they would have been very effective. They were, after all, "anti-aircraft" guns.

It was a superb example of "stickiness" in cultural evolution—failure to change in response to an altered environment. The Germans, as they industrialized, had moved more toward an industrial army and developed a greater capacity to learn from their mistakes. In the desert, the British ended up paying a high price for their conservatism. Fear of change, combined with the regimental system and old-boy control, killed large numbers of British soldiers needlessly in North Africa. Institutions such as that regimental system transcend individuals. They function to enforce rules of social order and mandate cooperative behavior. They embody the "rules of the game."

Plenty of cultural macroevolutionary factors lay beneath the panoply of cultural microevolution in action during the Mediterranean campaigns of World War II. For instance, Italy was no military powerhouse at the start of the war, a result of 1,500 years of complex historical interactions following the collapse of the Roman Empire in the fifth century AD but also of the limited resources at its command. It had too little good farmland and generally lacked enough key resources such as oil, coal, and iron ore to be an industrial power. England, by contrast, had taken good advantage of its fine harbors, its once abundant tall oaks, and later its rich deposits of coal in becoming the world's top naval power by 1900. During the Mediterranean campaigns, it often used that naval power to great advantage, as well as its possession of Malta, a key Mediterranean base astride Axis supply lines to North Africa.

Consideration of any significant aspect of history shows that numerous levels of analysis are usually necessary to get a reasonable grasp of causes and consequences. But most cultural evolutionary interpretations are nevertheless tentative and often controversial. The interacting natures of millions of people and their multitudinous institutions and diverse norms have produced an emergent complexity that frequently makes it difficult to identify a few driving forces. This is especially true because historians themselves are participants in the ongoing rewriting of history and bring their own cultural information to the task. As American historian Arthur M. Schlesinger Jr. put it, historians are prisoners of their personal experiences: "We bring to history the preconceptions of our personalities and of our age."[6] Nonetheless, there are broad evolutionary patterns in humanity's non-genetic information, just as there are in its genetic information. It could be a boon to humanity if a "cultural Mendel" appeared to start developing a unitary theory of the mechanisms of cultural change—how our non-genetic information has interacted with diverse social and biophysical environments to make us lords of Earth.

CHAPTER 9

Cycles of Life (and Death)

"Whenever we discover that humans have changed their global environment, we build upon our recognition that living organisms . . . can affect the conditions of an entire planet."
WILLIAM H. SCHLESINGER, 1997[1]

THE RISE OF *Homo sapiens* to dominance—indeed our very existence today—would have been impossible without the environments in which our ancestors flourished. Those environments can be thought of as everything external to people (or any other organisms) that can influence their lives. Your acquaintances and local mosquitoes are part of your environment; a planet circling a star 10 million light years away and a fish living in the deep ocean are not. As we have seen, both the biological and the physical environments with which all human beings interact have shaped both our biological and our cultural evolution, and they will continue to do so. To understand both how the dominant animal is reshaping its surroundings and what those transformations mean for the future of humanity, we first need to take a closer look at those environments and their workings.

A crucial aspect of the human physical environment that has been much in the news lately is the uneven distribution of mineral resources, so important to supplying materials and energy to modern societies. That unevenness traces to the uneven distribution of biophysical environments of the far distant past. That's when the plants (including, probably, algae), a few animals, and microorganisms whose remains were subsequently processed geologically by pressure and heat into fossil fuels—coal, oil, natural gas—were living and dying in swampy areas and shallow seas. Because the Middle East several hundred million years ago had the appropriate environment, it now has vast deposits of oil and gas in comparison with other regions of the world. Knowledge of that supply in turn has led most recently to the United States' invasion of Iraq and its attempts to assert control over Iraq's oil resource. And now competition is building among the United States, China, and India for access to the resources of the Caspian Sea basin in central Asia, underlying parts of Iran, Kazakhstan, Turkmenistan, Azerbaijan, and southern Russia.

Once itself a leading producer of oil and gas, the United States depleted its oil reserves over more than a century and around 1970 reached a peak of production (about 12 million barrels per day, which has slowly declined since then to a little over 8 million). Today some 60 percent of the oil used in the United States is imported. The vast majority of the world's remaining conventional oil reserves lie in the Middle East, but in 2006 that region supplied only about 12 percent of U.S. consumption. Other countries, such as Canada, Mexico, Venezuela, and Nigeria, supply significant amounts, though with smaller reserves to rely on in future years. Oil deposits, like those of other mineral resources, are indeed very unevenly distributed globally. To complicate the situation further, world and U.S. oil consumption have both been rising by well over 1 percent per year, and as long as demand continues to rise, so will the United States need to import oil from distant and increasingly remote and difficult-to-tap deposits.

Today the global supply of pumpable petroleum is also showing signs of reaching its production peak soon, if it hasn't already. For the past decade or so, annual oil production worldwide has exceeded the amount of oil found each year in newly discovered fields, despite enormous efforts of exploration, a telltale sign that the majority of accessible and more or less easily extracted conventional sources may reach exhaustion in the foreseeable future. Indeed, one of the potentially most significant oil discoveries in the early twenty-first century was under thousands of feet of water in the Gulf of Mexico, beneath further thousands of feet of submarine crust. Such discoveries will be costly, in both dollars and energy, to exploit.

Reaching the peak of global oil production was inevitable, sooner or later, because the resource is essentially finite (not renewed on a time scale of interest to society), but it appears to be occurring at a time when demand can be expected to accelerate as many developing nations rapidly industrialize. China in particular, with annual economic growth rates since 1990 of 8 to 10 percent, has depleted its modest oil resources and is seeking assured supplies elsewhere. India, whose population size will soon surpass China's, is following a similar path. The worldwide rise in oil and gasoline prices since 2005 is most likely related to tightening supplies as production increasingly lags behind rising demand. Supplies on the world market might have been even tighter if the rising prices hadn't forced the poorest countries to cut back their consumption. An ever larger global population competing with increasing intensity for a dwindling supply of oil does not bode well for a peaceful and prosperous future for civilization.

Oil is not the only mineral resource whose scarcity could constrain civilization's future, however. Sources of natural gas also are showing signs of depletion, and rich deposits of uranium (for nuclear power) and bauxite (the ore for aluminum), among others, are limited. These resources generally will not "run out" but will decline in extractability, concentration, or both until they become uneconomic to mobilize or their exploitation does more environmental damage than society will permit.

Living Resources

But minerals are not the only heterogeneously distributed resources that are significant to humanity. Over most of the hundreds of thousands of years of human history, the distribution of climates and the closely related distribution and abundance of fresh water have been even more important to the course of human evolution than mineral distributions have been. Both not only affect human populations directly but also are important factors in the patterns of abundance of the plants and animals that our hunter-gatherer ancestors depended upon for food, and upon which industrialized agriculture is based today. Just as the distribution of potential draft animals influenced the use of the wheel, the distribution of climates in which essential grains can be grown may have a huge influence on how civilization responds to rapid climate change.

Like Earth's non-living resources, its biological resources (biodiversity) are very unequally distributed as well. The most obvious large-scale pattern is the heavy concentration of species diversity in tropical areas, especially in tropical moist forests (tropical rain forests, "jungles") and on coral reefs. This means, among other things, that much of Earth's biological riches are in relatively poor nations.

There are also many important smaller-scale (secondary) patterns in the distribution of both populations and species. One of the most important of these secondary patterns depends on winds that produce nutrient-rich currents in the oceans by causing upwellings of cold water. The main examples are found along the western coasts of continents—including the Humboldt Current off South America, the California Current off North America, the Canary Current off northwestern Africa and the Iberian peninsula, and the Benguela Current off southern Africa. Those currents produce a substantial portion of the world's oceanic fisheries' catch. Thus some of our planet's extensive ocean areas with distinct interacting physicochemical characteristics and communities of organisms are shaped by currents.

You may not be familiar with those ecosystems, but you are familiar with others, even if you don't realize it. An aquarium containing some tropical fishes and aquatic plants is a miniature ecosystem. You are embedded in a vast ecosystem, the biosphere, which extends from a few miles above Earth's surface to the oceans' bottom and almost two miles underground, where a few bacteria live and grow slowly, using hydrogen or perhaps radioactivity in the hot rocks to power their lives. An ecosystem is simply the collection (community again) of interacting organisms—plants, animals, fungi, and microbes—that are found in an area, along with the physical environment that influences them and that they influence. Indeed, that little pool of water in a pitcher-plant leaf described in chapter 1 is an ecosystem, containing not just the pitcher-plant mosquitoes but also microorganisms, aquatic stages of midges, and other creatures that depend on the trapped prey for nourishment.

The boundaries of an ecosystem are simply a matter of definition, since every part of the world is bound into webs of interactions that include living and non-living elements. The aquarium, for instance, is also part of the global ecosystem, just as the underground bacteria and the pitcher-plant leaf are; if you are a farmer, your farm could be defined and studied as an ecosystem, which again is part of the larger ecosystem of the biosphere. Ecosystems, even the global one, are not "closed," at least not for long. If the sun's energy were blocked out for even a short time, the biosphere would collapse. That's because all ecosystems require inputs of energy.

Energy

Energy is a key concept, which we can define rather simply as the ability to effect change in our physical world. Two fundamental rules about energy help clarify how ecosystems work. The first is that, while energy can change its form, it can be neither created nor destroyed. If one does the accounting accurately, the sum of energy in the universe remains constant—but that energy can change from, say, energy stored in the chemical bonds of gasoline into the energy of heat and then into the energy of motion, as in the engines of automobiles. Even the energy released in an atomic explosion, or, more slowly, in a nuclear power plant, is not "new" energy. It is energy in the form of matter converted into (mostly) the energy of heat.

Matter is just another form of energy, and the relationship of that form of energy to more familiar forms is given by Einstein's famous formula, $E = mc^2$, where E is the amount of energy, m is the amount of matter (mass), and c is the speed of light in a vacuum. Since the square of the speed of light is a *very* large number, you can see that a tiny bit of matter can contain an enormous amount of energy. In a Hiroshima-sized atomic bomb with the explosive force of about 15,000 tons of TNT, enough to devastate any city's core and burn much of the rest, a total of about one-fortieth of an ounce of matter—that is, energy in the form of matter—is converted into heat energy to provide that blast, as well as other forms of energy such as X-rays and gamma rays (very high-energy penetrating radiation, some of which does not contribute to the explosion).

The rule that energy can be neither created nor destroyed is the famous "first law of thermodynamics." The second law of thermodynamics is trickier but also gives us a lot of information useful in understanding everyday events. One way to formulate the second law is that, although energy can be neither created nor destroyed, it can flow to places where it cannot do useful work. For instance, once gasoline is burned to drive a car, its energy has not disappeared—it still is present in the motion of the car and in the hot engine and exhaust. The heat energy is ultimately transferred into the atmosphere, and the energy of motion is eventually also transformed into heat and transferred into the atmosphere as the tires rub against the road and as the brakes are applied, heating the brake disks. All of that heat energy still exists, but it is dispersed in the form of (minutely) increased movement of molecules in the atmosphere, and it cannot then do useful work—that is, it cannot be recaptured somehow and employed to heat a cup of coffee.

Another way of stating the second law is that the universe as a whole is becoming ever more disordered—that there is a general tendency for structures to break down and for order to move toward chaos. Technically, the second law of thermodynamics says that the entropy—the randomness or uncertainty—of the universe is increasing. Order cannot be reestablished without the use of energy. For instance, if a drop of ink is placed in a glass of water, the ink-water structure, or order, breaks down into a murky chaos. The previous order can be restored only by using a substantial amount of energy—in this instance, the energy needed to distill the water and reconcentrate the ink. That a spontaneous reaction moves only in the direction of order to disorder is something everyone observes every day. Ice cubes in iced tea melt; they never appear spontaneously. Soles of shoes wear away; they never rebuild themselves. Dead squirrels on the highway rot and disappear; they never unsquash and run off.

The second law tells us a lot about ecosystems. First, they can't be closed for long because they must be open to fresh inputs of energy. Otherwise the inexorable workings of the second law would cause them to break down into disorder. For example, energy for the aquarium consists of light, which powers the plants, and food added for the fish, if they don't eat the plants. If you cover the aquarium with a dark cloth and don't add food, a gorgeous underwater display will quickly be transformed into a smelly, disordered mess.

Food Chains and Material Cycles

The fundamental source of energy for nearly all ecosystems is sunlight.[2] Sunlight drives the process of photosynthesis, whereby the energy derived from the sun is used by green plants, in combination with carbon dioxide from the atmosphere and water and mineral nutrients obtained from the soil, to build energy-rich molecules. Those molecules are combinations of carbon, hydrogen, and oxygen—carbohydrates—which the plants then use to power their life processes (metabolism). Plants extract the energy from the carbohydrates by an opposite process, a slow, low-temperature burning (oxidation) process called respiration.

Using that extracted energy, the plants synthesize other molecules, such as proteins, fats, and DNA, to build and repair their bodies and to reproduce.

Animals also use the energy of the sun, but they get it by eating the plants that captured the sun's energy through photosynthesis (these are plant-eaters, or herbivores), or they get it second-, third-, or even fourth-hand by eating other animals (flesh-eaters; carnivores or parasites). It is, then, basically correct to say that "all flesh is grass." If photosynthesis didn't exist, we wouldn't either. Furthermore, neither would fossil fuels—even they are ultimately derived from solar power.

The rate at which the sun's energy is fixed by photosynthesis, minus that used by the plants themselves, is the net primary production (NPP) of an ecosystem. NPP is an important quantity, since it is the fundamental food/energy supply for almost all non-photosynthetic organisms, from viruses and fungi to cockroaches and us. Starting with measures related to NPP, ecologists can calculate, for example, how much corn must be grown to provide a million people with corn-fed beef.

The eating sequences just described are called food chains. An example might start with a grass, which gets its energy directly from the sun (and so technically is a primary producer) and uses that energy to power its own life processes. Some of that energy is then obtained by a cow that eats the grass (herbivore, primary consumer), and a portion of that in turn goes to a woman who has a chunk of the cow, a steak, for dinner (carnivore, secondary consumer). Suppose the unlucky diner is devoured by a lion during the night (carnivore, tertiary consumer), which gets some of the energy but itself is plagued by fleas (parasites, quaternary consumers), which also get a little. Thus the sun's energy is passed in four steps from the producer to the last consumer in the chain—sunlight powers fleas! Each level in a food chain is often referred to as a trophic (feeding) level. In reality, food chains are generally interlinked in complex food webs (figure 9-1). After all, people eat apples and lettuce as well as steaks (technically, we are omnivores), and lions eat things other than people; food chains with single links are simply a useful abstraction.

Along the sides and at the ends of all food chains—in a sense, the last consumers—are the decomposers that feed on waste products such as feces and dandruff and on dead bodies—the ultimate trophic level. All non-living organic matter is ultimately decomposed (unless somehow sequestered from such activity, as was the matter that became fossil

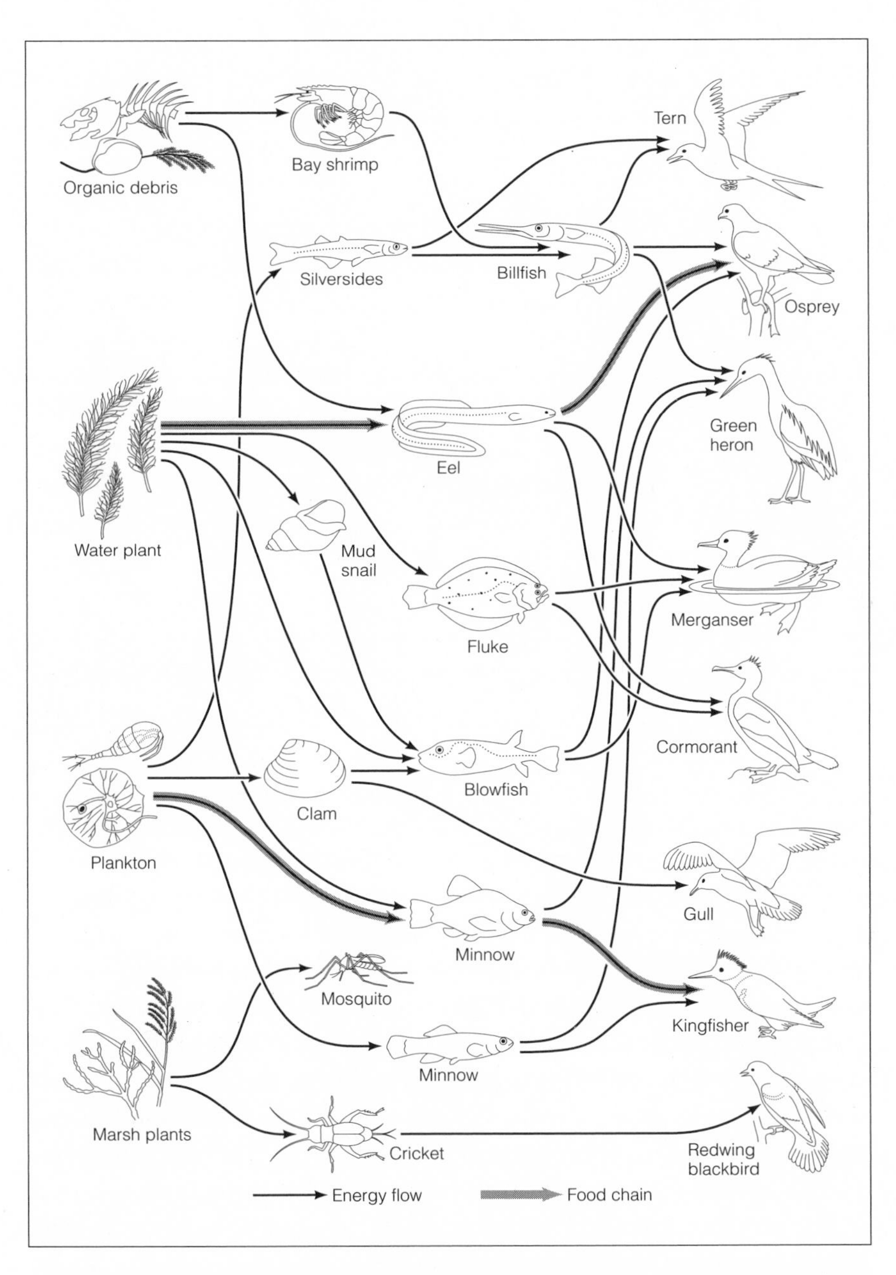

FIGURE 9-1. Portion of a food web in a Long Island estuary. Arrows indicate energy flows. Two food chains are indicated by the gray overlays: water plants → eel → osprey and plankton → minnow → kingfisher. Based on G. M. Woodwell, Toxic substances and ecological cycles, *Scientific American* 216 (1967): 24–31.

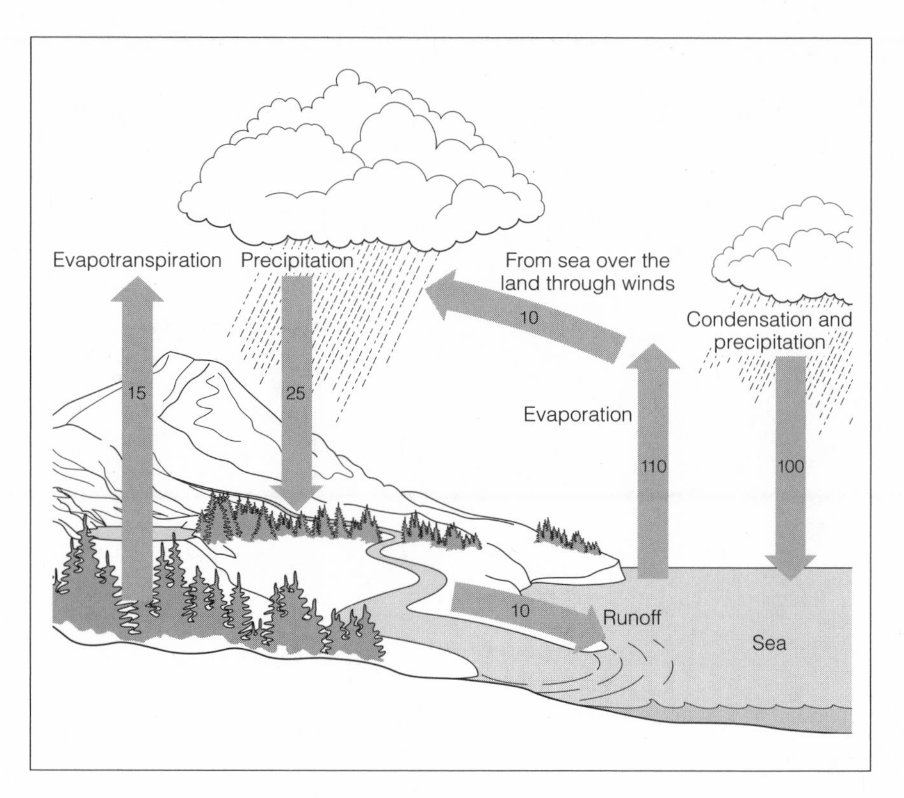

FIGURE 9-2. The hydrologic cycle. Evapotranspiration includes water vapor evaporation from the surface and that being transpired from plants. Approximate quantities are shown in cubic miles of water per year. Data from M. I. Budyko, *Climate and Life* (Academic Press, New York, 1974).

fuels), including dead leaves, dead vultures, and dead bacteria. The sun, of course, also ultimately powers the vultures, dust mites, dung beetles, and bacteria that do the decomposing. Scrounging hand-me-down solar energy from such sources may seem disgusting (to us), but it is the only way that complex molecules can be broken down into the simple mineral nutrients—carbon, phosphorus, nitrogen, sulfur, and other elements essential to life—that plants can absorb from soil or water bodies.

Decomposition keeps the nutrients in the ecosystem moving in circular paths from the physical environment to plants to consumers to decomposers and back to the physical environment and available to plants. These nutrient cycles—biogeochemical cycles, as they are called, because biological food webs interact with the geological and chemical properties of Earth—are important characteristics of ecosystems and of the entire planet. They are closely tied to the movement of water

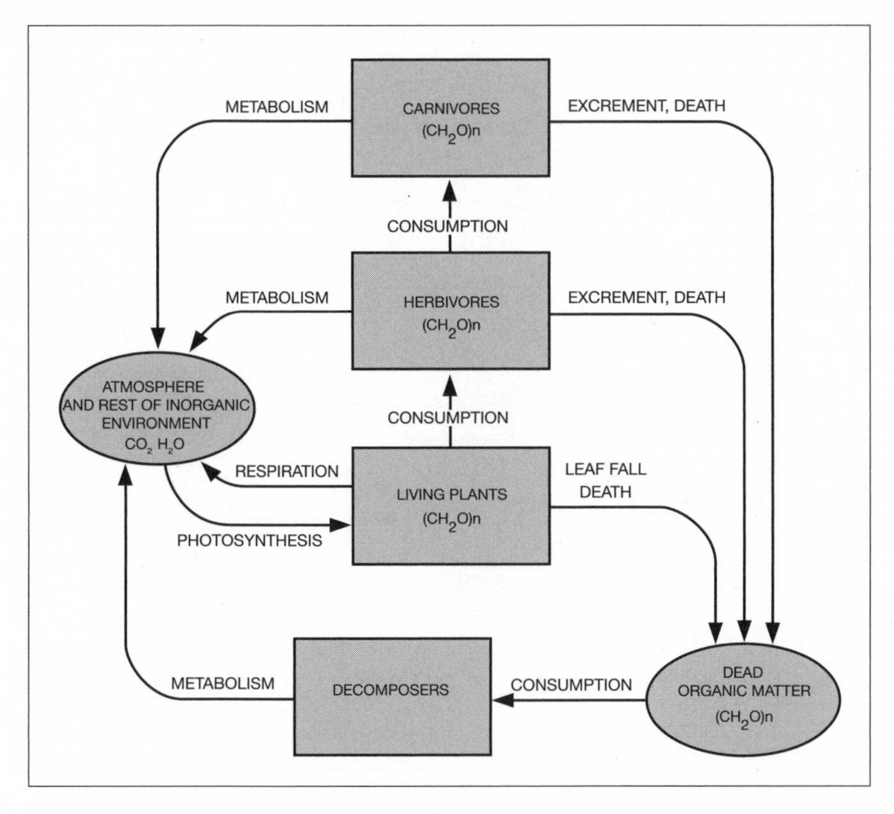

FIGURE 9-3. Global carbon cycle, simplified. "$(CH_2O)_n$" symbolizes carbohydrates and, in addition, amino acids and other carbon-containing organic molecules. Note that the whole system is driven by incoming solar energy, which powers photosynthesis.

through ecosystems, the critical hydrologic cycle (figure 9-2). The biogeochemical and hydrologic cycles together make life possible.

Among the elements most important to life are carbon and nitrogen; simplified diagrams of the global carbon and nitrogen cycles are in figures 9-3 and 9-4. The essence of the carbon cycle, as you can see, is centered in photosynthesis, respiration, and food webs. The nitrogen cycle illustrates the pattern of flows and pools typical of other biogeochemical cycles. The largest pool is molecular nitrogen (N_2), which makes up 78 percent of the atmosphere. But nitrogen in the form of N_2 cannot be used by most organisms, which require nitrogen for the manufacture of proteins—essential compounds in all life-forms. Some species of bacteria, however, can contribute to the critical and complicated process of nitrogen fixation, converting N_2 to ammonia (NH_3) or eventually to nitrates (NO_3), which can be used by plants.

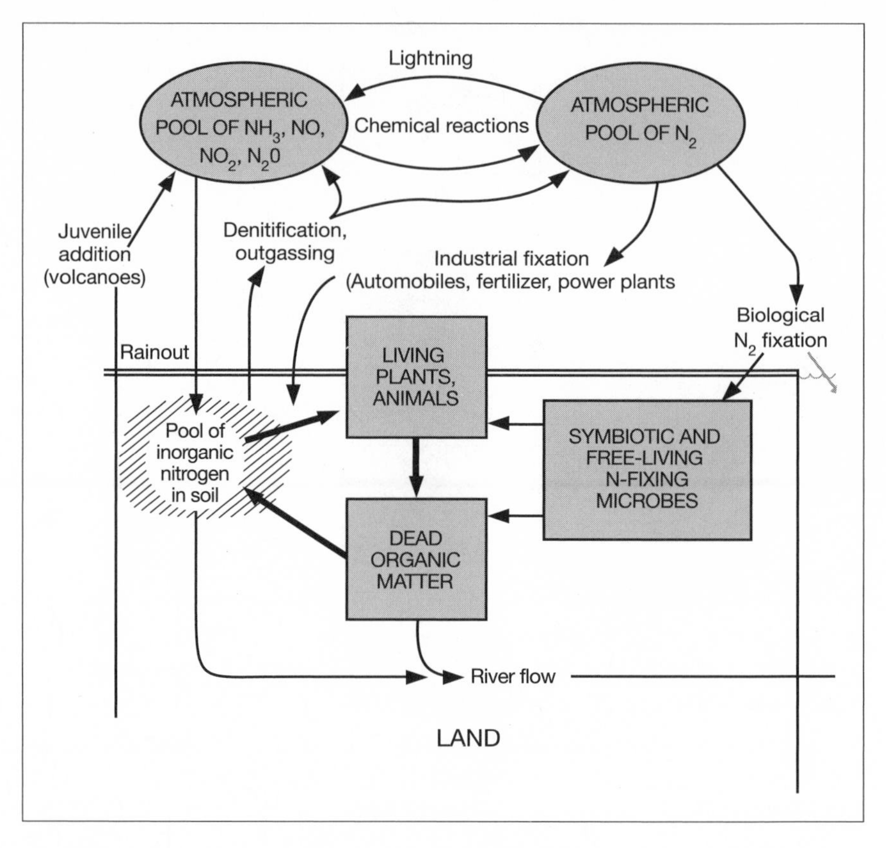

FIGURE 9-4. Global nitrogen cycle, simplified. Heavier arrows show the largest flows.

Thus, despite the vast quantity of nitrogen in the air we breathe, life on Earth as we know it is utterly dependent on a handful of kinds of tiny nitrogen-fixing microorganisms. The most famous of the nitrogen fixers are bacteria of the genus *Rhizobium*, which live in special nodules on the roots of legumes (plants of the pea and bean family). The bacteria get the energy they need for the nitrogen-fixing reaction from the legume and in turn supply the plant with vital ammonia to manufacture the amino acids that are assembled, under the direction of DNA, into proteins. This is a beautiful example of a coevolved relationship, one that humanity takes great advantage of by planting legumes in rotation with other crops because they produce excess fixed nitrogen that restores the ammonia and nitrate pools in the soil. Those nitrogen pools in soil are depleted by most other plants, such as wheat, which cannot fix nitrogen. Meanwhile other, denitrifying bacteria gain energy by converting nitrates to molecular nitrogen (N_2), which is then returned to the atmospheric pool.

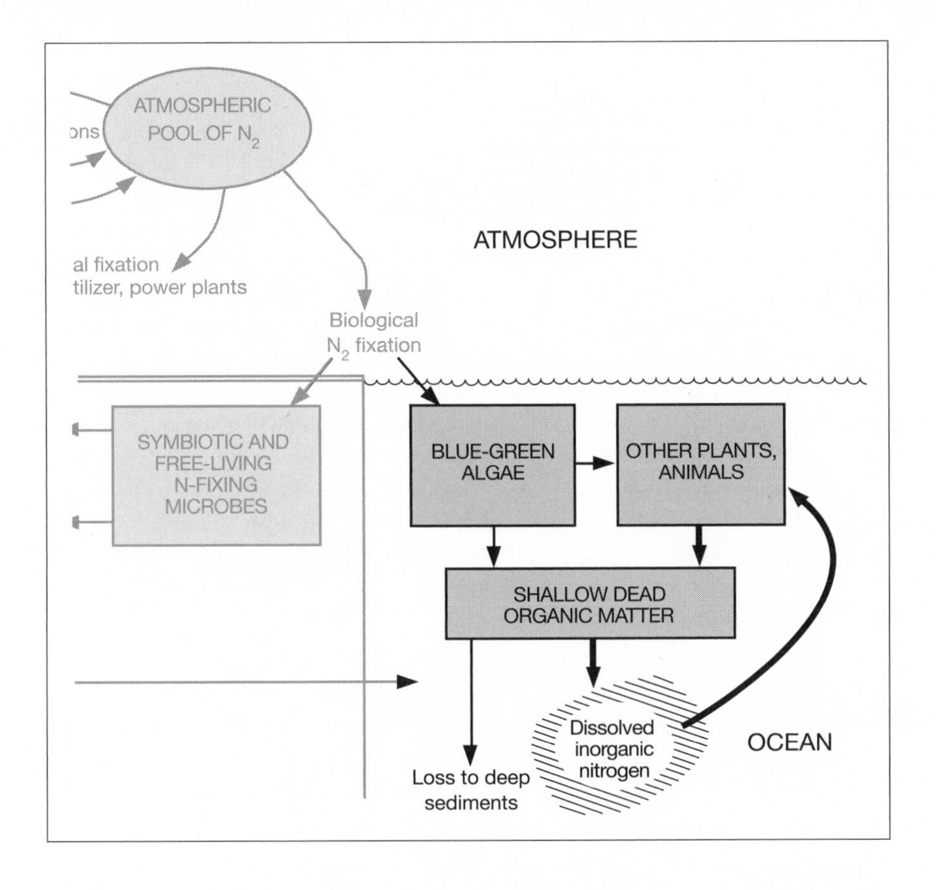

Industrial society deliberately adds fixed nitrogen, sometimes in large quantities, to the cycle in the form of fertilizers that make the growing of crops on nitrogen-poor soils economically feasible. Humanity also burns fossil fuels, wood, and natural vegetation, drains wetlands, and clears land, all of which free nitrogen and make it available to the global cycle. These human activities have doubled the rate at which nitrogen enters the land-based nitrogen cycle, which, among other things, can have nasty effects on aquatic ecosystems, which we'll discuss in chapter 12. And, by greatly changing their environments, such human activities alter the evolutionary trajectories of many organisms.

It's in the functioning of the food-chain parts of the cycles that the second law of thermodynamics has its best-known effect. Each time the sun's energy is used, less of it is available to do work. Thus fewer calories of the sun's energy can be captured by herbivores to drive their life processes than were originally captured by the plants. Even fewer can be obtained by carnivores, and fewer still by creatures that eat carnivores.

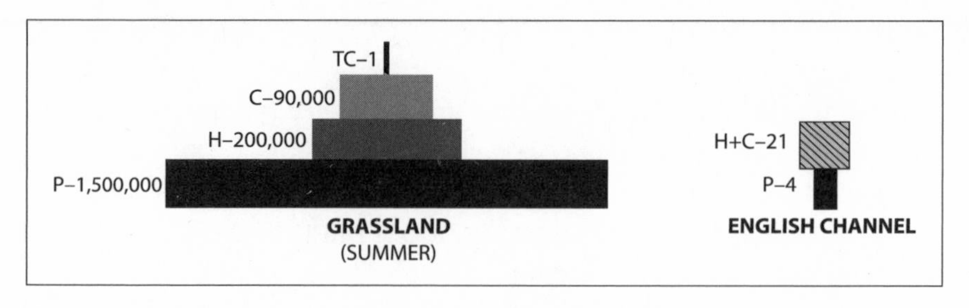

FIGURE 9-5. *Left*: biomass pyramid of a summer grassland. P = producers (photosynthesizing plants that use solar energy to generate living mass); H = herbivores; C = carnivores; TC = top carnivores—those that feed on other animals. Numbers represent estimated individuals per 1,000 square yards, a surrogate for their weight. *Right*: biomass pyramid for the English Channel. Numbers are approximate ounces of dry weight per square yard. Note that a smaller mass of producers can support a larger mass of herbivores and carnivores because of the rapid reproduction of the producers (plant plankton)—just as you don't need to have a mass of food in your refrigerator greater than your mass, since you "reproduce" the contents from the supermarket every few days. Adapted from E. P. Odum, *Fundamentals of Ecology*, 3rd ed. (W. B. Saunders, Philadelphia, 1971).

The poor decomposers, with their critical job of closing the cycles, as a group get the least energy of all. A rough rule of thumb, which may be too pessimistic, is that close to 90 percent of the useful energy becomes unavailable at each step of a food chain. Thus, all else being equal, in a given food chain there can be only about 10 percent of the weight of herbivores as of plants, and only 10 percent of the weight of carnivores as of herbivores. This produces the biomass (living weight) pyramids (figure 9-5) famous in biology. It also is the reason that insect carnivore populations are usually smaller than those of the herbivorous pests they feed on, why there are more wildebeests and zebras on the Serengeti Plain than there are lions, and so forth.

Of course, everything is always not equal. For instance, suppose the herbivores are tiny aquatic animals (zooplankton) that have a rapid turnover (say, a generation a week) and the carnivores are a fish with a generation time of a year. Then at any given time there can be a greater biomass of carnivores than of herbivores, producing an inverted biomass pyramid. But it takes a lot of energy to power all that zooplankton reproduction, and if one looks at the energy use at each level in the chain, the pyramidal structure is restored.

An important practical lesson from all this is simple: many more people can be fed directly on grains such as wheat, rice, and corn than if the grains are fed to cattle and then the beef is fed to people. An important basic principle that also follows from the second law of thermodynamics is that, while materials cycle in ecosystems, energy does not. The useful energy that enters an ecosystem eventually all ends up as heat energy too dissipated to do work.

The complexity of ecosystems is difficult to exaggerate, and the complexity starts at the level of individuals. Organisms are exchanging gases and materials with their environments; animals, for example, breathe in oxygen to use in burning their carbohydrates and exhale carbon dioxide (CO_2) as the exhaust of that process. When they digest food, they break down proteins and release key breakdown products in the form of ammonia, urea, or uric acid into the environment. The excretory product that evolves is primarily a function of how severe the organism's need is to conserve water. Ammonia requires relatively little energy to make but requires a lot of water to excrete, and it is the excretory product of fishes; uric acid is the opposite, and it is excreted by desert lizards. We're in between—the "active ingredient" in our urine is mostly urea, although we do excrete a little uric acid. People whose bodies, for reasons not well understood, either produce too much uric acid or can't excrete it readily enough deposit it in joints, where it causes the painful disease gout.

As already indicated, individual organisms can occupy any one of a variety of positions in ecosystems: producers, carnivores, parasites, and so forth. They also can play diverse roles in their communities, such as a shrub that has evolved to grow well in the shade or a bug that is specialized to detoxify the shrub's chemical defenses. Such roles, you will recall, are usually referred to as niches—the "big-beaked finch niche," "shade plant niches," or the "shade plant herbivore niche." More familiar niche names include "pollinator" (any critter that pollinates plants) and "insectivore" (anything that eats insects); more detailed descriptions such as "frugivorous bat" (any bat that eats fruit) are also used.

To make things a little more complicated, many organisms interact with their environments in ways that allow them in part to construct their own niches—as when a wasp builds a little mud nest as a nursery for its young. In some cases niche construction by one species creates or changes entire ecosystems. Obvious examples of the latter are beavers, which cut down trees and build dams that alter entire landscapes. On a smaller scale sapsuckers, a type of woodpecker, drill their own holes for nesting and sap wells for feeding, and in the process they create small ecosystems used by other birds, squirrels, and insects. And, of course, on a grand scale, human beings are modifying the entire biosphere into a niche for themselves. All three—people, sapsuckers, beavers—are examples of what have been termed ecosystem engineers, which, in the course of constructing their own niches, transform those of other organisms, and thereby influence the evolutionary trajectory of all.

Organisms play more than one role, and so their niches are multidimensional and may not be easy to define. Hummingbirds, for example, are nectarivores (they lap up nectar from flowers), and in the process they become pollinators: they get nectar from, and transfer pollen for, plants that have coevolved flowers of exactly the right shape to attach pollen grains to the hummers. And the hummers are also insectivores; they eat bugs, from which they get critical protein that typically is scarce in nectar. Grizzly bears also have multiple roles. They eat salmon from oceanic food chains, as well as rodents and berries. This is why ecosystems normally contain not a set of discrete food chains but instead a complex food web, one that is connected to the webs of other ecosystems. Niches, food chains and webs, and niche construction are all useful concepts; they are handy ways of thinking about the complexities of how living things interact—not concrete things we can see on a casual nature walk!

Soils and Sediments

Unnoticed complexity in ecosystems may be at its highest in the soil. Soil is not, as many think, just ground-up rock that helps prop up plants. It is a rich ecosystem in itself, although vitally connected to whatever is above it. Each square meter of rich soil may contain tens of thousands of tiny invertebrates, such as insects, mites, and nematodes, and trillions of algae, fungi, and bacteria, all of which are part of a vast, interconnected world of soils and sediments. In many respects, our knowledge of that world is at a stage similar to the understanding of the aboveground plants and animals possessed by Charles Darwin and Alfred Russel Wallace after their nineteenth-century explorations. They were familiar with the "entangled banks" Darwin wrote of, but not with entangled soils.

What can nevertheless be said with assurance is that ecosystems above- and belowground are tightly interconnected, and so are those between bodies of water and the sediments below—and that many activities belowground and within sediments are involved in supplying essential services to humanity. The most obvious example is dealing with the dead—the recycling of nutrients by soil organisms that break down plant and animal wastes and corpses into the essential nutrients that can again be utilized to build new plant and animal (including human) tissues. Without the activities of these unsung creatures, human life would be impossible, just as it would be in the absence of photosynthesis.

The Climate Connection

The linkage between the atmosphere and oceans is arguably the most important aspect of Earth's physical configuration from the standpoint of Earth's organisms. The atmosphere-ocean system influences virtually every ecosystem (the exceptions may be bacterial systems miles below Earth's surface). Most important, the atmosphere-ocean system controls not only the temperatures to which organisms are exposed but also atmospheric circulation, oceanic currents, precipitation and the hydrologic cycle, and so on, all of which profoundly influence the lives and evolution of virtually all organisms, people included.

The climate—the average temperature, wind, and precipitation conditions at any part of Earth's surface—is basically a product of the functioning of a heat engine: a device that changes thermal to mechanical energy. It is driven by energy from the sun, and its actions are mediated by Earth's rotation and the tilt of its axis, which create the seasons. It is this climatic system that is at the center of the issue of climate change, so much in the news today.

Energy arrives from the sun in the form of electromagnetic ultraviolet (UV), visible, and infrared (IR) radiation. Almost all the IR and the shorter wavelengths of UV (UVB and UVC) are absorbed by gases high in the atmosphere, but, unless they are reflected by clouds, longer wavelengths of UVA and visible light are transmitted to the surface almost undiminished. The stratospheric absorption of UVB is carried out almost exclusively by the unstable three-oxygen molecule ozone (O_3), which is critically important, since UVB radiation is poisonous to life. Indeed, before photosynthetic organisms in the sea produced enough oxygen to generate a constant supply of ozone in the stratosphere, life was restricted to the oceans.

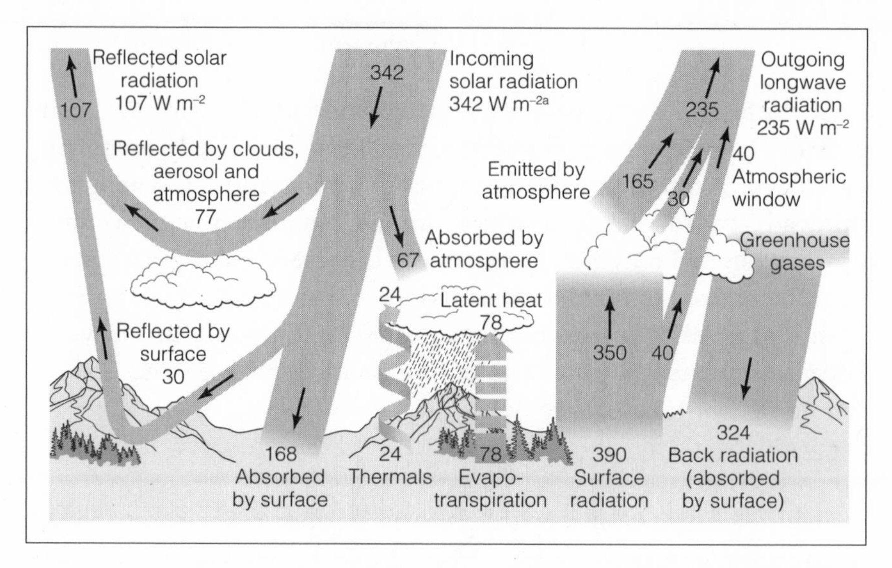

FIGURE 9-6. About a quarter of solar energy is reflected, and about half is absorbed at the surface by soil, oceans, vegetation, and human infrastructure, warming the surface, evaporating water, and powering photosynthesis. Numbers are rough estimates of watts per square meter (W / m^2) of Earth's surface. The total rate of energy leaving (235 W / m^2 as infrared and 107 W / m^2 as reflected sunlight) is the same total as the 342 W / m^2 of incoming sunlight, so Earth is in rough energy balance. Particles and greenhouse gases in the atmosphere absorb outgoing infrared radiation (except for wavelengths in the "atmospheric window") and reradiate it toward the surface. That back-radiation creates the "greenhouse effect." Some heat leaves the surface with packets of rising air (thermals), and some is released when rising water vapor (a gas) condenses into a cloud of liquid water molecules. Redrawn from J. T. Kiehl and K. E. Trenberth, Earth's annual global mean energy budget, *Bulletin of the American Meteorological Society* 78, no. 2 (1997).

Of the solar radiation that reaches the lower atmosphere, a little over a quarter is reflected by clouds, dust, and surface features, especially snow and ice. Clouds do by far the most reflecting. This reflectivity of the planet is known technically as its albedo. Overall, about half of the energy that strikes the top of the atmosphere penetrates to the surface and is not reflected. Rather, it is absorbed by soil, oceans, vegetation, buildings, people, and so forth, warming them, evaporating water, and so on (figure 9-6). A critical tiny fraction is captured by plants and converted to chemical energy by the process of photosynthesis.

But, as you will recall from the first law of thermodynamics, that can't be the end of the story. Since energy can be neither created nor destroyed, if Earth simply absorbed part of the incoming solar energy, it would long since have been burned to a cinder. What comes in must go out if the planet is to remain roughly in equilibrium. So energy is constantly leaving Earth, being reradiated to outer space in the form of long-wavelength infrared (IR) radiation. The flows among atmospheric gases, dust, and clouds and the surface are complex, but there is one overriding principle: the components of the atmosphere that were largely transparent to incoming short-wavelength visible and ultraviolet radiation are not all transparent to IR radiation. For instance, molecules of water vapor and carbon dioxide do not transmit IR but absorb it. They are themselves warmed and reradiate the energy, again as IR. But that reradiation is not all directed toward outer space—roughly half of it is returned toward the surface.

That reradiation of IR toward the surface creates what is often called the greenhouse effect—so named because it increases Earth's average surface temperature from –18°C (0°F) to 15°C (59°F).[3] This is fortunate, since without the greenhouse effect, Earth would be very inhospitable to most forms of life. The water vapor and other gases whose molecules intercept outgoing IR are collectively known as greenhouse gases, and they are central players in the story of climate change, which we tell in chapter 13.

More of the sun's energy reaches the equatorial regions than the poles, and the temperature difference between the tropics and the poles is a major driver of weather. Hot air rises and is cooled, and in the process moisture is squeezed out of it. That abundant rainfall is what makes forests of the Amazon River basin and other tropical rain forests flourish. The freshly dried cool air travels poleward at high altitude and then, at about latitudes 30° north and south, it descends and rewarms. That cascade of warm, dry air is what creates Earth's major deserts in those latitudes, for example, the Sahara and Chile's Atacama Desert. Further transport toward the poles is primarily accomplished by large cyclonic systems (areas of low pressure with winds spiraling inward), which operate counterclockwise in the Northern Hemisphere, transporting warm air northward in the eastern part of the circulation and carrying cool air southward in the western part. These are just some of the major features of the complex vertical and horizontal atmospheric circulation resulting from differential heating of the planet, which creates the climates that have been so crucial to human evolution, both genetic and cultural.

Differences in temperature, by creating pressure gradients that produce wind patterns, also drive global ocean currents. Those currents interact with winds to create weather patterns and diverse climatic conditions. For example, the Gulf Stream in the Atlantic Ocean carries heat northward and transfers it to winds that carry warm air over Europe, making England's climate much milder than that of Labrador, which is at the same latitude. Winds and oceanic circulation also perform such functions as carrying the dust that falls out on islands far out in the oceans, making them fertile, and causing the upwellings of nutrients that support the production of fishes that humanity harvests.

The Gaia Concept

The most comprehensive ecosystem is the biosphere—the thin envelope of atmosphere and even thinner portion of Earth's surface that supports life. This is a region about 10 miles thick around the surface of a planet that is almost 8,000 miles in diameter—roughly equivalent to the skin on an apple.

Not just humanity but life itself has had a dramatic impact on the planet. Among other things, life has injected enough oxygen into the atmosphere that it comprises one-fifth of its gases, has built huge coral reefs in the oceans (the joint work of coral animals and algae that leads to the large rocklike deposits), and has colonized and modified the land surface, in the process changing Earth's albedo. Life may even have played a role in creating the continents. Apparently the same photosynthesis that poured oxygen into the oceans and atmosphere may have (emphasis on "may," since this is recent speculation) supplied energy for the breakdown of basalt and production of the granite and granitelike materials that make up the continental crust.

There is thus no question that life has had, and continues to have, substantial effects on the atmosphere and physical features of Earth's surface. This has led to the "Gaia hypothesis," the brainchild of chemist James Lovelock—the idea that Earth itself is an organism-like, self-adjusting, evolving system whose component organisms are continually improving their physical environments. Those fascinated by this idea should contemplate the fate of the anaerobes, microorganisms that did not use oxygen to power their activities and for which oxygen was a poison. Increasing quantities of oxygen were added to the atmosphere and oceans by photosynthesizing organisms over billions of years. By 500 million years ago or so, the oxygen was poisoning large numbers of anaerobes. Some of those organisms still survive in a few oxygen-free habitats, but many must have been pushed to extinction. They would be unlikely to support the hypothesis that organisms were keeping Earth hospitable for life! Yes, life has certainly reshaped Earth (this has sometimes been called the "weak Gaia hypothesis"), but Earth in no way resembles an organism. There are no large populations of replicating Earths for selection to operate upon, no Earth-sized equivalents of speckled and melanic moths or DDT-resistant and DDT-susceptible fruit flies to differentially breed or survive. If Earth cannot replicate itself and does not react adaptively to its environment, that's because it is not a survivor of millions of generations of natural selection operating on many thousands of reproducing Earths. The idea that life is somehow "directing" the evolution of Earth for life's own benefit is a non-starter.

The Complexity of the Biosphere

Just how complex the biosphere is, and how impossible it is for humanity to "manage" it, was dramatically demonstrated by the "Biosphere 2" experiment. Starting in the mid-1980s, at a cost of some $200 million (this was to be no on-the-cheap operation), about three acres in Arizona were covered by a greenhouse that was to be hermetically sealed. It was supposed to become a self-supporting microcosmic representation of the entire biosphere ("Biosphere 1"). A series of small ecosystems were put together: a grassland, a marsh, a bit of ocean with a coral reef, all complete with appropriate plants, animals, and microorganisms. A good deal of the area was planned for intensive agriculture, which would supply food for four male and four female human "Biospherians," who were sealed into Biosphere 2 in 1991.

The results were quite predictable, at least in retrospect, precisely because of the unpredictability of many of the interactions in even such a simple ecosystem. There was not enough knowledge to design appropriately for them, so all of the component ecosystems quite rapidly collapsed. The oxygen level dropped to the equivalent of that 3,000 feet above the highest peaks in the contiguous forty-eight states (some 17,000 feet above sea level); concentrations of nitrous oxide rose to levels that could cause brain damage; nineteen of twenty-four vertebrate species went extinct almost immediately, as did all of the pollinators; and vines added to the experiment to absorb carbon dioxide went rampant. Since natural pest control services were absent, "crazy ants" (so called because they dash rapidly about), cockroaches, and katydids swarmed. Despite imports to Biosphere 2 of oxygen to breathe, energy to supplement the sunlight, and candy bars to prevent starvation, the whole endeavor collapsed and the Biospherians fled.

The experiment dramatically demonstrated that relatively stable ecosystem functioning depends on a vast diversity of organisms, and that human beings are unlikely to be capable of assembling reasonably stable large-scale ecosystems—especially ones able to adjust to constantly changing conditions (as have those in "Biosphere 1"). As we will see in the next few chapters, Earth's ecosystems—our life-support systems—are now being destroyed piecemeal. Biosphere 2 serves as a warning that today's demolition of the ecosystems of Biosphere 1 may portend consequences similar to those in Biosphere 2. And we have no Biosphere 3 to flee to.

CHAPTER 10

Ecosystems and Human Domination of Earth

"It is now generally admitted by plant ecologists, not only that vegetation is constantly undergoing various kinds of change, but that the increasing habit of concentrating attention on these changes instead of studying plant communities as if they were static entities is leading to a far deeper insight into the nature of vegetation and the parts it plays in the world."

ARTHUR TANSLEY, 1935[1]

THE BIOSPHERE is now being profoundly reshaped by human society, and the *pace* of that change now appears to be exceeding the highest rates seen in the past 65 million years, since collision with an extraterrestrial object led to extermination of the dinosaurs. But change, although ecologists were somewhat slow to recognize it, has always been a major feature of the biosphere and its component ecosystems—something noted by Arthur Tansley, who coined the term "ecosystem."

The human penchants to favor constancy and to categorize helped shape the conclusions of scientists in the 1930s, 1940s, and 1950s, the early decades of investigation of what was termed community ecology—the study of groups of organisms of different species, for instance, in entangled banks such as Darwin described (see chapter 1). Each part of the world was thought to have a "climax" community. That community was the assemblage of plants and animals to which any disturbed area tended to return eventually, determined primarily by the influence of physical factors on the vegetation. The process, known as succession, gradually restored nature to its "rightful" condition—rather like a geographic great chain of being—or so the idea went.

More recently, ecologists have realized that they were imagining climaxes as static conditions when in fact not only local areas but the entire planet is in a constant state of flux. Succession, in the sense of a sequence of plant replacement after disturbance in an area, is a common phenomenon. For example, when a glacier retreats in Alaska as the planet heats, the first organisms to colonize a given patch of newly exposed ground are usually mosses and a perennial herb, the dwarf fireweed. Other reeds and herbs follow, depending on slope and soil formation, and after that, prostrate willow shrubs. These are usually followed by erect willows and alders that form thickets mixed with other herbs and shrubs. Then in some areas what is generally called a "climax" gradually takes shape, which after a couple of centuries consists of a spruce forest with primarily mosses growing beneath it. In other areas a hemlock forest appears as the climax stage.

But the forest climax isn't the end—it will gradually disappear as the mosses accumulate water and kill the trees by starving their roots of needed oxygen. The result is another succession probably leading to open bog land interspersed with ponds (muskeg), and then very likely another succession as climate change or other factors influence the local ecosystem. Indeed, the successional process itself is always changing, often at rates slow enough to be almost undetectable during a single scientist's lifetime. Even before human beings began rapidly altering Earth, erosion wore away mountains, rivers changed course, climates shifted, and continents slowly moved, all of which had substantive effects on plant and animal ecosystems and triggered trillions (or more, depending on how you count them) of successional sequences.

How much of a "climax" community exists in any area is pretty much a matter of defining boundaries and time scale. Each year when Paul and his colleagues go up to Jasper Ridge Biological Preserve, the Stanford University campus home of the checkerspot butterflies we discussed in chapter 7, the vegetation looks pretty much the same in summer, fall, and winter. The checkerspots that lived on the grassland growing on serpentine soil (which has a special mineral mix with lots of heavy metals, making it inhospitable for plants not evolutionarily adapted to it) have been extinct for more than a decade. The species of plants that the checkerspot caterpillars fed on are still there, but in declining abundance, and the glorious display of spring flowers for which the serpentine is famous is fading. And deer and mountain lions may be more common than they were when Paul started research on Jasper Ridge in 1960. Furthermore, shrubbery is slowly invading some of the grassland areas that are not on serpentine soil and may be invading the serpentine itself. The main local "climax" vegetation is generally considered to be chaparral, dense stands of tough-leafed evergreen shrubs typical of Mediterranean dry-summer, wet-winter climates. But succession to chaparral has been delayed on the serpentine. Nitrogen pollution from automobile exhaust could change all this by enriching the soil and making it more hospitable to shrubs and Eurasian weeds, putting an end to the spring beauty in the relatively near future.

In the Rocky Mountains of Colorado, aspens are a fast-growing early successional tree species that took over after the climax spruce forests were logged; they are expected eventually to be replaced by slow-growing spruce. Aspens still dominate the slopes around the Rocky Mountain Biological Laboratory more than a century after logging ended. But now they are beginning to die out in much of the state, not because spruce are taking over their habitat but possibly because of some human alteration of the environment. Spruce forests are still considered the climax community, but climate heating and related insect attacks may decimate the spruce and set the stage for a new "climax" to be gradually established.

Although the idea of a climax community still makes some sense on a time scale of centuries, since within a few human lifetimes some communities seem permanent, climax communities certainly are not unchanging entities. Furthermore, that time scale of centuries for apparent permanence may be shrinking fast because of anthropogenic (human-caused) climate change. And as communities change, so do ecosystems, and so do the niches occupied by their component organisms.

Human beings, like beavers and many other organisms, as we saw in the preceding chapter, help to construct their own ecological niches. A burgeoning human population, perpetually trying to increase its consumption, is now reshaping the entire Earth to suit its own immediate needs—to be its niche. *Homo sapiens* has become the dominant animal on the planet, changing the climate, the land surface, the depths of the oceans, the global distribution of organisms, and the chemical composition of the biosphere at an accelerating rate. People grow crops on about 12 percent of Earth's land surface, have paved or built on another couple of percent, graze their livestock on 25 percent or so more, and in various ways exploit most of the roughly 30 percent that remains in forests or tree farms. It is only the remaining surface in high mountains, under ice, or in extreme desert that human beings have not extensively exploited.

People now mine many minerals—ecosystem components vital to industrial societies—at rates far above those of natural processes of weathering and erosion, and are changing the composition of the atmosphere, as we will see, in ways that could threaten the persistence of civilization. Indeed, with few exceptions, every cubic inch of the biosphere has been influenced directly or indirectly by our species. A 1986 study, for example, calculated that humanity had already destroyed or was currently co-opting for its own use almost half of Earth's net primary production (NPP), the basic food/energy supply of people and almost everything else but green plants, by farming, grazing, fishing, building, and so on. And in the process, society was interfering with the ways in which the biosphere supports people now and could support people in the future.

Ecosystem Goods, Ecosystem Services

Ecosystems are such complicated entities; is there a reason to bother to know them in any detail? The quick answer is that they support our lives—without them we would all be dead. As biologists have often pointed out, the human economy is a wholly owned subsidiary of Earth's ecosystems. To mention a few examples, without reserves of genetic variability in the wild relatives of crops, we would find it difficult or impossible to maintain much agricultural production. Without pollinators, our diets would change dramatically for the worse; without natural enemies of agricultural pests, we'd all soon starve to death. Without natural controls on the hydrologic cycle, we would be thirstier, hungrier, and more subject to death and destruction from flooding. In short, without ecosystems, there would be no economy at all.

Scientists often divide what humanity obtains from ecosystems into two categories, ecosystem goods and ecosystem services, as we do here (others just have one category and consider the delivery of goods as a principal service). Ecosystem goods include timber taken from natural forests and fish captured in the sea, among many others, all of which are key elements in providing well-being to human beings. In sub-Saharan Africa and parts of Asia, firewood and bush meat (wild animals harvested for food in tropical areas) are further examples of such critical ecosystem goods. Ecosystems everywhere have supplied people with natural ingredients for often very effective pharmaceuticals, such as morphine, aspirin, and digitalis, most of them based on those plant defensive chemicals we discussed in chapter 2.

Furthermore, scientists exploring the vast diversity of life are continually adding to benefits that can be extracted from humanity's natural capital—Earth's ecosystems, from which people draw a steady flow of "interest" in the form of goods and services. Those goods are not just medicines but everything from petroleum (derived from past ecosystems) and other minerals, timber, and fishes to novel structural materials inspired by the flexibility, toughness, and light weight of mollusk shells; organisms that can monitor the state of life-support systems for us; substances to help control pests; and more. For instance, some ant species that are threatened by fungal and bacterial diseases have a special gland that secretes a powerful antibiotic, which keeps their surface much cleaner than the skin of most people, and scientists are beginning to isolate chemicals from those glands to help humanity in its fight against pathogens.[2]

Valuable as ecosystem goods are, however, they are increasingly being replaced by goods from human-managed systems (tree plantations, fish farms, cattle feedlots) and industrial operations (kerosene substituting for firewood, synthetic fertilizers for natural ones, antibiotics for herbal remedies). While there is considerable discussion and debate about the costs and benefits of these substitutions—for instance, about the environmental impacts of fish and shrimp farming or of eucalyptus and pine plantations—it is clear that modern industry can more readily devise substitutes for the ecosystem goods flowing from our natural capital than for the ecosystem services.

Freshwater supply and flood control are two related and typical ecosystem services that are cases in point. Heavily vegetated highland ecosystems can help ensure that rainfall and snowmelt infiltrate into aquifers; groundwater provides a reliable base flow to streams and rivers. If changes to vegetation, such as by fire or logging, compact soils and reduce infiltration, excess surface water may cause flooding while reduced groundwater diminishes dry-season stream flow. In Rwanda's Parc National des Volcans, a home of the mountain gorillas, the forested mountains are the source of some crucial ecosystem services. About 25 percent of that desperately poor nation's water used in agriculture is metered out from those forests, and they supply the esthetic pleasure of viewing mountain gorillas, a source of a major portion of the nation's foreign exchange. About 40 percent of the montane forest was cleared for an ill-fated scheme to grow the flowers from which the natural pesticide pyrethrum (another plant defensive chemical) is extracted. In Rwanda it was claimed that one result of this clearing was the loss of about 10 percent of the water bound for farms, though many factors could have been involved in that loss.

Rwanda's deforestation and water problems are far from unique; without vegetation the provision of dependable water flows is often difficult or impossible. On the Philippine island of Luzon, deforestation is causing rapid siltation of the reservoirs that supply much of the water to irrigate rice. Globally, human beings make use of more than a quarter of the fresh water that returns to the atmosphere from the land—for instance, by passing it through corn, wheat, and rice plants—and over half of the accessible runoff (streams, rivers, replenishable groundwater). There appears to be no easy or cheap mechanism by which these proportions could be greatly increased without doing substantial environmental damage by diverting water from organisms involved in providing other services or by building dams in inappropriate places.

Wetlands are among the most productive ecosystems, but people frequently drain them in order to provide more cropland or living space for burgeoning populations. Over half of all original wetlands in North America and Europe have been drained, as have a quarter of those in Asia. Wetlands help control the flow rate of water, slowing it and absorbing floodwaters, thus providing time for the natural detoxifying activities of decomposing microorganisms to break down poisonous substances and for silt and other solid matter to settle out. The loss of this purification service results in either an increase in human waterborne disease or the high costs of building water purification facilities.

Increasingly the value of wetlands is being recognized and people are choosing to take steps to restore their natural services. Wetlands, for example, are now legally protected in the United States (though not always in practice). Some are also preserved internationally by the Ramsar Convention on Wetlands, which was signed in Ramsar, Iran, in 1971. The Ramsar Convention now has more than 150 signatory nations and protects more than 1,600 wetlands of international importance.

Perhaps the biggest success story in the area of the freshwater-supply service concerns the drinking water of New York City. Although the water was once highly prized and sold bottled to other cities, in 1989 its quality dropped below Environmental Protection Agency (EPA) standards. The city calculated that it would cost $6–$8 billion to build a water treatment plant and many millions a year to run it. New York chose instead to restore the watershed from which a network of streams and reservoirs provided the city's water. At a cost of some $1.5 billion, cattle were fenced out of stream courses, other farming activities that led to runoff of fertilizer and other farm chemicals were restricted, outhouses were replaced, property owners were required to collect and treat their own storm water, development was forbidden in some areas, areas under paving were limited, and so on. With these measures, the ecosystem service of water purification was restored at a fraction of the cost of replacing it industrially.

Generation and maintenance of soils is another critical ecosystem service. Soils are ecosystems in themselves, and their myriad organisms are essential to preserving soil fertility. That fertility is in turn essential to agriculture and forestry, supplying humanity with food and timber. All too often, fertile soils are now destroyed by carelessly removing vegetation cover and exposing them to wind and water erosion. Bad farming practices also contribute, for instance, by continually removing crops that carry nutrients with them without using environmentally sound techniques to return nutrients to the soil. This "mining" of the soil requires careful replacement of minerals—if, for example, crop residues are removed without taking other steps to replace mineral nutrients, soil fertility progressively degrades.

Ocean flood protection is a service vital to the huge proportion of humanity that lives near the sea. While most people viewed the December 2004 tsunami disaster in the Indian Ocean and Hurricane Katrina, which nearly destroyed New Orleans in August 2005, as "acts of God," both were, in their consequences, also partly acts of humankind. Most of the coastal mangrove forests in the Indian Ocean region had been cleared for resorts, shrimp culture, and other human activities.

FIGURE 10-1. If not properly vegetated, land is subject to rapid erosive loss of soil. Photograph courtesy of iStockphoto.

Dense mangrove stands can make excellent buffers against storm surges, as can coral reefs, another natural ecosystem type now being degraded by human activities. About 27 percent of Earth's coral reefs and 35 percent of its mangroves have been destroyed in just the past few decades. It seems likely that the 2004 tsunami was of a magnitude that largely overwhelmed natural defenses, but they certainly could provide some protection from smaller tsunamis and from the increasingly intense storms that are likely as Earth heats up. In New Orleans, human modifications of Mississippi River flows via dredging of navigation channels have led to loss of much of the coastal wetlands that once protected the city.

Another essential ecosystem service is *natural pest control*. It has been estimated that about 99 percent of potential crop pests are controlled naturally by climate (winter or dry seasons) and natural enemies. Populations of the herbivorous insects that attack our crops are presented with an abundant food supply and would destroy the crops were it not for their own enemies—predacious insects, spiders, birds, fungi, and so on. If those enemies did not exist or were greatly reduced in number and variety, it would be virtually impossible for crop agriculture to continue. That's because plant-eating insects already have a great deal of evolutionary experience with the poisons plants synthesize to protect themselves—remember the monarch butterflies (chapter 2)—and quickly evolve resistance to the poisons humanity manufactures to attack them. While more potent chemicals could be employed to battle the pests, there are limits because of threats that more toxic compounds pose to human life. The Cañete Valley pesticide disaster in which overuse of pesticides resulted in *more* cotton pests (chapter 2) suggested that it is not possible to use ever more toxic compounds without the risk of killing people.

The potency of "biological" pest controls was dramatically demonstrated when *Opuntia* cactus was imported from South America to Australia as an ornamental plant early in the twentieth century. The cactus took over some 100,000 square miles of Queensland and New South Wales, with half of that area so badly infested that the land was made useless. A natural enemy of the cactus, a tiny moth from Argentina called *Cactoblastis cactorum*, was introduced into Australia as a control measure. Its role in keeping *Opuntia* from being a pest in South America was suggested by the way it decimated the cactus in Australia. After suppressing the *Opuntia*, the moth also soon faded from prominence. Without knowing the history, no one would ever guess that the big cactus was being controlled by the tiny moth.

The use of exotic organisms as biological control agents carries substantial risks, however. For example, in 1935 cane toads (*Bufo marinus*) from Latin America were introduced into Australia to control cane beetles that were attacking sugarcane crops. From an original batch of some 100 individuals released in northern Queensland, the toads multiplied and spread rapidly across the landscape, devouring all manner of small animals and poisoning large animals that tried to eat them. They are thought to be responsible for a decline in native marsupial predators. The cane beetles have remained uncontrolled, but the toads spread over much of northeastern Australia and are still moving west at about a mile per year; they have become one of the most serious pests on the continent.

Another ecosystem service is *amelioration of the weather*. This is largely effected through moderation of the hydrologic cycle by land-based vegetation, which provides windbreaks and extensive shade and recycles water, thus preventing drought. Without the latter service provided by the dense vegetation, the Amazon River basin would be desertified, since most of the rainwater that falls on the rain forest is recycled many times over by the vegetation's transpiration—the evaporation of water by plants that draw water and nutrients into the roots.

FIGURE 10-2. *Top*: a dense growth of imported *Opuntia* (prickly pear) cactus making an area of eastern Australia virtually worthless. *Bottom*: same area cleared by the tiny moth *Cactoblastis cactorum*. The moth was introduced from South America, homeland of the *Opuntia*, where it naturally feeds on the cactus. Photo courtesy of the Alan Fletcher Research Station.

The *cycling of nutrients* is a service without which Earth would be lifeless. *Waste disposal* is a related service; both are accomplished in the process of decomposition. People believe that sewage disposal systems are entirely human made, but in fact they simply harness natural bacterial decomposers to perform their normal functions. In addition, both nutrient cycling and the hydrologic cycle are key factors in control of the gaseous composition of the atmosphere and thus help to regulate the strength of the greenhouse effect.

A more familiar, often observed service is *pollination of plants*, including many that are crops. This service is provided by the wind in the case of grains, and by numerous kinds of animals, including bees, butterflies, moths, hummingbirds, and bats, that move pollen from flower to flower. A great many plants are wholly dependent on animal pollination for their reproduction; certain crops may be able to reproduce without pollination, but their quality is improved by it. In Costa Rica, for instance, leaving 10 percent of a coffee farm as natural forest rather than clearing it substantially *increases* the yield and quality of the coffee. That is because bees that live in the forest pollinate the coffee, moving pollen around among the plants, which then produce more and better beans than if the pollen simply fertilizes the flower in which it is produced or another on the same plant.

Finally, ecosystem services include the *provision of entertainment and tranquility* to millions of bird-watchers, gardeners, hikers, scuba divers, butterfly enthusiasts, hunters and anglers, gorilla and elephant admirers, and others who derive pleasure from natural ecosystems, and, of course, all those human beings who simply appreciate natural beauty and often travel to attractive places for vacations. That appreciation amounts to a significant part of the global economy, although it is often ignored by interests that profit from exploiting and depleting natural systems such as forests, wetlands, and fishery stocks.

So, we repeat, ecosystem services (and goods) support the human economy—they are absolutely essential. They often operate on a scale far larger than artificially contrived human replacements could possibly match. When you understand the overwhelming importance of all those ecosystem services, and you hear a politician say, "We've got to focus on the economy and not the ecology," you will know she doesn't understand very basic facts of human existence.

The State of Ecosystem Services

The Millennium Ecosystem Assessment was a three-year study, published in 2005, in which 1,360 scientists from ninety-five nations participated. The primary intent was to evaluate the suppliers of these vital services (table 10-1). The report is intended to be useful at global, regional, and local levels. It includes not only an assessment of the current state of the world's ecosystems but also projections of alternative future trends and consideration of related policy choices. The report summarizes the condition of our life-support systems as follows (*our comments are added in parentheses*):[3]

- The supply of certain ecosystem services has increased at the expense of others. Significant gains in the provision of food and fiber have been achieved through habitat conversion, increased abstraction and degradation of inland waters, and reduced biodiversity. (With almost 7 billion people to feed and care for, this focus on food, clothing, and structural materials is a natural human response, but one with side effects that may compromise the human future.)
- Fish cannot continue to be harvested from wild populations at the present rate. Deep-ocean and coastal fish stocks have changed substantially in most parts of the world, and the harvests have begun to decline and will continue to do so. (Fishery harvests increased to around 80 million metric tons in the 1980s, and since then they have declined somewhat, replaced in part by fish farming, which was producing one-third of the seafood harvest by 2000. The most recent expert opinion is that the seas will be essentially devoid of harvestable wild fish in a few decades.)

TABLE 10-1 **Classification of Ecosystem Services in the Millennium Ecosystem Assessment, with Examples**

Supporting Services	*Provisioning Services*
Nutrient cycling	Food
Soil formation	Timber and fiber
Primary production	Fresh water
Preservation of genetic resources	Fuel
Regulating Services	*Cultural Services*
Climate amelioration	Esthetic
Flood control	Spiritual
Agricultural pest control	Educational
Water purification	Recreational

- The supply of fresh water to people is already inadequate to meet human and ecosystem needs in large areas of the world, and the gap between supply and demand will continue to widen if current patterns of water use continue. About one-fifth of the entire human population has no access to dependable supplies today. This situation seems likely to worsen, as climate change may lead to desertification and reductions of precipitation in many regions, generally altering patterns of water availability continuously and making many facilities such as dams and aqueducts less useful.
- Declining trends in the capacity of ecosystems to render pollutants harmless, keep nutrient levels in balance, give protection from natural disasters, and control the outbreaks of pests, diseases, and invasive organisms are apparent in many places. (Dilution and buffering against disaster get tougher and tougher as the numbers of people grow and the demand of each person for consumables also grows. This is especially serious because many pollutants may cause damage in very small quantities, such as human-made chemicals that function in a manner similar to the hormones—chemical messengers—that control human development.)

The Three Horsemen of Environmental Impact

The magnitude of the effects of human activities on our life-support systems and the services and goods they supply can be visualized as the product of three factors, summarized in the *I = PAT* identity. The overall ***I***mpact is equal to the product of ***P***opulation size times per capita ***A***ffluence times the ***T***echnologies and socioeconomic-political systems used to generate the affluence. "Affluence" in this context is simply per capita consumption. And inclusion of the "socioeconomic-political systems used" is to emphasize, for example, that automobile commuters in a community with HOV (high-occupancy vehicle) lanes and flextime work schedules will have a smaller impact, ceteris paribus, than would an otherwise identical community without such measures (among other things, traffic jams waste gasoline).

The *I = PAT* equation was developed by John Holdren and us as a simple aid to analysis and was not intended for detailed calculation. It is primarily useful in making comparisons. Its original intent was to point out such things as that the population problem in the United States was more severe than that in many "overpopulated" poor nations because of the high level of affluence and wasteful consumption in the United States. The average American consumes dozens of times more resources than the average citizen of a very poor nation. In general, a population's impact on the environment could be reduced by their shrinking the population's size or by taking numerous other actions, such as using trains rather than cars to get to work, settling in high-density apartments rather than in suburban sprawl developments, or reducing their consumption of meat.

The condition of human life-support systems has long been of great concern to many scientists. They believe that, with present levels of population size and aggregate consumption, humanity cannot be supported long on our single, finite Earth. The potential consequences of trying were emphasized in the 1992 *World Scientists' Warning to Humanity*, which was endorsed by more than 1,500 leading scientists, including more than half of the living Nobel laureates in science. It stated in part:

> Human beings and the natural world are on a collision course. Human activities inflict harsh and often irreversible damage on the environment and on critical resources. If not checked, many of our current practices put at serious risk the future that we wish for human society and the plant and animal kingdoms, and may so alter the living world that it will be unable to sustain life in the manner that we know.[4]

That humanity may be ever more rapidly sawing off the branch on which it is perched was echoed in a 1993 report, issued by fifty-eight of the world's academies of science,[5] and has been a conclusion reached in numerous reports issued in the United States, some as early as the 1940s.[6] In the chapters ahead we will explore what lies behind these conclusions and what can be done to avert the unhappy outcomes they portend.

CHAPTER 11

Consumption and Its Costs

> "There is no human circumstance more tragic than the persisting existence of a harmful condition for which a remedy is readily available. Family planning, to relate population to world resources, is possible, practical and necessary. Unlike plagues of the dark ages or contemporary diseases we do not yet understand, the modern plague of overpopulation is soluble by means we have discovered and with resources we possess."
>
> MARTIN LUTHER KING JR., 1966[1]

TWO KEY related factors we've seen again and again in patterns of human cultural evolution, and in the patterns of environmental impacts caused by the consumption activities of human societies, are the number of people involved and how that number is changing. The United States, of course, would need far less oil today—and its contribution to climate change would be far less—if its population had not more than doubled in size since 1950. If the population of India were declining rather than rising, Indians would be competing less heavily with other animals for that nation's net primary production, the basic food supply of all animals.

Even though virtually every environmental and social problem nationally and globally is exacerbated today by continued population growth, the topic of human numbers is absent not only from policy discussions in the United States but also largely from public discourse. You don't need to be a rocket scientist to see that increased population size, all else being equal, means more greenhouse gases released into the atmosphere and thus more rapid climate change, more tropical forests cut down, more traffic jams, and more extensive and intensive agriculture.

But all is *not* equal. Human beings are bright apes—they brought the richest land under agriculture first; took water from the nearest, most convenient sources first; mined the most concentrated ores before tackling those that had only traces of the desired metal; and exploited the shallowest, most extensive oil deposits that would flow naturally before drilling down thousands of feet, opening fields beneath shallow seas, and forcing recalcitrant thick oil from wells. So, on average and with a given set of technologies, the addition now of each person to the population has a disproportionately negative effect on the environment as poorer soils are cultivated to feed her, water is brought from more distant, more polluted sources to supply her, coal mines are dug deeper (and made more dangerous as a result) to generate electricity for her, and oil is transported farther to power her car, should she have one.

The growing human population, having passed 6.7 billion, is heading for a projected 9 billion or more in this century, as we saw in chapter 7. Yet worldwide household size (the number of people living together) has on average been shrinking as a result of lifestyle changes—more divorce, less likelihood of elderly people living with their children and grandchildren, and so on. For example, average household size between 1985 and 2000 so declined that 155 million more households were added in countries with high levels of biodiversity than would have been had household size remained constant. In turn, growth in numbers of households requires more materials for construction, uses more fuel for heating (additional space per person to warm), and covers more land with structures that therefore can't be covered with productive farms or natural habitat to shelter other organisms.

The disproportionate environmental costs associated with further population growth are seldom brought into public discussion anywhere. Indeed, political commentators typically focus instead on increases in wealth and health over the past century, which occurred even though the population has quadrupled in size. They either ignore or don't realize that this has been achieved by depleting humanity's natural capital—using up vital resources far more rapidly than they can be replaced—and thus mortgaging our future.

Still less do political observers focus on the role that curbing population growth has played in underpinning the increase so far in average prosperity. Could the "Asian Tigers" (e.g., South Korea, Taiwan, Hong Kong, Singapore) have reached their present level of affluence if they had not first reduced their once high birthrates? Would China be achieving such remarkable economic expansion without its prior success in curbing population growth? Might India have achieved more if its family planning program had been more successful? No country with a high birthrate today, other than a few oil-rich oligarchies, has attained a high level of well-being for the majority of its population.

Moreover, not everyone has enjoyed the global increase in average prosperity over the past few decades. Three billion people are living in poverty today—half again as many as the entire world population in the 1930s. Nearly half of the world's population today survives on an income of less than two U.S. dollars per day. More than 850 million are severely or chronically undernourished, and some 2 billion people suffer from "hidden" hunger: micronutrient deficiencies leading to lost vision, disabled immune systems, and, often indirectly, death. Perhaps worst of all, malnutrition contributes to the deaths of some 6 million children annually. True, the proportion of people in poverty has declined over the past half-century, but the absolute number in poverty has increased enormously.

The total human population has been increasing almost continually since the agricultural revolution, multiplying roughly a thousandfold since our ancestors took up farming, although there have been occasional temporary local and regional declines associated with epidemics and wars. But the most rapid phase of growth, often called the "population explosion," began after World War II, and its roots lay in a humanitarian triumph, as mentioned in chapter 7. That triumph resulted from the export from developed to non-industrial nations in Asia, Africa, and Latin America of "death control" technologies, especially pesticides for use against crop pests and disease vectors; vaccines; and antibiotics for fighting bacterial diseases. Death rates plummeted in these regions, yet birthrates remained high. High birthrates persisted for a few decades after death rates fell in most countries, and the global population shot up from 2.5 billion in 1950 to 3.5 billion by 1970.

Although the global rate of growth slowly declined after 1970, the actual number of people added to the population each year continued to increase until 1990, the result of a still expanding population base. With a worldwide population of 3.5 billion in 1970, a 2.1 percent annual growth rate added 73.5 million people that year, for example. The population passed 6 billion just before the turn of the century, with a peak annual increment at that time of 86 million. By 2007 the growth rate was down to 1.2 percent, but the annual increment was still over 75 million, and the world population passed 6.6 billion. There still is great truth in an anonymous slogan from the 1960s, "Whatever your cause, it's a lost cause without population control," as the United Nations Environment Programme and a number of non-governmental organizations (NGOs) are rediscovering.[2]

Calculating Consumption

Concern about population has begun to reawaken, especially because the enormous increase in numbers in the past half-century has been accompanied by a great surge in per capita consumption. Environmental scientists often employ statistics on energy use as a surrogate for all consumption activities. Total human energy use (e.g., all the petroleum burned as gasoline, jet fuel, or fuel oil; coal for electricity generation and home heating; wood burned for cooking in poor countries) has increased about twentyfold since 1850, while the population has increased almost sevenfold (6.5 times).[3] Among large nations, the United States has the highest level of consumption per person. While China has more than four times as many people, by a conservative measure, each on average consumes about a sixth as much as the average American. India has more than three and one-half times as many people, but each consumes on average about a fourteenth as much. Not only does the United States consume a lot of energy (which can also be viewed as a surrogate for A times T in the $I = PAT$ equation); the resources and technologies it uses in order to produce useful energy also have massive greenhouse gas and land-use impacts. Strip-mined coal, a vast array of pipelines for natural gas and oil, and large coal-fired power plants are defining characteristics of our national technology infrastructure (much of the T in $I = PAT$), and China and India are moving in similar directions.

The good news is that we could make significant headway in reducing environmental impacts by either reducing our consumption or switching to cleaner resources and technologies such as wind and solar power—

or, ideally, by doing both. Although such changes might not be easy to implement on the scale needed to reduce the impacts to a safe level (think of the subsidies for fossil fuels, the potential costs of transforming the energy system, and the political access that major energy companies have), it is easy to estimate the environmental benefits.

With its large and growing numbers and its huge energy use, is the United States the front-runner in overpopulation? To what extent is the world as a whole already overpopulated? Social scientists sometimes say that overpopulation occurs when human numbers press on human values, such as when they produce highway congestion. But this leads to questions such as "How much pressure?" and "Whose values?" One claim is that there is overpopulation when the density of people is so great as to impair the quality of life, cause serious environmental degradation, or create long-term shortages of essential goods and services. But how do we measure impairment, degradation, and long-term shortages, some of which might be solved by substitution of one good or service for another?

Biologically, less subjective criteria are available. With non-human animals, overpopulation occurs when the number of individuals in a given habitat is so large that the population's resource base is being depleted to the extent that it will not be able to support as many animals in the future. With human beings, we can look at what is happening to a population's resource base of natural capital, a key source of the flows of often irreplaceable "income" that support a population.

Of the three principal kinds of capital, human-made, human, and natural,[4] human-made capital (i.e., machinery, roads, houses, airplanes) can normally be produced and replaced by people's efforts. Human capital—the knowledge, experience, and energy embodied in the labor force—can be maintained and restored by education and retraining. Economists try to keep careful track of society's human-made and human capital, and accountants routinely assess the value of and depreciate human-made capital for tax and other purposes. Unhappily, though, little attempt is made to measure and value natural capital—Earth's ecosystems—and only very recently has its depreciation started to be taken into account in evaluating a society's wealth and its capacity to support a growing population.

Natural Capital

Many forms of natural capital are irreplaceable on a time scale of interest to humanity. This is true of coral reef ecosystems, which are key sources of seafood for many societies. Damaged reefs may take hundreds of years or more to return to their previous productivity. The same can be said of ancient forests and, of course, fossil fuel resources such as coal and oil, which would take tens of millions of years to regenerate. Thus if a population—in this case, the global human population—is being supported not by the income from natural capital but by steady depletion of the capital itself, there is overpopulation.

The idea of "irreplaceable" resources often brings to mind the depletion of non-renewable resource capital such as fossil fuels or mineral ores, but supplies of those are generally adequate for the foreseeable future. While petroleum and natural gas may reach limits of economically viable extraction within a few decades, there are still gigantic amounts of other fossil fuels, such as coal and tar sands, although the environmental liabilities of their extraction and use make them problematic. The rich deposits humanity is now exhausting cannot be replaced; instead we must consider replacing their functions, that is, substitutability. For instance, various forms of solar energy or nuclear power could replace fossil fuels in some uses.

Ironically, it is the rapid consumption of theoretically more readily renewed natural capital that shows clearly that Earth is overpopulated. Three key types of that capital are being expended at a rapid rate: agricultural soils, groundwater, and biodiversity.

Agricultural soils are usually generated on a scale of inches per millennium as rocks are worn down to fragments and gradually colonized by myriad plants, animals, and microbes. In many parts of the world today, they are being eroded away at rates of inches per decade. Soils are a renewable resource, essential for agriculture and forestry, but once they are severely depleted, they cannot be renewed on a time scale of interest to society. Soils also can be ruined for practical purposes by accumulations of salts, as often results from poorly managed irrigation or from overcultivation or poor land-use practices. With careful husbandry, good soils can remain productive seemingly indefinitely; without such husbandry, they can be converted to a non-renewable resource. Unfortunately, not all soils are robust enough to support continuous cultivation.

FIGURE 11-1. Mining the tar sands in Alberta, Canada, an environmentally destructive process. Like strip-mining for coal or oil shale, it not only contributes to global heating but also destroys habitat for biodiversity. Photograph courtesy of Greenpeace / Ray Giguere.

Nutrient depletion, erosion, overgrazing, and poorly designed irrigation have all contributed to degradation of agricultural lands. Consequently, it has been estimated that each year an additional 20 million hectares (some 75,000 square miles) of the world's agricultural land either becomes too degraded for crop production or is lost to urban sprawl. Sometimes this proceeds to the point of land becoming desertlike as farmers abandon exhausted soils and put other areas under cultivation.

Most groundwater is found not in underground streams but in relatively small pores in rock within aquifers (underground water-bearing formations of sand or porous rock). The water-saturated sands of beaches make a good aquifer model: dig a hole and water flows out of the tiny spaces between sand grains and fills the hole to the level of the ocean—the "water table." Many aquifers around the world are being drastically overpumped today. Water that accumulated during the last ice age (often referred to as fossil groundwater) is being removed so fast that water tables are dropping a yard or more per year. This is a serious problem in all three of the world's most populous countries: China, India, and the United States.

In the United States, for instance, the Ogallala Aquifer, located under the High Plains, is being rapidly emptied, especially in its southern portions, to maintain irrigated agriculture. On average, water levels have dropped about 10 feet, but in some areas as much as 200 feet since pumping began. The Ogallala is one of the biggest underground storehouses of water anywhere. It extends from South Dakota to Texas, and before electric pumping began, it contained more water than Lake Huron. In much of the world, groundwater is needed for use in homes and industry as well as being critical for agriculture, and almost everywhere groundwater is being overexploited. The results can be striking; so much water has been drawn from aquifers in southern and western parts of the Central Valley of California that the water table has dropped as much as 400 feet and the land has sunk several feet since its conversion to farming in the nineteenth century.

In addition, recharge areas are being paved over to the extent that, in places such as Long Island, much rainfall runs off into the sea instead of percolating through the soil to recharge an aquifer. Rapid pumping can cause an aquifer to collapse, and rising sea levels in response to global warming can cause salt to intrude into coastal aquifers, rendering them useless. In many areas, we will have to wait for another ice age to fully recharge aquifers. In some places it will never happen.

Groundwater depletion has serious consequences for human health, since groundwater is less likely than surface water to be polluted. An estimated 1.2 billion people, most of them in poor countries, do not have a source of pure water—and waterborne pathogens now cause more human deaths, on the order of 5 to 10 million per year, mostly of children, than any other environmental factor. Furthermore, depletion of ice-age groundwater may put farmers in very difficult positions if, as predicted, more areas suffer drought as a result of climate change. Indeed, depletion of this vital form of natural capital could lead to a struggle between Canada and the United States over the water in the Great Lakes.

A third critical form of natural capital that is being diminished by increased human numbers and expanding human consumption is biodiversity: the populations and species that are working parts of human life-support systems. Biodiversity is also the most overlooked of natural capital forms, as its connections to human economic systems are complicated and often indirect. This can be demonstrated with a thought experiment from our colleague Gretchen Daily. Consider the challenge of trying to build an ecosystem on the moon to support people, assuming that crushed rock with which to form soils, an Earth-like atmosphere, abundant water, and so forth were readily available. Which species would you bring with you? Beyond the obvious choice of crops, you would need to consider trees for shade and wood, ornamental and medicinal plants, pollinators and other organisms they need to maintain their populations, predators that would regulate their populations, soil bacteria and fungi, especially to serve as decomposers and nutrient recyclers, and so on. In addition, you would want the genetic diversity of individual species to be broad enough to adapt to unforeseen changes in the ecosystem, and enough species to stabilize interactions between predators and prey.

The point is that, to be safe, you'd need to take with you a vast array of different organisms to be sure your lunar economic system would work. Unfortunately, biology isn't as deterministic as the laws of Newtonian mechanics; we simply cannot predict, with sufficient precision, what characteristics different combinations of species and environments will produce in an ecosystem, as the demise of the Biosphere 2 experiment (described in chapter 9) drove home. As our knowledge of biology increases, it becomes ever clearer that the strategies and complexities of organisms and ecosystems are more sophisticated than even the most advanced human technologies.

Biodiversity everywhere on Earth now is threatened as growing numbers of people harvest organisms, transform habitats, and change the climate as they attempt to support themselves and increase their consumption. The wealth of species and populations is the least appreciated and the most slowly restored of all renewable resources, but it is critical for agricultural ecosystem services such as maintenance of soil fertility, pollination, and pest control. Forest ecosystems, among the richest homes of biodiversity, as well as savannas, river valleys, lowland plains, and prairies, have been giving way to the production of food for human consumption ever since agriculture was invented.

A great expansion worldwide of grazing by cattle, sheep, and other ruminants, requiring a roughly 600 percent increase in area over the past 300 years, has, along with clearing of lands for crop agriculture, also been a major engine of global change. The impacts of grazing animals on other organisms vary, particularly in response to different stocking rates (number of animals per unit area) and soil and climate conditions. The greatest effects are seen in arid and in very humid conditions. In arid areas, overgrazing of grasses by cattle and sheep, destruction of shrubs and saplings by goats, and compacting of soils by too many hooves cause desertification or, especially in the absence of goats, encroachment of shrubs into grasslands. In the moist tropics, deforestation often precedes short-lived cultivation, followed by overgrazing, which leads to serious land degradation and loss of the land's original biodiversity.

In response to the effects of ever more people and ever more consumption, innumerable populations of organisms are disappearing at a high rate. No one knows exactly how many populations and species are being exterminated annually as well, but there is general agreement among scientists that numerous kinds of organisms are following the dodo and passenger pigeon into oblivion at a rate far faster than any seen in the past 65 million years. In some cases populations can be replaced, but not in many. And the loss of species is forever (despite *Jurassic Park*–style science fiction); the waiting time until suitable functional replacements might evolve is generally measured in millions of years.

Furthermore, evolutionary pathways are unique. Although we sometimes observe convergence—that is, evolution among separate species of similar morphologies or behaviors—the paths these species take to get there, and hence their genetic histories, are different. That biologists now agree that humanity has entrained a massive extinction crisis, the sixth in the history of our planet, is perhaps the clearest sign that a single species, *Homo sapiens*, now dominates the planet and has reached a population level that seriously jeopardizes its own well-being. By wiping out populations and species at a rate far beyond that at which evolution can replace them, humanity is limiting both the potential size of the human population and the amount that each individual, on average, might consume.

Patterns of Consumption

The problems associated with the human rise to planetary dominance, as we have indicated, are not just a consequence of how many of us there are. And the contributions of population growth to the human predicament certainly can't be judged simply by population density, the number of people per square mile. For example, people sometimes claim that there is no global population problem because Earth has only about 130 people per square mile of land area on average, whereas Holland has over 1,000, and the Dutch are big-time consumers. The error (known as the "Netherlands Fallacy") is that the people of the Netherlands don't live on the resources of the Netherlands any more than the people of densely populated New York City live on the resources of their tiny land area.

The Dutch and the New Yorkers can maintain such dense and high-consuming populations only *because* the rest of the planet is relatively thinly populated. They both must import resources, the Dutch to maintain their lifestyle and the New Yorkers in order to survive. The Dutch probably could *survive* in the Netherlands (if at a much lower standard of living) because it is a relatively resource-rich area. There are, however, many places on Earth the size of the Netherlands (16,000 square miles) that couldn't support 16 million people (the current Dutch population) no matter how mean their lifestyle. And there is *no* area on Earth of some 320 square miles that could support the 8 million or more people of New York (picture a square mile anywhere housing and feeding 25,000 people).

The amount that people consume, of course, is a critical factor in whether humanity remains on a collision course with disaster. The United States, because of its large population size, continued growth, and high per capita level of consumption, is the champion consumer of the world. With only 4.5 percent of the global population, the United States accounts for more than 20 percent of the world's energy use (and energy use is closely related to environmental damage). Only western Europe, with more than 400 million people and a comparable level of affluence, can rival the United States today in disproportionate consumption and attendant environmental destruction, yet Europeans are much more conserving of resources, especially of energy. Each person in the United States on average uses over twice as much energy as a European, and 15 to 150 times more energy than a citizen of a very poor country.

Further population-consumption problems and related environmental dilemmas are traceable to the unevenly distributed ability to consume. In the United States, for example, more than in most industrialized nations, a large and growing disparity between income groups exists. In the early twenty-first century, the most affluent fifth of the U.S. population received more than 50 percent of the nation's income while the poorest fifth received barely 3.5 percent. The disparity is even greater when the incomes of the richest 1 percent and 0.1 percent are considered: they receive 16 and 7 percent, respectively, of total income, proportions that have more than doubled since 1980. The income of the average corporate chief executive officer was 300 times the average annual wage in 2005, a tenfold increase in twenty-five years. By contrast, in Sweden, the richest fifth of the population received less than 35 percent of the income and the poorest fifth received nearly 10 percent.

America's skewed income distribution leads to power imbalances, failures of education, feelings of social and political impotence, and distress at being unable to keep up in the consumption rat race, reflecting a perception of wage stagnation in the middle class over the past decade or so. The skew also reflects a growing lack of good jobs available for the less affluent (and generally less educated) segments of society. The resultant atmosphere of frustration in much of the American populace does not seem conducive to the cooperative actions needed to make the public increasingly a part of the environmental solution rather than a central part of the problem. The distributions of income and power have complex environmental ramifications, such as helping to determine who consumes how much and where. They also determine who is forced to suffer most directly from poor environmental practice—for instance, it is generally the poor and powerless who live near toxic waste dumps and downwind of the coal-fired power plants that supply electricity used profligately by the rich.

Organizations to counter overpopulation, such as Planned Parenthood, have had a global reach and substantial success, but on curbing overconsumption there's been far less activity. There have been no "planned consumption" or "zero consumption growth" movements developed in parallel. Nobody is passing out "consumption condoms" or "morning-after-shopping pills." Despite various environmental voices pointing out the effects of overconsumption and a scattering of individuals following "voluntary simplicity" principles, most people in both rich and poor countries still view growth in consumption as an unalloyed good.

During the twentieth century, the industrialized world became a world of triumphant consumerism. The dominant belief was "that goods give meaning to individuals and their roles in society," as historian Gary Cross put it in his fascinating book *An All-Consuming Century*. In the United States after World War II, consumption was believed to hold the key to the economic growth necessary to avoid a slide back into the Great Depression. The key worked, and twentieth-century consumerism, in partnership with capitalism, was largely victorious over the rival ideologies of fascism, communism, and socialism.

But there is little reason to believe that the American style of consumerism can perpetuate its triumph through the twenty-first century without substantial modification. Clearly, physical growth of the economies of the globe must soon be constrained if essential ecosystem services—humanity's life-support systems—are to be maintained. Some countries are so poor that their consumption must be increased if their populations' basic needs are to be met, which indicates a need for redistribution from rich to poor countries if global sustainability is to be achieved—something not politically palatable in the United States and some other countries.

To assume that economic growth in its present form can simply continue indefinitely, however, is to ignore civilization's progressive malady, of which the most prominent symptom is the loss of natural capital. Unless the ability of that capital to deliver essential ecosystem goods and services can be maintained, economic growth will come to an end. The conventional focus on using up resources to keep consumption climbing, furthermore, ignores many other human values—such as peace, equity, tranquility, spirituality, and survival as a civilization. It also highlights a common present-day view of markets that would have made Adam Smith, the father of modern market economics, cringe: the notion that unrestrained and unregulated markets should be the instruments that run the world. Smith's *Inquiry into the Nature and Causes of the Wealth of Nations* (1776) is often cited, but also important is an earlier book he wrote, *The Theory of Moral Sentiments* (1759). In that book Smith makes it clear that financial interests should be pursued to acquire basic necessities and that other, non-financial, values should take precedence after that. Today he would quite likely see markets as a fine tool for producing certain economic goods but not a be-all and end-all in themselves.

The situation in other developed countries is broadly similar to that in the United States. In western Europe and Japan, however, the sizes of homes and the extent of suburban sprawl are constrained by the already high densities of populations (respectively about five and ten times that of the United States) and generally stricter zoning codes. Similarly, narrow city streets and shorter intercity distances have helped dampen (but have far from extinguished) enthusiasm for large, gas-guzzling automobiles. Both Europe and Japan also have fast, efficient, and convenient public transportation systems both within and between cities. The higher taxes that make fuel prices roughly twice as high as those in the United States are accepted by Europeans and Japanese, whose cultures may not tie self-image and freedom as closely to the personal automobile as does American culture.

Nevertheless, car cultures seem to be developing in parts of the world as diverse as Australia, Mexico, and China, and zoning restraints once found in places such as England and Japan are breaking down as more and more land is devoted to automobiles rather than to people, agriculture, or nature. The case of China is special because of its enormous population—1.3 billion as of 2007—as well as its government's attitude. Massive public works projects are common in China, where a high national savings rate facilitates huge investments. In anticipation of the growth of its budding car culture, China is building new freeways on an incredible scale, which will presumably accelerate the demand for personal vehicles.

Potentially one of the greatest engines for expansion of per capita consumption is what Norman Myers and Jennifer Kent refer to as the "new consumers." These are the more than a billion people in twenty leading developing and transitional countries such as China, India, South Korea, Malaysia, Brazil, Argentina, Mexico, Russia, and Turkey who now have a purchasing power equivalent in the United States to at least $2,500 per person annually, a huge advance over the situation thirty years ago. In 2001 the new consumers around the world collectively had a purchasing power about three-fifths of that of the American population, and the proportion has increased significantly since then.[5] New consumers by 2001 were driving some 125 million cars, almost one-fourth of the world fleet, and by 2010 that could increase to more than 245 million.

In one sense, that's a great part of the human triumph—more and more people are getting access to the good things of life. But the usual downside of neglected social costs (those captured by the market plus those borne by society as a whole)—climate disruption, resource wars over petroleum, and pollution and its health effects associated with automobile use—will need to be dealt with. Most of the countries in which new consumers live can ill afford the local environmental effects that, for instance, a transition to a car-dominated transportation system like that of the United States would cause. Increasing consumption by the formerly poor will also exacerbate global environmental and resource problems—unless, of course, those of us in the rich countries find ways to compensate by lessening our own negative effects and developing countries find ways to avoid the environmentally damaging processes the now-overdeveloped nations used in the course of the Victorian industrial revolution and its sequelae.

How Others Pay the Price for Us

In wealthy nations, it's easy to remain unaware of how a consumer lifestyle feeds into the destruction of natural capital in distant, less affluent countries. Much of the damage traces to consumption of mundane items ranging from bananas and coffee to magazines, furniture, and materials for housing construction. The degree of destruction caused in poor countries in the service of consumption in rich countries varies from place to place. The vast banana plantations of the Sarapiquí River region of northeastern Costa Rica, where we and some colleagues have done research, now harbor only a handful of the hundreds of bird species that once occupied the forests the banana plants replaced, for example. The plantations, however, produce great profits for big corporations as huge quantities of bananas are shipped to the residents of industrial countries to slice over their breakfast cereal. It's worth observing in this context that human demand typically isn't some exogenous and inevitable force. Spread of the banana-eating habit itself, for example, was created originally by vast advertising campaigns, such as that featuring a singing, dancing "Chiquita Banana."

In the Coto Brus area in the far south of Costa Rica, the landscape is dominated by small coffee plantations and degraded pastures. It is typical of the habitat that now covers a great deal of that country's previously forest-clad hill country, much of it devoted to growing coffee for rich-world consumption. A single 600-acre forest remnant and scattered smaller degraded patches are all that remain of what four decades ago was a continuous cover of magnificent midelevation rain forest. But this mixed landscape still supports more biodiversity than the intensively cultivated lowland areas.

A great deal of the destruction of rain forests over much of the world can also be traced to activities to support consumption in the world's rich nations and the rich minorities in many developing nations. The escalating global demand for sugar, coffee, tea, rubber, beef, tropical fruits, timber, and pulpwood has caused enormous but little-appreciated damage to biodiversity and human cultures. We have personally seen the march of biologically impoverished oil palm monocultures replacing species-rich tropical moist forests in Malaysian Borneo, New Britain, the Chocó region of Ecuador, and even Costa Rica. These forests are the most complex and diverse terrestrial ecosystems and are vanishing at an alarming rate. They are being destroyed to make room for the production of some 40 million tons of palm oil annually, about a third of which is used in food despite the unhealthy fat it contains.

Oil palm plantations are typical engines of rain forest destruction. While some poor people may be helped by giant palm operations that need workers and supply them with cheap cooking oil, others are dispossessed. So some social benefit is derived, but it only marginally compensates for the environmental costs incurred in the creation of the biological deserts of palm plantations, not to mention the social costs of dispossession of local inhabitants. And in the background is the looming threat of further environmental disruption from greatly expanded plantations to supply palm oil as a biofuel for consumption by an increasingly automobile-oriented world.

The lowland tropical forests of the Malay Peninsula, Java, Sumatra, Borneo, Sulawesi, and the Lesser Sunda Islands—collectively, the Sundaic lowland tropical forests—may house more plant species than any equivalent area on Earth, and they are the tallest and perhaps most beautiful of all tropical forests. In addition, they support an extraordinary array of mammals, including such charismatic species as the tiger, Asian elephant, orangutan, Malaysian tapir, clouded leopard, gaur, banteng, and proboscis monkey. The bird community of those forests is no less exciting and includes nine species of hornbill, several pheasants (including the spectacularly ornate Bulwer's pheasant of Borneo), large numbers of attractive woodpeckers, and a mass of fascinating babblers. Those forests have been almost completely destroyed in recent decades, largely by local people paid by exploiters and entrepreneurs in service of rich-world consumption, threatening a regional extinction episode of tragic proportions. The tragedy, of course, extends to the masses of poor people who gain little as their homelands are decimated.

The lowland forests of the trans–Fly River area of southeastern Papua New Guinea, the third largest remaining expanse of lowland tropical forest in the world (those of the Amazon and Congo river basins are first and second), are now threatened with similar destruction. When we were in the Kiunga region of the Fly River drainage in 2003, we learned that Malaysian corporations were planning a massive timber-harvesting campaign there that was to begin with the clear-cutting of more than 650,000 acres. If the plan is carried out, the New Guineans will have both their forests and their culture destroyed for the short-term gain of a pittance. The bright, traditionally very political New Guineans are very likely no match for the globalized, Chinese-backed Malaysian steamroller bearing down on them. They are too naïve about the ways of the outside world; national politicians in Port Moresby and local headmen are easily bribed with small amounts of money, alcohol, and access to prostitutes.

Exploitation of indigenous peoples for the benefit of faraway populations is, of course, not new. Nor is its recognition as such. As long ago as 1948, conservationist William Vogt, who was deeply concerned about Central American rural development, vividly described in his pioneering book *Road to Survival* how population growth, consumption, and destruction of natural capital are linked:

> By excessive breeding and abuse of the land mankind has backed itself into an ecological trap. . . . I do not mean the other fellow. I mean every person who reads a newspaper printed on pulp from vanishing forests . . . who eats a meal drawn from steadily shrinking lands . . . puts on a wool garment derived from overgrazed ranges that have been cut by the little hoofs and gullied by the rains, sending runoff and topsoil into rivers downstream, flooding cities hundreds of miles away.[6]

From Caribbean and Hawaiian lowlands to the hill country of Brazil and the forests of central Africa, Southeast Asia, and the Philippines, deforestation, erosion, and squalor have been generated by firms that bought governments and cared nothing for sustainability. It is not a pretty story, but it is a supremely important one. The process of "development" as conceived after World War II envisioned the industrialization of less developed nations in Asia, Africa, and Latin America, and international agencies such as development banks and the assistance programs of the industrialized world were dedicated to that end. The result was that hundreds of millions of people, especially in rural areas, were left out of the development process for several decades. Agricultural development was concentrated on establishing the "green revolution," which, as we'll see in the next chapter, helped to create disparities of wealth among farmers. Growing crops for export to rich countries was also encouraged, often squeezing poor farmers off the most productive land and into marginal areas. Colonialism and subsequent policies that have resulted in a skewed distribution of resources have left billions of very poor people in their wake.

Today's poor people endure a different and more complicated kind of poverty from that of a hundred years ago. Previously, the poor were relatively (compared with the then "wealthy") well off and often found ways to sustain themselves and lead satisfactory lives, enriched with cultural practices. But broader awareness of today's much greater inequity, supplied through TV and pictures in magazines, naturally spurs unhappiness. Now hundreds of millions of people in developing nations aspire to an American lifestyle, but they have been given a model of "development" that too often deepens the misery of the majority, even though an elite minority may enjoy Western-style affluence.

The capitalist economic system has turned out to be miraculously efficient at mobilizing resources and producing goods. The question is, however, for whom has the system been efficient? In addition, important ethical and economic questions arise about what exactly an economic system should maximize. "Utility" is what most economists would reply—which is roughly equivalent to "happiness" or "satisfaction," but generally with regard to consumption. The question of whether, after basic needs are met, consumption should in some sense be maximized has been given little public discussion.

But capitalism's biggest failure so far has been in not paying more than lip service to valuing environmental "externalities"—costs of production not captured in market prices (i.e., those "external" to the market). For example, the costs of the loss of tropical biodiversity are not captured in the price of palm oil. That loss could deprive future generations of a wonder drug for use against cancer, given that many of our most important pharmaceuticals are derived from plant defensive compounds. When you buy a gallon of gasoline, you contribute to the costs of climate change and lung disease generated by smog, not to mention the military costs of securing oil in unstable countries and guarding oil-shipping routes. None of these factors is captured in the price at the pump—they are external to the market price.[7] It has been said that socialism collapsed because it did not allow the market to tell economic truth, and that capitalism may collapse because it doesn't let the market tell ecological truth.

How the Rich Pay the Price Along with the Poor

Ironically, the process of producing consumer goods, which are disproportionately enjoyed by the rich, is the prime driver of the toxification of Earth, and consumption is the main route by which people in the rich countries are exposed to toxic substances. Industrial society releases tens of thousands of chemical compounds, many of them toxic, into the environment in huge quantities—billions of tons annually. The most obvious manifestation of this is visible, irritating, and sometimes lethal air pollution—smog. "Smog" was a term coined for the combination of coal smoke and fog that created hideous air pollution in London in the early twentieth century. The more widespread current phenomenon is photochemical smog. It is caused by the interaction of emissions of many chemicals from automobile exhausts, factory smokestacks, backyard barbecues, and the like with sunlight, hence the "photo" in its name.

Air pollution can cause severe direct health effects, especially among the very young, the elderly, and people with heart and lung conditions. This is made clear by the correlation of rising hospital admissions and death rates with the severity of air pollution, in particular the small particles that result from the burning of fuels. A large-scale global risk assessment managed by the World Health Organization indicated that combustion particles (outdoor, indoor, and occupational, including passive tobacco smoke) are responsible globally for nearly 3 million premature deaths per year (2.4 million from indoor and outdoor pollution, which is nearly all caused by fuel combustion). Including active smoking, the number of deaths approaches 8 million, larger than the annual number of deaths from malnutrition. Air pollution is classically a problem of big cities such as Los Angeles, São Paulo, Tokyo, and Beijing, but air currents spread air pollution far and wide. Some ingredients, such as sulfur and nitrogen oxides and ground-level ozone, also damage crops and forests in many areas.

Ironically, however, in light of the prominence of discussions of rich-world urban air pollution, Kirk Smith, professor of public health at the University of California, Berkeley, has shown that the most serious health effects of air pollution result from the burning of fuelwood and other biomass fuels (such as charcoal or dried cow dung) for cooking and space heating in Third World huts. This use of biomass amounts to almost 10 percent of today's energy demand, more than that supplied worldwide by nuclear and hydroelectric power combined. When these fuels are not burned thoroughly (to carbon dioxide), they generate primarily carbon monoxide but also a variety of small particles and organic compounds. All of these cause health damage—especially to the hearts and lungs of women and young children, who are most likely to be present during cooking. Among other problems, many women develop chronic bronchitis or emphysema, and many of their children get pneumonia.

Pollution from cooking and space heating in the Third World could be greatly ameliorated by relatively inexpensive changes in stoves—simply providing ones that burn fuels more completely, putting chimneys on them, and using properly dried wood, for example. In Guatemala, Smith showed, such changes could greatly reduce the incidence of childhood pneumonia, which kills more children and generates more lost years of life than any other disease. So both rich and poor pay a health cost imposed by the side effects of consumption—although the cost for the poor could be greatly reduced at small expense.

The mobility of the toxins humanity lets loose in the biosphere is legendary. Many known poisons, carcinogens, and compounds that mimic human hormones have been tracked to the farthest reaches of our planet. DDT, the use of which was long ago banned in the United States, and its poisonous breakdown products have been found in the recent past in rain over the midwestern United States, in California cities, and in polar regions. Numerous pesticides; PCBs (polychlorinated biphenyls, persistent toxic compounds used in insulators and various industrial processes); prescription drugs; industrial solvents; mine wastes; various components of plastics; and mercury, cadmium, lead, and other heavy metals—the list is nearly endless—now also have nearly global distributions.

Can humanity adjust to living with small amounts of toxins in each individual, or will some unexpected pattern emerge? Could, as some fear, hormone-mimicking chemicals cross a threshold concentration in males and drop human sperm counts to the point where the survival of *Homo sapiens* is threatened? Recent reports of a decline in male births to half the number of female births in heavily polluted Arctic and subarctic locations are worrying. Or does the continuing rise of life expectancies in much of the world indicate that most people can easily handle the toxic load, with increases in general health more than balancing an increase in environmentally induced cancers? And what harmful effects are toxins having on other organisms and on ecosystem services?

These are difficult questions to answer. It is very hard to estimate the toxic effects of a single compound, let alone potential synergisms among two or more, or effects that will show up in later generations. Remember the long struggle to show the hazards of cigarette smoking when, among other things, it was quite easy to judge the exposures of individuals—something that is nearly impossible for most environmental toxins. Humanity is running a vast chemical experiment in which all of us are the lab rats. The results are still in doubt.

Consumption and Agriculture

The most important activity of humanity is growing and distributing food for ourselves, and that certainly is the enterprise most critical to human health and happiness. In our heavily urbanized society, most of us are far removed from the systems that supply our food, and so we take them all for granted. But many of the processes involved in providing the variety of foods now demanded by consumers in developed countries carry substantial environmental costs. They include the energy costs of growing, harvesting, storing, and processing food and transporting it around the world; the health costs of pesticide use and antibiotic resistance promoted by the extensive use of antibiotics in livestock production; and the environmental costs of unnecessary packaging used to help promote sales. All that is on top of the ecological costs, some of which we discussed above, of converting so much land to agriculture in the first place, and still more for the growth of export crops in developing countries in place of subsistence crops, plus the degradation of many countryside areas that has followed.

One reason for the growing concern of agricultural experts about future global food production is the prospect of feeding approximately 40 percent more mouths by the middle of this century. In addition to population growth, demand is rising fast for higher-quality foods, especially animal products, by the new consumers in middle-income developing nations. In China, more people could be classed as new consumers (well over 300 million) by 2004 than the United States had consumers (slightly fewer than 300 million). Per capita meat consumption has doubled since 1990, making China the world's largest carnivorous nation. By 2010, with its economy expanding at 8 to 10 percent per year, there could be almost *twice* the number of new consumers in China as there are consumers in the United States, and their collective purchasing power could approach half that of the United States.

As the Chinese use more and more grain to feed livestock, they will put ever more pressure on their already stretched water supplies and agricultural system—indeed, on a marginal world food economy. The Chinese, however, are just the most prominent example of rising consumption in developing countries; new consumers in many nations are pushing to emulate the consumption patterns of the world's rich. Demand for meat in developing nations, generated primarily by new consumers, is projected to nearly double between 2000 and 2020.

In terms of environmental damage, meat is much more costly to produce than are staple food crops such as rice, wheat, maize, and potatoes. A 2006 report from the Food and Agriculture Organization of the United Nations (FAO), *Livestock's Long Shadow*, pointed out that as much as 30 percent of the world's land surface is used for livestock production, amounting to 70 percent of all agricultural land. This includes the land used to grow feed crops, that for intensive livestock production facilities (concentrated animal feeding operations, or CAFOs), and grazing land for cattle, sheep, and goats. A major cause of deforestation, especially in the tropics, is clearance for pasture. The world's herds include some 1.5 billion cattle and buffalo and 1.7 billion sheep and goats—a total biomass exceeding that of the human population.

Furthermore, as much as 40 percent of today's grain production and most of the soybean crop are destined to feed livestock (including billions of pigs and poultry)—enough vegetable food to nourish hundreds of millions of people. The growing of feed grains in rich countries is a leading cause of farm-based pollution, especially water pollution, and increasingly so in developing regions as livestock production becomes more intensive. Huge numbers of animals are fed in small areas, and their wastes, rather than being returned as nutrients to the soil over the areas where feed grains are produced, often pour into watercourses. Given that the potential for expanding grazing land is limited, the rising demand for meat for consumption in developing countries may well be accompanied by as much as an 85 percent rise by 2020 in global demand for feed grains to produce the meat.

Moreover, grazing modifies the planet's climate. It does so in part by influencing the amount of solar energy that is reflected from the surface into space (more from grasslands and deserts than from forests) and by determining whether water percolates into the soil, runs off in floods, or is recirculated by plants, creating cloud cover. But more important, as the FAO notes, some 18 percent of anthropogenic greenhouse gas emissions can be ascribed to livestock production. Roughly 9 percent of global carbon dioxide emissions are caused by livestock production, mainly from land-use changes, especially deforestation, caused by expansion of pastures and land for feed crops. "Chewing of the cud" and fermentation by microorganisms of the resultant mash in the guts of ruminants is responsible for a significant portion of human-caused methane emissions. An even larger portion of nitrous oxide emissions, mostly from manure, are from livestock production. Both gases contribute significantly to global heating. The exact proportions from livestock are at present unknown, as there are many unevaluated sources. In addition, livestock emit almost two-thirds of anthropogenic ammonia, which is a significant factor in causing acid precipitation.

Migration's Effects

An influx of immigrants can functionally increase the human impact on a region's life-support systems just as population growth by natural increase does, while the departure of emigrants also can ease environmental and resource pressures in their home country. Human migration is a difficult population-consumption issue that has its roots in Earth's uneven distribution of resources and in cultural evolution. It generally consists of a movement of people from poor areas toward wealthier ones—the same direction as the general flow of resources—and toward the prospect of jobs and a better life. To the degree that migrants are successful and acquire the consumption habits of their adopted country, they add to the global pressure on human life-support systems. Migrants also often depart from areas under relatively oppressive governments and move toward ones with more freedom, which in most cases also translates as moving from poor toward rich. The United

Nations estimates that, during the years 2000–2010, nearly 3 million people annually are moving, mostly from developing to developed nations; overall, some 200 million people are estimated to live in different countries from those in which they were born. About half of the migrants move to North America; millions of others move to Europe, Russia, Australia, and a few wealthy enclaves in Asia. Strangely, the United Nations' 2006 projections anticipate a marked decline in international migration after 2010, perhaps because they underestimate the number of likely environmental refugees.

In recent decades, migration increasingly has been a matter of people escaping resource wars—for example, the millions of Iraqis who fled their homes in the aftermath of the United States' attempt to secure control of the world's second largest oil reserve. Millions more flee environmental degradation, which is a major source (and often a result) of poverty. An early example of the latter was the some 3 million people who, during the dust bowl era of the 1930s, fled desertification in the southern U.S. Great Plains but nonetheless remained in the United States. More recently, people trying to escape the ecologically devastated island of Haiti for a life in the United States have been much in the American media. Today, from Africa to Iran to China, people are being forced to move in the face of spreading deserts. And rising sea levels caused by global heating are likely to displace growing numbers of people on low-lying islands and coasts. In coming decades the majority of migrating people may well be environmental refugees—either a trickle moving out of ecologically degraded areas or from coastal areas threatened by increases in giant storms and rising sea levels, or a flood escaping from future resource wars.

Density and Disease

One final way in which the size of the human population influences its fate is in its contribution to our vulnerability to epidemic disease, what is sometimes referred to as the decay of our epidemiological environment. Scientists are very concerned as a growing, ever denser human population, containing hundreds of millions of undernourished and thus immune-compromised people, is pushed into closer contact with animal reservoirs of infectious diseases. Such contacts greatly increase the chances of epidemics of novel diseases occurring, such as that

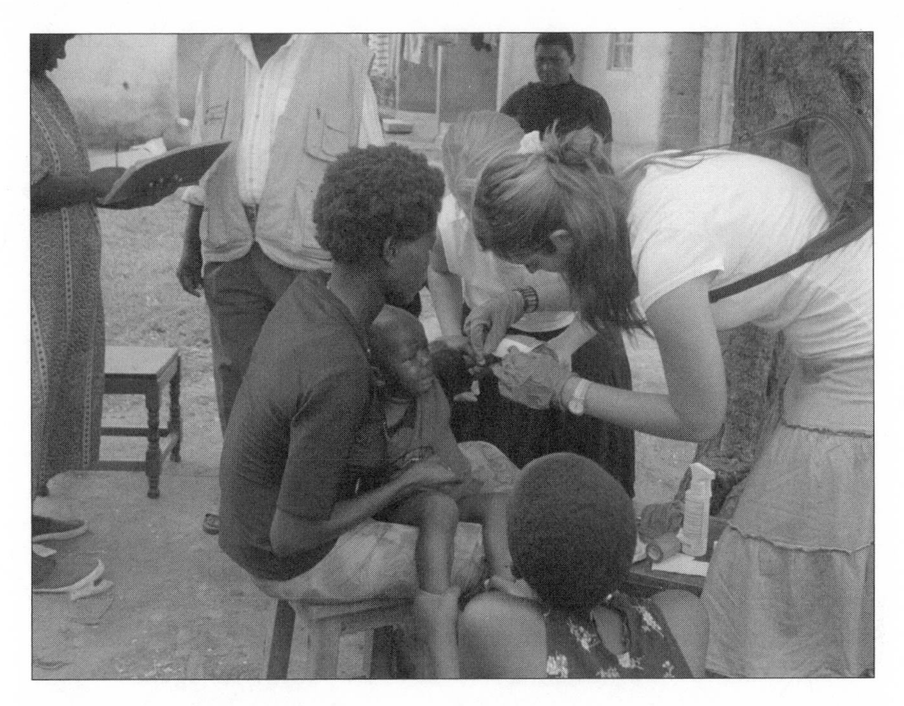

FIGURE 11-2. Uganda Rural Fund (www.ugandaruralfund.org) volunteer students from College of St. Benedict and St. Johns University, MN, dressing a wound of a child on an HIV / AIDS family outreach visit in rural Kenya.

of the largest modern novel epidemic, HIV / AIDS. The epidemic of AIDS, transferred from African apes, has caused untold human suffering; in Kenya alone in 2006, there were nearly a million AIDS orphans in a population of 35 million people. By 2007 some 33 million people worldwide were infected with HIV / AIDS, and nearly 20 million had died from it. In the absence of treatment, tens of millions more will die prematurely. No cure exists as yet, and treatment is expensive. Thus millions in poor nations who are infected, if they have even been diagnosed, may never receive help in spite of recently expanded efforts to provide it.

Not only do larger human populations increase the chance of a disease transferring from animals to humans by providing more "targets" for the transfer, but larger numbers of people also increase the chance of the disease persisting. A town of 2,000 people cannot sustain a measles epidemic: people either die of it or become immune, the pool of susceptible individuals is exhausted, transmission fails, and the disease disappears. A city of 500,000 easily can support such an epidemic, though, as births continually introduce new potential victims into the pool, allowing transmission to continue. Population size also influences the odds of new strains of known pathogens developing and spreading, such as the deadly Ebola and Marburg viruses, which have occasionally been transferred from other African primates into *Homo sapiens*. Ebola has recently killed dozens of people in western Africa, where it is decimating remaining small populations of gorillas and chimpanzees.

Especially worrisome is the potential emergence of new lethal flu strains. The World Health Organization (WHO) has been trying to mount an early warning system for avian (bird) flu, which has killed millions of wild and domestic birds (including many on farms, which were killed in attempts to prevent the spread of the disease) and infected several dozen people in Asia. Masses of people there are in close touch with pigs and ducks, which serve as reservoirs in which flu viruses might evolve new strains deadly to people. These are emergent viruses, viruses or virus strains that evolve a greater ability to cause disease or, especially, the ability to cross from an animal host to human beings.

The chances of horrendous epidemics are greatly enhanced by high-speed transportation systems. Ten generations ago, the fastest any human being could travel was the speed of a galloping horse; today an infected person can fly to another continent in a few hours. In the modern era an airline attendant, Gaëtan Dugas (who died in 1984), played a key role in the spread of HIV in North America. He acquired the disease in Europe through sexual contact with men from Africa. Forty or more of the first 248 AIDS cases in the United States by April 1982, it turned out, were in men who had had sexual relations with Dugas or one of his other sex partners.

Meanwhile, we are busily disarming ourselves for the continuing war with bacteria by encouraging the evolution of more antibiotic-resistant strains. This is done not only through overuse by the medical profession, which at least is an error committed in a good cause, but also through massive use of antibiotics to treat entire populations of domesticated animals to enhance their growth and the bank accounts of the owners of "factory farms." Those antibiotics escape into water supplies and result in more resistant pathogens. As the *New York Times* put it in late 2007, "The trouble with factory farms is that they are raising more than pigs. They are raising drug-resistant bugs as well."[8]

The now-dominant animal, then, faces a series of intertwined dilemmas surrounding its numbers, its patterns of consumption, its susceptibility to epidemic diseases, and the distribution of resources, wealth, and people. The continued failure to ask what human beings really want, to address issues of overpopulation and overconsumption, and to ensure for the poor a better level of well-being by their own standards could endanger the well-being of even the rich through increased social and political instability and the debilitating costs of continuous resource wars.

CHAPTER 12 A New Imperative

"Over the last two decades a new imperative has come to dominate environmental concerns with the growing awareness that human activities have an increasing influence on Earth System functioning, upon which human welfare and the future of human societies depend."

WILL STEFFEN AND COLLEAGUES, 2004[1]

"We have grown in number to the point where our presence is perceptibly disabling the planet like a disease."

JAMES LOVELOCK, 2007[2]

THE ROOTS of global change go far back in time; the configuration of continents, climates, the array of living organisms, and so on have been continuously changing since the beginning of the biosphere. Large-scale anthropogenic (human-caused) environmental change also is an old story. Primarily through hunting and setting of fires, people have been altering the environmental milieu (and thus the evolutionary trajectory) of their own species and many other organisms for hundreds of thousands of years. They began to outstrip other large animals in this regard only more recently, although overharvesting of resources, at least with respect to large mammals and some coastal marine organisms, is a human tradition that goes back at least to the Pleistocene epoch.

In Africa, the average size of mollusks collected by our distant ancestors declined as harvesting intensified. Some 45,000–30,000 years ago, human beings invaded Australia and apparently helped to kill off most of the large marsupials. About 14,000 years ago, perhaps earlier, our species entered the Western Hemisphere, and most scientists believe that those invading human beings were mainly responsible for the extermination of woolly mammoths, giant ground sloths, and other indigenous large herbivores. Whether climate change played a significant role is controversial, at least in North America for that period. The animal extermination in turn altered selection pressures on plants there. Although the actual changes have not been well documented, the disappearance of giant browsers (shrub-eaters) such as the sloths would undoubtedly have affected the physical structure and populations of shrubs, just as browsing by modern-day elephants affects plant communities in Africa. In much of the Old World, many large mammal species had disappeared earlier, probably first reduced in numbers by climate change and then given the coup de grâce by modern hunters (who hadn't been present during previous climatic disruptions). Most scientists agree that the very rapid "Pleistocene overkill" of Australia and the Western Hemisphere happened because the large animals there, unlike the big mammals of Africa and Eurasia, lacked a long evolutionary experience with human hunters.

By exterminating many of the large mammals and setting numerous fires, hunters and gatherers may have caused small ripples in Earth's capacity to sustain life, but those were minor compared with the truly global-scale modifications human beings caused as agriculture spread and populations expanded. Millennia later, the industrial revolution carried the whole global change enterprise to a new level, and truly dramatic transformations were wrought in the past century. When scientists talk of global change today, they are usually referring to human-caused changes in the environment of sufficient magnitude eventually to *reduce* the capacity of much of Earth (as opposed to local or regional areas) to support organisms, especially *Homo sapiens*. This is a very recent development—a result of the rise of *Homo sapiens* to planetary dominance.

Those environmental changes include the transformation of more than half of Earth's ice-free land surface in the service of farming, grazing, forestry, mining, and human settlements; emissions of various substances into the atmosphere, leading to ozone depletion, acid precipitation, and global heating; homogenization of Earth's biota through transport of plants and animals around the globe; redirection and disruption of many freshwater flows by damming, groundwater depletion, rechanneling of rivers, and so on; global dissemination of DDT, lead, mercury, dangerous plastics, hormone-mimicking chemicals, and other toxic substances; and dramatic modification of global nutrient cycles, especially of carbon and nitrogen.

This chapter focuses on the changes humanity is making to Earth's land surface and oceans and their linked life-support systems; the next one will explore changes to the atmosphere and their effects on climate and natural systems, among other topics.

Land-Use Change

Probably the single most important development in the history of our species began about 10,000 years ago when some human groups first settled down and began to farm, as we suggested in previous chapters. Because people have continually altered land cover over thousands of years, and because this process still occurs piecemeal across the planet, summary statistics on recent land-cover changes are often difficult to assemble. One might imagine, for example, that in the case of the most obvious major change, deforestation, it would be simple to calculate its extent from satellite imagery. But unhappily no such imagery exists for the years before 1960, and today, furthermore, it is often difficult to distinguish in those images tree plantations or degraded (e.g., partially harvested) forests from relatively undisturbed virgin (never cut) forest. Therefore, we have to depend on ground-based surveys in combination with satellite photo analysis to arrive at an estimate.

Land-cover change was dramatic during the twentieth century and is still proceeding rapidly. More than 70 percent of the world's temperate broadleaf forests, the kind that turn gorgeous colors in the fall before losing their leaves for winter, had been chopped down by 1950. The same percentage of Mediterranean-type forests, the sort made up of evergreen trees such as those also found in coastal California, and temperate grasslands, such as existed naturally in Kansas when the bison roamed, also had been lost by the middle of the twentieth century. At present, the most rapid land-cover changes are occurring in tropical moist forests ("rain forests" or "jungles") and in the world's remaining temperate, tropical, and flooded grasslands. It is estimated that, of what was left of each of those ecosystem types in 1950, more than 15 percent has already been converted for human purposes.

FIGURE 12-1. Deforestation in the Amazon, still an all too common sight. Photograph courtesy of iStockphoto.

Such land-cover disturbance is usually also accompanied by fragmentation of the remaining habitat. For instance, when farmers displaced by industrial agriculture move into the Amazonian forest to set up small agricultural plots, they use roads created by loggers, then clear space for homes, small communities, and connecting roads (figure 12-1). The small, disconnected patches of forest that remain are often substantially modified by "edge effects"—for example, the interiors of tropical forest patches may dry out because the entire patch is relatively near a forest border, and adjacent cleared land is open to heating by sunlight and winds that can penetrate the patch if it does not become overgrown with vines.

Much has been learned of such edge effects from a large-scale experiment begun by ecologist Thomas Lovejoy in 1980, called the Biological Dynamics of Forest Fragments Project. In the Amazon River basin near Manaus, Brazil, in cooperation with landowners who were clearing land for cattle, forest fragments of various sizes (1, 10, 100, and 200 hectares, a hectare being equal to about 2.5 acres) were created between 1980 and 1990, and the fates of their biological communities were studied. The studies showed how complicated the changes were in the fragments, how much they depended on other factors besides fragment size alone, and how they varied with the kinds of organisms found in the original patch. Relative to the original fauna of a patch, the numbers of species of primates, birds, and some insect groups tended to decline, while the species richness of small mammals, butterflies, and amphibians tended to increase. The reasons in some cases were clear—for example, butterflies depend on a variety of food plants for their caterpillars to feed upon and others from which adults gather nectar and pollen. The diversity of such plants increased along the sunny patch edges. Also, the species of mammals and butterflies showed some "turnover," with some kinds disappearing and being replaced by others.

From Lovejoy's project and other studies, it is clear that small size and edge effects make many patches unsuitable for numerous plant and animal species that would live quite happily in large, unbroken stretches of habitat. Not only drying out but also higher temperatures and exposure to non-forest pests and predators are among the hazards of edges to plants and animals not adapted to them. Even normally long-lived rain forest trees die sooner if they're located within 100 yards of the edges of the fragments they stand in; a change in tree species composition then has cascading effects on other organisms in forest fragments. Habitat destruction and fragmentation (which dilutes the conservation value of patches) are thus principal causes of the extinction of populations and species of organisms that are essential to maintaining ecosystem services.

Deforestation, accompanied by massive biodiversity loss, is concentrated today mainly in the tropics. Of the roughly 65 million hectares of land deforested worldwide between 2000 and 2005, more than 42 million were in Africa and South America. Major deforestation also continued in Southeast Asia, where from 1985 to 2001 in Kalimantan (Borneo), "protected" lowland forests declined by more than 56 percent (more than 11,000 square kilometers). In some areas, up to 80 percent of the lowland biodiversity-rich dipterocarp forests (named after a valuable characteristic tree group) have been cleared.

Overall, the world has lost nearly half of its post–ice age forest cover in the past 8,000 years, most of it since 1970. In many temperate areas, such as the eastern United States and parts of Europe, there has been extensive natural regrowth of forests, as well as the creation of large tree plantations. For instance, more than a third of the total worldwide deforested area from 2000 to 2005 was partly offset by natural forest regeneration and establishment of tree farms. Areas of natural regrowth differ in many significant respects from the original forests, though, since there usually is no reestablishment of key elements of biodiversity, especially large predators. Nevertheless, many of the previous attributes of the forests will be restored if wildlife gradually returns. Not so in tree plantations, which are essentially monocultures and quite inhospitable to most wildlife and other plants.

Land-use changes can have dramatic impacts on aquatic systems as well. Deforestation, for example, can radically alter the temperature and the fauna of streams that once ran under a canopy of trees. And the application of inorganic fertilizers to farm fields overfertilizes freshwater streams and coastal waters. The result is eutrophication: water bodies receive excess nutrients—typically nitrogen, phosphorus, or both—which leads to excessive growth of plants and often algae, whose dense growth creates what are called "blooms." Dissolved oxygen in the water may be greatly reduced, especially at night, when algae are not photosynthesizing or when dead algae are decaying, which in turn suffocates valued fishes and other aquatic organisms. This effect often creates "dead zones" in the oceans close to the shore. The oceans now have hundreds of such zones, resulting not just from overfertilization but also from the results of many land-use changes leading to flows of animal wastes from feedlots and farmlands and of human wastes from privies and sewage systems. As oceanic dead zones proliferate, blooms that result in breezes carrying toxins to people on the shore in seawater droplets are also becoming more frequent.[3]

The giant dead zones that appear annually in bays and coastal waters around the world (most famously in the Gulf of Mexico) are traceable to nutrients running into rivers such as the Mississippi drainage. Several other dead zones have appeared off the coasts of the United States, including in the Chesapeake and Delaware bays and along the Washington and Oregon coasts; others have been seen off European and eastern Asian coasts. Globally, the number of dead zones seems to be doubling about every decade. In addition, eutrophication has contributed to the decimation of diverse freshwater faunas in many inland waters, such as Africa's Lake Victoria.

Coral reefs are also affected by changes in land cover. Resort development on tropical islands, for example, can lead to silt moving in streams from disturbed lands, entering bays, and killing silt-sensitive corals. Corals are also sensitive to eutrophication, which affects them in offshore areas of Hawai'i and other islands in the Pacific and the Caribbean.

Agriculture and Its Challenges

Homo sapiens could never have become the highly technological, environment-dominating and environment-transforming animal of today if agriculture had not been invented. Remember, it was farming that first produced enough surplus food that significant numbers of people in early societies could be freed from subsistence activities and dedicated to other tasks as administrators, priests, merchants, artisans, and so on. Now at least 20 percent of Earth's ice-free land surface is covered by crops, and more than another 25 percent is devoted to grazing domestic animals. The agricultural enterprise is a major driver of global change, intimately connected with the decline of three crucial kinds of natural capital, as described in the preceding chapter: fertile agricultural soils, "fossil" groundwater, and biodiversity.

As the twenty-first century unfolds, the human agricultural enterprise is in an especially precarious position. As agronomist Ken Cassman has neatly summarized,[4] the multiple demands being placed on it will be very difficult to meet without substantial luck with the climate and massive new investments to increase harvests. Climate change is likely to alter not only temperatures but also the amount and the temporal and spatial distribution of precipitation, all of which are critical for growing crops. Besides appropriate temperatures for crop growth, water must be supplied in approximately the right amounts and at the right time. While most agricultural areas are rain-fed, the roughly 18 percent of agricultural land that is irrigated (often by groundwater) supplies some 40 percent of the world's food.

Almost all the world's land that is suitable for agriculture, including what can be reasonably irrigated, is already being cultivated, and what still could be converted to cropland is generally of inferior quality. In recent decades, a significant portion—on the order of 10 percent or more—of the world's agricultural land has become too degraded for productive farming and has been taken out of production. In many cases, the cause has been poorly managed irrigation; in others, soil erosion and loss of fertility. In cases of serious degradation, restoration would not be possible at a reasonable cost.

Because farming is so spatially extensive, transport is a key part of any modern agricultural system. Food must be transported to markets, and inputs (seeds, fertilizers, and pesticides) must be supplied to farmers. In addition, much food undergoes processing and packaging before it reaches your local supermarket, all of which involves further transport of food and materials and expenditure of fossil fuels. A key component of industrial agriculture thus is energy, which is needed for operating agricultural machinery, powering the necessary transportation, processing, and storage. In addition, fossil fuels are essential for producing fertilizers and pesticides. As one ecologist put it long ago, we eat potatoes partly made from petroleum.

Today, as domestic shortages of oil threaten, the stability of its import flow seems uncertain, and prices rise, people are looking to agriculture as a potential major source of commercial energy. More and more effort is now being put into developing liquid fuels—so-called biomass fuels—from grains and other plant materials by converting them by fermentation into ethanol, an alcohol. As early as 2006, rapid expansion of ethanol production (due largely to its use as a gasoline additive) was raising maize prices in the United States, despite recent record and near-record harvests. Thus the need for energy is beginning to compete with food production for agricultural land. In the United States, the domestic food supply is unlikely to be seriously affected even if much of the maize crop is diverted to energy production, although prices of feed grains have already risen, followed by higher meat prices. But since a large portion of the grain crop is customarily exported, developing countries could find themselves unable to buy the grain they need to feed their populations. Thus, for example, instability in the Middle East threatening oil supplies for the West could also threaten the food security of poor people everywhere as food prices on the world market rise with oil prices. The prospects seem to be for food prices to keep rising as the competition escalates between feeding fuel to Hummers and feeding increasing numbers of people. By the end of 2007, not only higher food and fuel prices but also food shortages and signs of related hunger and political instability had already appeared.[5]

One major aspect of recent global change has been shifts in the geography of crop production. Food security in many areas greatly depends on patterns of international trade. For instance, Costa Rica, following the economic principle of "comparative advantage" (basically, doing what you can do best relative to what trading partners can do best), some years ago concentrated on growing coffee. The nation could make lots of money exporting coffee and then buy from China the black beans people needed for subsistence. But when the coffee market collapsed because of new coffee production in Vietnam and a recovery of yields in Brazil, Costa Rica's farmers and agricultural laborers faced trouble that could have been avoided had the nation had a more balanced agricultural economy. A greater mixture of crops would not only lessen the dangers of economic overdependence on one or a few but also reduce the risks associated with giant monocultures—those huge tracts of land planted in a single crop strain. The latter make especially attractive evolutionary "targets" for insect pests, which can quickly evolve ways to overwhelm the natural or pesticide defenses of the crop. Smaller plantings of mixed strains are much less likely to be subject to catastrophic events such as Peru's Cañete Valley disaster, described in chapter 2, in which overuse of pesticides greatly reduced cotton production.

In the late 1960s and the 1970s, a "green revolution" in many developing countries substantially increased the productivity of land already under cultivation and did increase the food security of millions of poor people. The surge in grain harvests was based on the introduction of high-yielding strains, primarily of wheat and rice, that respond well to inputs of fertilizers and abundant water—a technology already established in developed countries but largely lacking elsewhere until then.

The large increases in food supplies, however, came with sizable costs. A few strains of the grains replaced a wide diversity of traditional strains, creating monocultures that were much more susceptible to diseases and pests and contained much less genetic variability. The disappearance of traditional varieties made it more difficult to selectively breed new strains adapted to changed conditions. Furthermore, the greater need for water gave an advantage, especially in India, to wealthier farmers, who had the money to drill deeper tube wells as well as to buy fertilizers and pesticides. This exacerbated social inequities, but the rewards in alleviating famine seemed worth it to many observers, at least in the short run. But today India's water situation is perilous, roughly half its farmers are in debt, a few rich farmers are getting richer, and the soils are becoming exhausted of nutrients.

In the past decade or two, expansion of global food production has failed to match the growing global demand, and reserve supplies by 2007 were at a worrisomely low level, enough to meet demand for barely eighty days. Increases in grain yields (amount produced per unit area) have lagged globally, despite the need for such increases as populations in

FIGURE 12-2. Farming vast areas of the same crop (monocultures) can be very efficient in the short term, but the genetic uniformity of the crops makes them susceptible to pests, and they provide poor habitat for pollinators and organisms that act as natural pest controls. Photograph courtesy of iStockphoto.

developing countries continue to grow. At least with present knowledge of plant genetics and physiology, in many areas limits to the potential yields of some important crops seem to be approaching. No new "magic bullets" are known today that could continue to expand crop harvests to meet the needs of growing populations.

Indeed, there are fears that in India a much-advertised "second green revolution" today is simply a U.S.-based effort to promote privately owned agricultural biotechnology that will negatively affect Indian farmers and consumers. While biotech improvements to crops have gained most of the research attention and funding in recent years, they do little to increase yields except in special situations, such as to increase salt tolerance in crops grown in arid areas. Attempts to increase pest resistance in crops by genetic engineering have already begun to encounter the evolution of ways around that resistance by the pests themselves.

To meet the growing need for food, and to compensate for deteriorating farmland in many areas, large investments are needed to support research explicitly directed toward increasing yields, not only of the major grains but also of a variety of neglected traditional foods. Many such foods are important regionally and provide essential subsistence for poor rural societies with rapidly growing populations, who are most in need of expanded food supplies. Yet the agricultural research establishment until relatively recently has largely ignored the potential for higher yields of these traditional crops and the possibility that such strains might be grown in ecologically less damaging ways.

The relationship between productivity and the extent of plant diversity of crop and grazing lands is now a hot topic of ecological research, and an important one. There is considerable evidence, especially in long-term field experiments in grasslands carried out by ecologist David Tilman and his colleagues, that mixes of species can be much more productive than monocultures, generating more than double the biomass. Similar results have been obtained by agricultural ecologists in experiments with mixed-crop cultivation in tropical areas. Some such systems were practiced traditionally, and in the best of them two or three crops planted together can produce yields of each nearly equal to those obtained by growing them alone. Moreover, they are inherently more resistant to pests and protective of soils. Such findings could have major implications for increasing agricultural production in many developing countries. There are several obstacles to implementation, however. One is that finding the best mix of species and appropriate harvesting and pest control strategies in any given area is relatively complex and requires some experimentation. And such systems in practice are likely to be labor-intensive, although that could be an advantage in many poor countries with high unemployment. They also often are not readily amenable to farming with American-style heavy machinery—another plus in a world already damaged by the burning of so much petroleum.

Moreover, mixed-crop farming and mixed-species pastures could provide a form of insurance for farmers and ranchers in a time of increasingly changeable and uncertain weather. If conditions are adverse for some crops or grasses, others may do well. Finally, a focus on mixed-species regimes in grazing areas and forests may help sequester carbon dioxide in long-lived plants, in a world where global change in various forms is steadily *reducing* the diversity of ecosystems. And, perhaps most important, it can help preserve precious biodiversity.

Oceans: An Open Access Resource

One general problem that is tied to both loss of biodiversity and climate change is that many critical resources are "open access" or nearly so. An open access resource is a service or good that no one owns—one over which nobody exercises property rights. Such resources, without some form of group regulation, are available for anyone and everyone to exploit, and the result often is overexploitation or even destruction of the resource. Economists call them "common pool resources" when it is difficult or impossible for some actors to prevent other actors from exploiting them. They are distinguished from "public goods," which are resources that because of their nature or abundance cannot be overexploited. The latter are sometimes called "non-rival goods" because one person's use does not reduce their availability to others. Until the industrial revolution got under way, the atmosphere as a sink for waste was primarily a public good. Since then, it has increasingly become a common pool resource that is considerably at risk of overexploitation and abuse.

A classic open access resource has been oceanic fishes, which ironically were once viewed as inexhaustible—virtually a public good. And that resource has suffered from the textbook fate of open access resources: overexploitation. Today, many oceanic fish stocks are overfished by industrialized fleets, and the total yield has been essentially constant in weight but declining in quality as more desirable species are overharvested. It's a process that fisheries biologist Daniel Pauly calls "fishing down the food web." First to be exploited are the largest and most desirable fishes—usually top predators such as grouper, tuna, swordfish, and the like. When the big fishes are depleted, the fishers switch to smaller or less valuable prey, ending up in some cases catching jellyfishes. Hauling in smaller fishes or bycatch—fishes not sought after—along with the intended quarry reduces the prey that large predators eat, further impoverishing the food web. Furthermore, even where fish stocks are protected, the taking of larger animals amounts to selection favoring individuals that reproduce early. Thus the average size of individuals declines and their reproductive output suffers (big female fishes generally produce disproportionately more eggs).

Indeed, human additions to the oceans, overfishing of desirable animals, and ocean heating and acidification (the seas become more acidic as they absorb excess carbon dioxide released into the atmosphere by human activities) seem to be favoring a return to more primitive conditions in the oceans. Fishes, shellfish, corals, and marine mammals are dying out and plankton with calcite shells are threatened by acidification,[6] while populations of bacteria, algae, and jellyfish, given the appropriate physical conditions and a diminution of predators, explode. Marine ecologist Jeremy B. C. Jackson calls it "the rise of slime." In Japan, jellyfish have become so numerous that they clog intakes to nuclear power plants; in the Gulf of Mexico, they interfere with shrimping operations and are sometimes so abundant that people claim you can almost walk on them. Comb jellies introduced from North America have colonized the Black Sea, destroying the local fishing industries. In Moreton Bay, Australia, a toxic seaweed (a strain of cyanobacteria) is spreading rapidly, making fishers' bodies break out in rashes and boils, and killing their livelihoods. Only six of what were once forty commercial shrimp trawlers and crab boats still work the area, and some of those operate only part-time. The oceans are being transformed, in Daniel Pauly's words, into a "microbial soup." "My kids," Pauly said, "will tell their children: Eat your jellyfish."

Overharvesting, pollution, and dead zones are not the only problems plaguing marine fisheries. Another is the method by which fishes are harvested. Particularly damaging is bottom trawling, which destroys critical ocean-floor habitat. Draining and development (destruction) of coastal marshes and estuaries that serve as nurseries for many commercially exploited oceanic fishes also reduce the fisheries harvest. And there are the side effects of land-use practices along shorelines that we discussed earlier. The situation is sufficiently serious that recently a group of leading marine scientists concluded that current trends will lead to the global collapse of all organisms currently fished by the mid-twenty-first century.[7] To some degree, the lack of wild seafood will, for the rich, probably be somewhat compensated by the products of seafood farming (marine aquaculture), but the environmental and financial costs promise to be very high and the quality, from a flavor and health viewpoint, probably low. As marine ecologist Steve Palumbi said, "This century is the last century of wild seafood."[8] The seas as humanity knew them for more than 100,000 years, a source of physical and mental nourishment and honorable (if frequently dangerous) jobs, are rapidly dying.

Homogenization of Nature

Human beings were the invaders in the case of Pleistocene overkill. But subsequently, *Homo sapiens* has vastly altered Earth's flora and fauna by moving organisms from the ecosystems in which they evolved and inserting them in others, sometimes purposely. Remember the case of *Opuntia*, the cactus that ate Australia (chapter 10), which was imported to that island continent on purpose. Likewise, in the 1950s the predatory Nile perch (*Lates niloticus*) was deliberately (and secretly) introduced as a sport fish into Africa's Lake Victoria, where it contributed to the decimation of what had been an extraordinary diversity of cichlid fishes. More often, however, disasters are entrained accidentally, as when zebra mussels were carried from Europe to the Great Lakes of North America in ships' ballast. Zebra mussels, without the natural controls that kept them in check in Europe, spread widely throughout the Great Lakes. They filter vast amounts of the microscopic food from lake waters, starving native fishes and other aquatic animals dependent on it. They form big colonies that can clog the intakes of power plant cooling systems, drinking water supplies, and so on.

All over the world, invasive species are causing problems for agriculture in particular and for people in general. For instance, the Hessian fly (*Mayetiola destructor*) was introduced into North America, probably from Europe, at the time of the American Revolution. It is now the most serious wheat pest in the United States, causing as much as $100 million in damage in some years. Plant geneticists struggle in a coevolutionary race with the fly, continually working to produce Hessian fly–resistant strains of wheat. But it usually takes the fly only six to ten years to evolve its way around any new defenses. On the island of Guam, to take another example, poisonous brown tree snakes (*Boiga irregularis*) imported from New Guinea have killed off most of the native birds, decimated mammal populations, and even come up through toilets and attacked people. The little *Cactoblastis* moth that saved Australia from *Opuntia* has now invaded the United States from the Caribbean and is moving toward Mexico, threatening the *Opuntia* grown there as a commercial food crop. In a different place, the savior of part of Australia becomes a serious potential pest. Over much of the planet, exotic earthworms, spread by anglers and horticulturalists, are battling with native ones, with variable and sometimes distressing results for local ecosystem services.

Examples of problems with invasive species could be multiplied ad infinitum. Much of this occurred before nations, states, and international organizations began to institute controls, but many of those instituted have proven inadequate, and more are badly needed.

Toxic Substances

One of the most difficult to evaluate of all the global changes is the toxification of the planet. Yes, we know that toxic pollutants can kill us when we're exposed to substantial doses—as smoking demonstrates to more than 5 million people every year. But the impact of global low-dose pollution of multiple substances on both human health and the health of ecosystems is more difficult to evaluate. For instance, it remains unclear whether a widespread drop in human sperm counts really has occurred, and if it has occurred, whether chemical pollution is responsible. The same can be said for reports of a deficit of male births, especially in the far north. Many signs are ominous. For instance, the fat of polar bears, which live very far from most sources, is loaded with pollutants. Concentrations of some of those pollutants are correlated with declining size and weight of the bears' ovaries and the bone that supports their penises.

Many of the synthetic chemicals we add to the environment—ones, for example, that leak out of plastics used in food containers—can function as hormones, chemical messengers that control many bodily functions in people and other animals. A compound called bisphenol A (BPA), which leaks out of ubiquitous polycarbonate plastics, has properties in common with estrogen, a steroid that functions as a human female sex hormone. BPA is present in the milk that mothers feed their babies from polycarbonate baby bottles, and it enters virtually all of us from the linings of many "tin" cans, hard plastic sports bottles and food containers, and other sources. Compounds such as BPA occur mostly in doses so low that they were formerly considered harmless. But as we learn more about the complexities of gene regulation, it is becoming clearer that they may in fact be quite potent, especially in the very young, in changing patterns of development.

One reason that polar bears carry such heavy pollutant loads is that they are long-lived animals living at the top of food chains. Chemicals such as residues of DDT that easily dissolve in fat are not lost at high rates, as is useful energy, when they move up food chains. Rather, they concentrate; more and more of the poison is found in every step upward. Much of the DDT residue in a fish is added to the DDT load in a seal when the seal eats the fish, and much of the seal's load is passed on to the polar bear that eats it. The process, called biomagnification, is complicated, but a basic message is that you should beware of eating too much flesh of species, such as tuna, that live high on food chains. After all, when you do, you are feeding even higher on the chain!

Another trend that may be related to toxification is a worldwide decline in amphibians (frogs and salamanders). Like other organisms, they are suffering from habitat loss and, probably to a much lesser degree, overharvesting (for instance, for frogs' legs for the restaurant trade and, at least at one point, for use in biology classes). But about half of the species' declines are in seemingly pristine habitats and thus seem to be tied to undetected factors. Diseases or parasites may sometimes be responsible, and many scientists suspect that chemical pollution is also involved, in some cases perhaps in combination with disease. Professor Tyrone Hayes of the University of California, Berkeley, has in fact produced persuasive evidence that the weed killer atrazine escaping from farm fields is involved in causing reproductive difficulties in leopard frogs. Most amphibians require streams or ponds in which to breed, and they need moist environments to keep their skins wet. They have not evolved the eggshells and watertight skins that allowed most reptiles to escape the need for standing water or high humidity and become fully terrestrial. The simple lungs of amphibians cannot supply enough oxygen for their activities, so their skins must be kept wet and permeable so they can take in oxygen. But that permeability also provides a gateway for toxic substances. Increased ultraviolet radiation has been hypothesized as another detrimental effect, and fungal diseases have been shown to be a factor, although what drives the fungal outbreaks remains unclear. Most recently, some studies have implicated climate change, suggesting that global warming is causing fungal outbreaks that are wiping out entire frog populations and pushing many species toward extinction.

It is not surprising that scientists have been unable to determine clearly the factors that have caused amphibian declines, or other phenomena such as the apparent increase in potentially lethal beak deformities in Alaskan birds. Although many people seem to think that the shifting population dynamics (size changes), ecological relationships (including responses to toxins), and evolutionary history of every animal or plant population they have heard of are known in detail, the truth is that we have substantial (and still very incomplete) information on only a relative handful of species. This is partly due to a lack of funds and personnel, as well as being a reflection of the vast diversity of life. It also is partly a result of much research on scattered groups of organisms, for which the results are fragmentary and difficult to put into context, and too little focus on what could be a few dozen carefully selected test systems.

Ecologists also were initially quite slow to begin investigating human-generated global change, including the distribution of toxic substances, and its consequences. A few still complain, as one did recently, that ecologists speaking out on their diagnoses of the human predicament were "undermining the discipline of ecology." (Curiously, physicians alerting people to the toxic dangers of smoking or the cardiovascular risks of high blood cholesterol levels don't seem to be perceived as having undermined the discipline of medicine.) Denial occurs not just among politicians and the general public but also among scientists, who are human beings just like everyone else.

Regardless of effort, however, it is extremely difficult to assay the effects of individual low-dose chemical pollutants, and even more difficult to quantify their impacts in combination. Although tens of thousands of potential toxins have been released into our environment, we are ignorant of most of those chemicals' toxicity to people and to non-human organisms that are crucial to human lives. Worse yet, scientists know virtually nothing of the possibly more serious dangers of any combination of a few or thousands of pollutants acting in concert. If we discover that some combinations of the pollutants people are adding to their environment are extremely damaging or lethal, it most likely will be very difficult to solve the problem. The second law of thermodynamics tells us that the world generally moves from order to disorder—from a concentrated drum of poison to tiny amounts dispersed around the planet. Rarely would humanity have the resources (energy) required to reconcentrate a toxin, as nature does when it reconcentrates DDT residues in a food chain. It's always easier to avoid dispersing it in the first place, just as it's easier to keep the salt in its container than to pick it out of the rug once it's been spilled.

The Ecological Footprint

Even if it weren't for toxification and the other environmental problems we have been discussing, our species would still be in trouble. There is considerable evidence that the enormous expansion of the human enterprise in recent decades has already caused *Homo sapiens* to overshoot Earth's long-term human carrying capacity—the number of people that could be sustained for many generations without jeopardizing the planet's capacity to support a similar-sized population in the future.

In 2002, a large and diverse team of scientists used existing data to determine how much of the biosphere would be required to support today's human population sustainably—that is, "to translate human demand on the environment into the area required for the production of food and other goods, together with the absorption of wastes."[9] The scientists considered the needs for croplands and grazing lands, forests for timber, productive fishing grounds, space for infrastructure (housing, transportation, industry, hydroelectric power, etc.), and carbon sequestration (to retard the atmospheric buildup of carbon dioxide) to support the population.

This study, while preliminary, estimated that humanity's "load" was equal to about 70 percent of the biosphere's regenerative capacity in 1961, when the human population was about 3 billion and per capita consumption of resources and energy was far less than today's. The scientists further estimated that we have exceeded that capacity since the 1980s as the population and per capita consumption both soared. They estimated that, in 2002, when the population had passed 6 billion, civilization's demand on the biosphere had reached more than 120 or 140 percent of capacity, depending on the assumptions made. The scientists were basically saying that things could not long continue as they were—that either human behavior toward the environment had to change or the number of people would have to be reduced, whether through lowered birthrates or an unhappy increase in death rates.

Much can be communicated about the human overshoot of Earth's carrying capacity through the tool of what is termed "ecological footprint" analysis. This work was begun decades ago by economist Georg Borgstrom and brought to fruition in the 1990s by William Rees and Mathis Wackernagel. The eco-footprint of a designated population is "the total area of land and water ecosystems required to produce the resources that the population consumes, and to assimilate the wastes that the population produces, wherever on earth the land/water are located."[10] Eco-footprint analysis, as we saw, suggests we may well have already exceeded the long-term carrying capacity of our planet by as much as 40 percent. Such estimates are, to say the least, deeply dependent on the assumptions that go into them, such as how successfully underground carbon dioxide sequestration can augment its absorption by natural systems. But our own estimate is that optimistic and pessimistic assumptions are very likely mostly to balance each other, and that footprint analysis showing that people have substantially overshot the long-term carrying capacity of Earth is probably accurate.

That and other analyses, and even common sense, suggest that the human enterprise is already unsustainable—human demand is outstripping what nature can provide—even though the great majority of human beings have not begun to approach the extraordinary American level of resource consumption. The eco-footprint of an average American is roughly four times the global average, and as much as ten times larger than those of the citizens of very poor countries such as Bangladesh and Chad. That difference does not simply reflect different consumptive desires and incomes; it reflects the great disparity in power between the United States and the poorest nations.

Tying It All Together

The various current trends that are generating global change and population overshoot are, as one might expect, not independent. For instance, it is commonplace to believe that hunger in the world is caused by poverty and a malevolent economic system, but that population growth is not connected to the problem. In a recent article about genetically modified foods in an important anthropological journal, any mention of population size in connection with malnourishment was repeatedly called "playing the Malthus card."[11]

Certainly, factors such as agricultural subsidies in rich countries and pressures to produce cash crops for export in poor ones have contributed considerably to problems of hunger and famine. So have the gross inequities in income and power that plague humanity. Those 850 million people in the world who are significantly undernourished and the 6 million children dying needlessly each year because of malnutrition serve as a stark index of extreme poverty. Poverty is one of the worst environmental problems because it requires that people wrest a living from their land without regard for environmental consequences. It also largely deprives close to a third of the global population (those living on less than US$2.00 per day) of any power to improve the world their descendants will inhabit. Yet if everyone were willing and able to share more equitably—for example, if the rich were willing to modify their current diets, significantly reducing their consumption of beef and other animal products—then hunger might be largely eliminated, and as a bonus the flux of greenhouse gases into the atmosphere could be significantly reduced. Ample food is produced today to feed more than 6.7 billion people a reasonable diet, but a disproportionate share goes to the already overfed.

Hunger *is* connected to population size and growth, of course, and it may well be more so in the future. The need to feed ever more people has led to an expansion in areas supporting crops (especially in poor countries) and an intensification of agriculture (first in rich and then in poor countries). Further intensification, especially in developing regions, will be necessary in order to keep increasing food production, at least until the world population has passed its peak size and a basic level of nutrition for all societies has been achieved. If the human population were not already huge and still growing, providing that basic diet would be much easier and would carry many fewer environmental risks.

Interactions among population pressures, consumption, and inequitable political and economic power can exacerbate the consequences of "natural" disasters, sometimes making the population component of the problem difficult to detect at first glance. Too many people and economic inequality in Honduras, combined with overharvesting of forests and planting of crops on steep slopes, turned a nasty hurricane into a catastrophe in 1998. Many of the poor in that country had no alternative to living in precarious situations. Hurricane Mitch dealt them a devastating blow, triggering floods and mudslides that would not have occurred if fewer people had needed to clear land for living space, if forests had not been overexploited, and if the distribution of land had been more equitable so that the poor were not crowded into vulnerable areas. Mitch killed thousands of people and made tens of thousands homeless in Honduras and Nicaragua. Most of the victims were squatters who were crowded onto riverbanks or into villages on steep mountainsides.

In India, large extents of the coastal mangroves in the Bay of Bengal have been destroyed to make way for shrimp farms designed to supply global trade with the rich countries. In 2000 a giant cyclone struck the coast, devastating communities never before affected by a cyclone because the coastal buffering of the mangrove swamps was gone. As a result, 30,000 people and perhaps 100,000 cattle were killed.

Even in such rich countries as the United States, one can see the results of such interactions. The effects of the infamous Hurricane Katrina, which devastated New Orleans in 2005, were worsened by the presence of large numbers of poor people living in vulnerable areas below sea level, by previous erosion of buffering land along the shore, by diversion of the Mississippi River, and by failure of the political system to ensure adequate protective measures. A combination of fewer people and more economic equity could have ameliorated all of these disasters.

Global change not only modifies the environment in which organisms are evolving genetically; it can have a major impact on human cultural evolution as well. We humans started our planet-altering rise to dominance by killing and eating big mammals. Now our hard-earned dominance threatens our own civilization, and it will therefore greatly influence future trajectories in cultural evolution—and, in turn, what they portend for humanity's life-support systems.

CHAPTER 13

Altering the Global Atmosphere

"Climate change is already occurring far faster than most scientists predicted."

JOHN P. HOLDREN, 2006[1]

FOR MOST OF human history, *Homo sapiens'* environment-altering activities did not significantly affect the functioning of the atmospheric system. The acidity of rainfall in the water cycle was not noticeably increased by human activities, the stratospheric ozone that screens out deadly ultraviolet B (UVB) radiation was produced and destroyed at a rate people did not influence, and the carbon cycle hummed along without significant human interference. Climates marched through the seasons unperturbed by human inputs of greenhouse gases. Now all that has changed as human beings have become so dominant a force that we are significantly modifying the thin layer of gases that envelops Earth, changing it so that, as examples, rain is more acid and the planet is heating up, both of which have enormous implications for us and other life.

ACID PRECIPITATION

One of the first human-caused impacts on the atmosphere to attract public attention was the rise in the acidity of precipitation in many regions. The increase was caused by the injection into the atmosphere of nitrogen oxides (NO_X) and sulfur dioxide (SO_2) principally by automobile exhausts and power plants that burn fossil fuels. In the atmosphere, the oxides of nitrogen and sulfur are chemically converted into nitric and sulfuric acids, respectively, which then combine with water droplets and make rain, snow, and fog more acidic than normal, sometimes extremely so—approaching the acidity of vinegar.

The connection between air pollution and the acidity of rain was first made by a Scottish chemist in 1852, but the effects were not really noted until acidity began to kill fishes in Scandinavian lakes after 1950. Fishes and other aquatic animals have evolved to live within certain limits of acidity; when those limits are exceeded, eggs often won't hatch, and if the change is even greater, adults die. The recognition that lakes were losing their fish populations was followed in the late 1970s and early 1980s by claims that acid rain was damaging forests in Europe. It was so serious in Germany that the damage was christened *Waldsterben*—forest death. By the 1990s acid precipitation had rendered about 20,000 of the 90,000 lakes in Norway and Sweden "dead" or "dying," and similar lake effects had appeared in some areas of the United States, especially in the Adirondack Mountains. Adverse effects on forests at higher altitudes along the Appalachian Mountains chain, slowing the growth of some types of trees and reducing their resistance to disease, had also become evident by that time.

Acid precipitation has also been suspected as a factor in a worldwide decline of amphibian populations, and it has been implicated in the decline of some bird populations. Diving birds such as loons have had difficulty raising their young on fish-poor acidified lakes, and some songbirds may have suffered from acid-caused eggshell thinning or reduced availability of prey. Acidification can alter soil chemistry as well, shifting the balance of organisms in the soil so that fungi are replaced by bacteria. This can harm trees and other plants that depend on a mutualistic (mutually advantageous) relationship with certain fungi called mycorrhizae. The fungi transport essential nutrients into the plants and receive energy-rich carbohydrates in return. Acid rain also threatens the pitcher-plant ecosystem—remember, the home of those evolving mosquitoes. While the bogs in which the plants live are naturally acidic, they are exposed to excess nitrogen from the nitric acid in acid precipitation. This causes an imbalance in the plant's nutrients—too much nitrogen relative to phosphorus—and the plant's growth is stunted. This threatens the survival of pitcher-plant populations and the tiny aquatic ecosystems dependent upon them. Those ecosystems can serve as microcosmic warnings—systems brought to the brink of ruin by global change.

People have tried to counter the negative effects of acid precipitation in various ways. Putting lime into lakes was one of the most successful remedies in Scandinavia, a treatment that was emulated in Adirondack lakes, with mixed results. But the most effective measures in North America were a result of the 1990 amendments to the Clean Air Act, which introduced a tradable emissions permit system to reduce emissions of SO_2. In order to lower the level of acid precipitation gradually, government permits were issued, allowing each firm a portion of the permissible release. Firms that could maintain or increase production while reducing their emissions (by, for instance, investing in scrubbers for smokestacks to remove sulfur from the emissions) could sell the unused portion of their permits to other firms that could not, usually because of older equipment and an inability to meet the costs of upgrading to new technology. This "cap-and-trade" system, in which the total emissions were held under a shrinking cap, and rights to emit could be traded, provided financial incentives for firms to reduce pollution, and it seems to have been very effective. Altogether, including controls on tailpipe emissions of NO_X from automobiles, regulation has had substantial success in reducing the acidity of rain and snow, although complex effects on ecosystems are still occurring.

The same success has been seen in Europe, where the acidity of both precipitation and lakes has been dramatically reduced. There, control has had its biggest effect on sulfur emissions, while nitrogen emissions are still increasing in many industrialized areas, for instance in Wales and parts of the Mediterranean basin.

The extent of acid precipitation in places such as China, Brazil, and South Africa is less well documented, but in large parts of China it clearly is quite serious. Acid rain is reported to bathe one-third of that country's land surface, damaging croplands and threatening aquatic food chains. Chinese officials are worried about social stability in acid-affected areas as farmers and fishers struggle to survive in the face of a deteriorating environment. Sulfur dioxide emissions in China in 2005 caused an estimated $65 billion in damages, and this number is expected to increase, as China has embarked on the biggest and fastest expansion of coal-burning power plants in history. Damage from these plants is not limited to Chinese health and air quality; significant amounts of pollution are transported across the Pacific Ocean and even affect the air in the western United States.

Acid precipitation (and the parallel problem of ocean acidification) remains an important part of global change and will continue to be as long as fossil fuels, especially coal, the main source of SO_2 emissions, are heavily used. The long-term effects of even reduced acidification, which will depend on many factors, such as the acid-buffering capacity of local soils, are poorly known. In North America, for example, acidification may continue to have deleterious effects on breeding bird populations, especially at high elevations, where soils are already somewhat acidic. While the problem of global heating (often called "global warming," which sounds much too benign and cozy) is currently commanding the most attention, acidification of much of Earth's land, fresh water, and oceanic systems could become extremely serious.

Ozone Depletion

Denial and ultraconservatism among politicians and the general public were a problem when scientists first tried to bring acid precipitation to the attention of policy makers—a situation that has often prevailed with respect to large-scale environmental issues. They also played a major role in the way society addressed the issue of stratospheric ozone depletion. What could have become the most critical global change was entrained by one of *Homo sapiens'* notable technological triumphs: the creation of Freon, a trade name for several chlorofluorocarbon (CFC) compounds. When first synthesized in the early 1930s, CFCs were heralded as one of the great inventions of the chemical industry—stable, non-toxic, non-flammable compounds that seemingly were ideal replacements for the toxic operating fluids then used in refrigerators. Later, the spread of air-conditioning and similar technologies caused CFCs to be produced in ever larger quantities. Then British chemist James Lovelock invented a device, the electron capture detector (used in gas chromatography), that permitted detection of the presence of CFCs as amazingly long-lived contaminants in the atmosphere. An atmospheric chemist, F. Sherwood Rowland, became curious about their persistence and encouraged his postdoctoral associate Mario Molina to help him look into it.

Rowland and Molina concluded that the CFCs could destroy the stratosphere's ozone layer, a shield extending between about eight and thirty miles above Earth's surface that protects all life on land from damage or destruction by the sun's dangerous UVB radiation. The ozone layer was produced over billions of years by an accumulation in the atmosphere of oxygen generated by photosynthetic organisms in the oceans. Life was able to leave the sea and colonize dry land only some 500 million years ago, after a sufficiently protective layer of ozone had been formed.

In the atmosphere, oxygen molecules, consisting of two oxygen atoms (O_2), are split by ultraviolet radiation, and single atoms combine with a complete oxygen molecule to form the unstable molecule of ozone, consisting of three oxygen atoms (O_3). In turn, UV radiation breaks down ozone, in a continuing cycle that normally maintains a steady concentration of ozone. But when CFCs reach the stratosphere above the ozone shield, they too are decomposed by UV, releasing chlorine. The chlorine, unfortunately, catalyzes a chain reaction of ozone destruction, pushing the oxygen-ozone cycle toward diminishing ozone.

Substantial depletion of the ozone layer today would be calamitous for terrestrial life. Although most popular press accounts have focused on human skin cancers, far more serious problems would attend severe ozone depletion, since many kinds of plants and some marine organisms are susceptible to damage from UVB. There could be widespread disruption of agriculture and damage to forests and other ecosystems. So Rowland and Molina's conclusion was one of the most unpleasant surprises of theoretical science of all time. When Rowland came home from his office after his discovery, he told his wife: "The work is going well. But it looks like the end of the world."

Rowland and Molina were subjected to much abuse by spokespeople of corporations that were producing CFCs for use in refrigeration and air-conditioning and as propellants in aerosol cans, who claimed the scientists' conclusions were faulty. The use of CFCs as aerosol propellants was banned in many countries in the late 1970s, but then the issue faded from public attention, even though that was only a partial solution. Eventually, however, empirical science came to the rescue of theoretical science. British scientist Joe Farman and his colleagues with the British Antarctic Survey, using "old-fashioned" but well-tested instruments, documented the thinning of the ozone shield over Halley Bay and the Argentine Islands in the Antarctic. That thinning had previously been missed by a sophisticated satellite launched by the National Aeronautics and Space Administration (NASA) because of a series of errors in computer programming and data analysis.

There is a cautionary tale about technology here; when we pull technological rabbits out of the hat to "solve" our problems, we must be very careful that they don't produce nasty droppings. The NASA satellite in question, Nimbus-7, carried TOMS (Total Ozone Mapping Spectrometer), an instrument that sent back to Earth several hundred thousand ozone measurements every day, covering essentially all of the sunlit planet. The first satellite measurements in October, the southern spring and the time of year that atmospheric conditions are best for ozone destruction over Antarctica, were made in 1979. For several years, they showed consistent ozone values of around 250 Dobson units (the standard measure); 300 is the approximate global average value at any time. But this turned out to be an artifact of the measurement system. Apparently TOMS was programmed in its early years to reject any reading below 180 Dobson units as obviously erroneous because such low values had never been known to occur.

By 1983, reexamination of the data showed that values below 175 were appearing, but a deep plunge during 1980–82 still had not been detected by analysts. The way TOMS was programmed, combined with NASA's failure to make any provisions for adequately handling the flood of data that came from it, gave Farman and his team the opportunity to be the first to publish the news of the ozone hole (defined as an area with readings below 220 Dobson units). When the problems were corrected, reexamination of raw TOMS data quickly confirmed and expanded the findings of the Farman team. Meanwhile, the size of the ozone hole, which was about 6 million square miles in the late 1980s, grew to 11 million square miles at its largest in 2000—about three times the area of the United States and about twice the size of Antarctica itself.

In 1986, a brilliant young atmospheric scientist, Susan Solomon, led a team to the Antarctic. Chemical data they gathered in high-altitude flights showed beyond reasonable doubt that CFCs were the culprit causing the thinning. The next year, the international community took action at a meeting in Montreal and banned production of CFCs and some related chemicals in industrialized countries, with a grace period for compliance and financial support provided for developing countries.

The agreement, known as the Montreal Protocol on Substances That Deplete the Ozone Layer, has been more or less successful, although some smuggling of the banned products still occurs, and developing countries are still producing too much of them. Because of the latter activities, along with the continued use of some ozone-depleting chemicals (especially bromine) at levels permitted under the agreement, and the long residence time of CFCs in the stratosphere, the Antarctic ozone hole has persisted. The concentration of ozone-depleting chemicals in the stratosphere seems to have peaked in 2001, but the ozone hole is not expected to be fully healed until late in this century. Farman himself is so concerned about a continuing threat that he has called for a faster phase-out of ozone-destroying chemicals and a comprehensive reexamination of the Montreal Protocol.[2]

In 1995, Sherwood Rowland and Mario Molina shared, with Paul Crutzen, another fine atmospheric scientist, the Nobel Prize in Chemistry. Their work helped save humanity from untold misery, and at the same time they revealed a story of unintended consequences that should be engraved in the memories of those who think technology will solve all human problems. Yes, technology *can* help to save us, as it did in detecting the fate of the ozone shield. Technological research by the chemical companies also found relatively ozone-safe synthetic substitutes for the ozone-destroying CFCs. But it was actually political action, mobilizing international cooperation, that ultimately fixed the problem. After all, it was technology, in the course of providing humanity with great benefits, that nonetheless put humanity in peril in the first place. And the possibility of unintended consequences persists. Some new substitutes for CFCs, for example, have turned out to be potent greenhouse gases (as are CFCs themselves) and thus may exacerbate global heating even though they are no threat to the ozone layer.

Humanity can't do without technology, but some technologies can stress human capacities so severely that errors—including potentially lethal ones—can occur at the technological-human interface, as they did in the NASA Nimbus-7 satellite case. Fortunately, that should alert NASA to such dangers and make even more effective its critical Earth satellite program. Nonetheless, despite the delays in recognizing and addressing the problem, the Montreal Protocol is counted by most as an international success story, one that could be a partial model for negotiating agreements on the far more complex problem of climate change. But it can be only a partial model, because of its own shortcomings and especially because of several vast differences between the two problems.

While both issues were subjected to public deception and lobbying from polluting companies, the political and industrial circumstances are very different. In the case of CFCs, it was necessary only to create new chemicals to replace the offending ones. This was a much smaller challenge than reinventing the way we use energy to fuel our modern lives, which now requires the use of billions of tons of coal, oil, and natural gas each year. The enormous volumes of fossil fuels produced as well as the diversity of both the sources and applications of energy are simply incomparable to the situation with CFCs. Furthermore, the invention and production of CFC replacements, hydrochlorofluorocarbons (HCFCs), weren't too costly, and after a short time the CFC market switched to the far less ozone-damaging substances. It took only a few innovations by leading companies to make the change; the costs of transferring the new technology were minimal, and the new manufacturing process was hardly any different.

Thus the solution to ozone depletion provides only limited guidance in dealing with the energy-climate dilemma. All countries have infrastructure to supply electricity, often massive and built primarily around the use of particular energy resources (coal in the United States, natural gas in Pakistan, nuclear power in France), in addition to transportation systems dependent mainly on petroleum. These systems are not easily changed. They certainly cannot be changed over a handful of years, as was the case with CFCs.

Another huge difference is that the industrialized countries accounted for most of the CFC production and use, and so collective action there could greatly reduce CFC releases. That is not so much the case with greenhouse gases. Although the United States was still the biggest emitter of greenhouse gases in 2006, and Japan and Europe had been the next largest for decades, China's emissions exceeded both soon after 2000 and were reported to have overtaken those of the United States in 2007. China and India, along with other transitional nations, are on track to dominate the global emissions profile within a decade or so. Whatever strategy the world eventually follows to curb greenhouse gas emissions must be designed to take those contributions into account.

Finally, the compliance of developing countries in the Montreal

Protocol was gained by providing a direct subsidy to cover the costs of adopting the new technology. A similar program was developed for curbing greenhouse gas emissions, called the Clean Development Mechanism, but participation is voluntary, and its auditing standards are weak. The policy has not achieved significant progress in slowing the rise in greenhouse gas emissions. Politically, the absence since 2001 of climate policy in the United States, the world's largest economy and historical leader in greenhouse gas emissions, has given developing countries a viable rationale for not taking emissions more seriously. But even with universal participation in a solution, the scale of the climate crisis dwarfs that of the ozone problem. In the long run, developed countries might need to transfer hundreds of billions of dollars to poorer countries to help them afford low-carbon energy and to adjust to the negative impacts of climate change.

Climate Change

Changes in climate are not a new phenomenon, of course. In its history, our species has lived through four major ice ages spread over the past 650,000 years or so, with the most recent extending from roughly 90,000 to 12,000 years ago. And, as recently as 12,000 years ago, people went through a sudden (in terms of geologic time) 1,300-year cooling period over substantial portions of the globe as the last glaciation ended. The cooling was named the Younger Dryas after a tundra flower that flourished as treeless plains replaced forests in Scandinavia. The onset and the end of the Younger Dryas were very sudden, each on the order of a mere decade, with changes of 5°C–10°C (9°F–18°F) in many areas. It is believed that the cooling was caused by an ice dam giving way and emptying a huge North American glacial lake into the North Atlantic Ocean. That was Lake Agassiz, northwest of today's Great Lakes and larger in area than any lake on Earth today. The influx of cold fresh water, in turn, upset ocean circulation, including the Gulf Stream, causing a significant cooling in the Northern Hemisphere, especially in Europe. Gradual remixing then restored the previous circulation and allowed the post-glacial warming to continue. The effects of these rapid changes on the sparse human populations at the time can only be conjectured, but in some areas they may have been disastrous.

Today the situation is very different. The climate change we now face is the inadvertent result of human actions. In the minds of many, if not most, global heating is the single most critical environmental threat to civilization. Even if, say, the release of toxic chemicals into the environment ultimately proves to be a more serious, if more subtle, threat to the human enterprise, global heating almost certainly poses an array of gigantic problems.

The climate is largely controlled by the heat balance of Earth—how much the sun warms the atmosphere and the surface—and human beings have been changing that balance in two important and related ways. One is by land-cover change, modifying the albedo of the planet—that is, the amount of sunlight it reflects directly back to space. Forests, for instance, absorb more solar energy than cleared areas, and they often generate clouds, so removing forests tends to alter local or regional climate, although the direction and amount of change depend on many factors. When dark boreal forests are clear-cut, the resultant increase in albedo may reflect roughly enough solar radiation to compensate for the carbon dioxide greenhouse gas added by the burning or decay of the cleared wood. In the Amazon, the loss of large, reflective clouds generated by the forests more likely *reduces* the albedo and allows more absorption of incoming radiation, which reinforces the warming caused by forest burning and decay.

The other way we are altering climate is by adding greenhouse gases to the atmosphere. Remember, greenhouse gases are transparent to short-wavelength incoming solar radiation, but they trap heat by absorbing outgoing infrared (IR) radiation and reradiating about half of it back toward the surface. Most of the natural greenhouse effect (on the order of 60 percent) is caused by one gas, water vapor. Trends in the atmospheric concentration of water vapor are strongly coupled to those of temperature: more water evaporates as it gets hotter. The role of water in the atmosphere is complicated, however, because of its involvement in cloud formation, since clouds can have either a warming (blanketing) or a cooling (albedo) effect, depending on their thickness and altitude.

The most important greenhouse gas being added to the atmosphere by human activities is carbon dioxide (CO_2), the concentration of which has been more or less steadily rising for many decades. Before the industrial revolution, which began in the late eighteenth century, the atmospheric concentration of CO_2 was about 280 parts per million (ppm). By 2007 it had risen to 383 ppm, an increase of about 37 percent, and was climbing by about 1.9 ppm per year (up from an average of 1.4 ppm per year between 1960 and 2005). The most recent evidence suggests that the CO_2 flow into the atmosphere is actually accelerating.

The increase in CO_2 concentration is due mainly to the burning of fossil fuels and, to a lesser extent, to the clearing and burning of forests, chiefly in the tropics. In the nineteenth century, land-use change, especially deforestation, was the principal contributor of anthropogenic greenhouse gases to the atmosphere by releasing CO_2 as vegetation and wood decayed or was burned. Deforestation continues to make a significant contribution, although considerably less than the CO_2 now released from combustion of fossil fuels; deforestation accounts for roughly 20 percent of human-caused emissions and is declining in relative importance as fossil fuel emissions continue rising rapidly.[3]

While correlation is not necessarily causation, there is ample evidence to connect the increasing release of CO_2 from industrialization with rising atmospheric concentrations and to conclude that industry and its products (such as automobiles) are releasing the CO_2 and causing the rise. Not only do the quantities of emissions match the observed atmospheric buildup of CO_2, but also the specific contributions from fossil fuels can be measured chemically because their carbon has been sequestered from the environment for millions of years.[4]

Other important anthropogenic greenhouse gases include methane (CH_4), nitrous oxide (N_2O), and several chlorofluorocarbons (CFCs). Methane is emitted naturally by chemical reactions in wetlands; human-caused emissions come from rice paddies, landfills, and the flatulence of cattle and other ruminant domestic mammals. Land-cover change affects more than the atmospheric CO_2 balance; it also is a major source of rising atmospheric concentrations of both CH_4 and N_2O, it turns out. Deforestation, by changing chemical processes driven by soil microorganisms, can be a source of either CH_4 or N_2O, augmenting natural emissions; and changes in farming practices can also alter the emissions of either gas. Part of the atmospheric buildup of N_2O can be traced to intensive use of inorganic nitrogen fertilizers, some of which escape from the intended targets in farm fields.

While CO_2 is projected to account for about 70 percent of the anthropogenic warming over the next century, CH_4 will produce almost 20 percent and N_2O almost 10 percent, with a small (roughly 2 percent) contribution from CFCs. Both CH_4 and N_2O are more potent, molecule for molecule, than CO_2, but they are present in much smaller concentrations. While there are natural sources of CO_2, N_2O, and CH_4, they are mostly balanced by absorption in the oceans, destruction in atmospheric chemical reactions, and natural "sinks" on land.

In addition, there is a small contribution from hydrochlorofluorocarbons (HCFCs), which, despite being less ozone-dangerous substitutes for the CFCs, also happen to be greenhouse gases. Unlike the other principal greenhouse gases, these chemicals are entirely human-made, and there are no natural sinks for them. The volume of these chemicals' emissions is tiny compared with that of CO_2, but their potency and centuries-long persistence in the atmosphere make them important. Finally, ozone in the lower atmosphere, produced as a by-product of air pollution, is a significant greenhouse gas in many areas (as opposed to its beneficial global service in the upper atmosphere, shielding life from UVB).

All this is complicated because different greenhouse gases have differing abilities to absorb outgoing IR thermal radiation and they persist in the atmosphere for different lengths of time; hence their effects cannot simply be added together. But the basic message is simple: people are adding greenhouse gases to the atmosphere and changing the climate in the direction of increasing the *average* surface temperature—global heating. The emphasis on "average" helps us keep in mind that the complexities of climate make it possible that some areas will cool while others heat much more than the average.

The human contributions to the greenhouse gas content of the atmosphere are usually measured as CO_2 equivalents; in aggregate, the non-CO_2 greenhouse gases approximately double the effects of CO_2 alone. Remember, not all greenhouse gases are the result of human activities; the single most important natural greenhouse gas is water vapor. The warming effect of accumulating greenhouse gases, however, increases both the evaporation of water and the capacity of the atmosphere to hold water vapor.

The net result of the anthropogenic addition of greenhouse gases to the atmosphere is to augment the *natural* greenhouse effect (which keeps the planet habitable), thereby causing a gradual heating of Earth's surface and the lower atmosphere as the gases accumulate. This will cause Earth's climate to change, and it could do so rapidly. Half a century ago, scientists believed that climate change perceptible to human beings would take centuries to occur. The more recent view is that significant changes have started already.

The Extent and Consequences of Global Heating

Among the difficult issues facing humanity is determination of both the climatic and social consequences of the inevitable further loading of the atmosphere with greenhouse gases in coming decades. Assessment of and predictions about global heating have been topics of continuing investigation by the Nobel Prize–winning Intergovernmental Panel on Climate Change (IPCC), an open forum sponsored by world governments, through the United Nations Environment Programme and the World Meteorological Organization, that seeks to find consensus on these vexing problems. In its deliberations, the IPCC involves essentially all the world's leading atmospheric scientists, as well as hundreds of other scientists from nearly every nation representing diverse disciplines, from physics and biology to economics and other social sciences.

In addition, representatives of more than 100 governments (including fossil fuel producers such as Saudi Arabia), the fossil fuel industry, and environmental organizations participate as observers in the open and transparent sessions. All participating governments must accept the language of an IPCC assessment before it is published. The scientists, lacking the ability to run vast experiments of the climate on hundreds of replicate Earths, use computer models of the global climate to predict what the effects will be. These models are far from perfect, but they are well validated by their ability, when fed data from the past, to reconstruct trends that are now history.

The IPCC's most recent report, the *IPCC Fourth Assessment Report: Climate Change 2007*, makes it clear that there is no credible argument for further debate on whether humanity is making a significant contribution to global heating. It states that evidence of global heating is unequivocal and that humanity's release of greenhouse gases since the middle of the twentieth century is "very likely" the cause of most of the warming. By 2006, the IPCC reports, the atmospheric buildup of CO_2 had exceeded by far the natural range of concentration during the past

650,000 years; eleven of the twelve years between 1995 and 2006 were among the warmest dozen ever recorded; and over the past century the global average temperature had risen some 0.76°C (about 1.2°F). Moreover, the rate of greenhouse gas buildup had accelerated since 1995.

The IPCC's scientists must deal with myriad uncertainties when it comes to evaluating the likely future course of regional and global climate change, given the inevitable continued buildup of greenhouse gases in the atmosphere. Changes, they know, will not be uniform; in addition to variable temperature effects, changes in circulation will affect precipitation patterns, with important implications for agriculture and many other human activities. Such trends can't be specifically predicted in terms of time and place. The greatest uncertainty, of course, is in the actual levels of future greenhouse gas emissions.

Global heating effects will also depend heavily on the balance of little-known feedback mechanisms. Negative feedbacks are ones in which the output of a system is fed back into the input in a way that reduces the output. For instance, if warming of the oceans caused by the buildup of greenhouse gases in the atmosphere leads to more evaporation and creation of more low clouds that tend to cool the planet, that would be a negative feedback. So negative feedbacks tend to be stabilizing. A thermostat on a combined heater and air conditioner is a familiar negative feedback device—when the temperature goes above the set point, cooling is called on; when it drops below the set point, a heater is turned on.

On the other hand, if warming caused by greenhouse gas emissions speeds the melting of permafrost (soil at high latitudes or high altitudes that previously remained frozen from year to year) and rotting of tundra vegetation so that even more CH_4 and CO_2 are added to the atmosphere, or if it creates high-level clouds that tend to warm the planet, those would be positive feedbacks. Such feedbacks would be destabilizing and would further increase the temperature rise caused by the original warming. A thermostat sensor that turned the heat up further, for example, as the temperature rose would be a positive feedback device. A clearly recognized positive feedback from global heating is the enhanced warming caused by increased water vapor in the atmosphere. Another is the decrease in albedo as glaciers, ice caps, ice sheets, floating ice in the Arctic, and winter snow coverage shrink. One of those positive feedbacks that already seems well advanced is the unexpectedly rapid albedo-reducing shrinkage of the area covered by Arctic sea ice.

Overall, there remains substantial uncertainty about the nature and magnitude of the feedbacks. Some scientists, for example, combining theory with ice core temperature records from the past 360,000 years, have concluded that it is likely that future heating has been underestimated.[5] The most recent CO_2 emissions data, as we indicated, also suggest that IPCC estimates of potential heating are conservative.

What kinds of problems for humanity are likely to be caused by global heating? The most certain threat is sea level rise and coastline retreat—which already appears to be wreaking havoc for some farmers on the British coast, whose fields are dropping into the sea as the beaches erode. Indeed, the beach ridge in front of Shishmaref, Alaska, on which Paul landed in a light plane in 1955, no longer exists, because of greater wave action associated with reduced sea ice and melting of the coast's rocklike shield of permafrost. The Inuit have been forced to move farther inland.

The upward creep of the oceans is caused by the expansion that water undergoes when it is heated and by the melting of glaciers and polar ice sheets and caps (ice sheets are huge areas of terrestrial ice covering over 20,000 square miles; caps are dome-shaped ice bodies covering less area). Sea level rise, which was roughly 6 to 9 inches in the twentieth century, is already threatening the habitability of some countries such as Tuvalu and the Maldives, which occupy low-lying atolls in the Pacific and Indian oceans. Conservative estimates, such as those that appear in the IPCC's *Fourth Assessment*, indicate a sea level rise in the twenty-first century ranging from about 7 to 24 inches (18 to 60 centimeters). But these estimates don't include the possibility of significant melting in this century of the Greenland or Antarctic ice sheets, which recent research suggests may be happening.

With the recent pace of increasing greenhouse gas concentrations, it had been estimated that the Greenland ice sheet could melt off in the next thousand years. That alone would raise sea level about 22 feet (7 meters), an extraordinary amount. Recent observations have indicated that melted ice is finding its way through fissures to the bottom of the ice sheet, lubricating it and causing it to slip faster toward the sea. As a result of this finding, scientists are revising estimates upward for the rate of both ice sheet melting and sea level rise (one should note that even at higher rates of melting, a complete loss of the Greenland or West Antarctic ice sheet would probably take 500 to 1,000 years). Since the albedo of Greenland would be much lower without the ice sheet, given that

FIGURE 13-1. Melted ice flowing across the ice cap in Greenland. Photograph courtesy of Alberto Behar/JPL/NASA.

rock reflects much less radiation than do ice and snow, loss of the entire ice sheet might result in a new, stable equilibrium—one that might prevail in perpetuity even if greenhouse gases returned to preindustrial levels in the atmosphere.

Of course, Greenland is not alone. There already has been a considerable loss of ice in montane glaciers around the world, posing a threat to freshwater supplies in many regions. And Antarctica, which has by far the largest stock of ice, has undergone significant loss of fringing ice shelves. In 2002, for example, a large Antarctic ice shelf suddenly broke apart and launched hundreds of icebergs into the surrounding sea. Ice shelves are already floating on the ocean, so their disintegration cannot cause a sea level rise, but they play an important role in stabilizing adjacent land-based ice sheets. Motion toward the ocean of the now-exposed Antarctic ice sheet has been observed to speed up since the ice shelf disappeared.[6]

The actual amount by which sea level is likely to rise by 2100 remains unclear, but a very conservative estimate would be a foot and a half (50 centimeters), with further rises continuing over the next several centuries. Eventually, if excess greenhouse gases continue to accumulate in the atmosphere, the sea level could rise dozens of feet, which would significantly change the shapes of continents and flood all coastal cities. Major coastal flooding could happen much sooner if there were unexpected "discontinuities" such as rapid slippage into the sea of the West Antarctic ice sheet. Such an event could increase sea level by as much as 16 feet (about 5 meters), in some estimates even more, catastrophically flooding Earth's coastlines and causing untold devastation.

But even a rise of a foot or so would cause serious problems, especially in low-lying poor countries such as Bangladesh. The result would be salinization of aquifers near coastlines, flooding of agricultural fields, destruction of natural fish nurseries in salt marshes, and more acute vulnerability to storm surges. The Netherlands and Florida are other low-lying coastal areas that are highly vulnerable to even modest rises in sea level for the same reasons. Rich nations can more readily compensate for such changes, but the effects would still be serious—especially where there is costly infrastructure near the coasts, such as expensive homes, resorts, marinas, aquaculture facilities, and the like. Timely and strategic preventive action in coastal development could forestall much of the future costs, and insurance companies and some local authorities in the United States are already trying to discourage home building in vulnerable areas.

One effect of global heating, related to warmer oceans and higher sea level, quite likely will be an intensification of giant storms. Global climate models indicate that severe weather events—large, intense cyclones (hurricanes and typhoons), as well as severe floods and

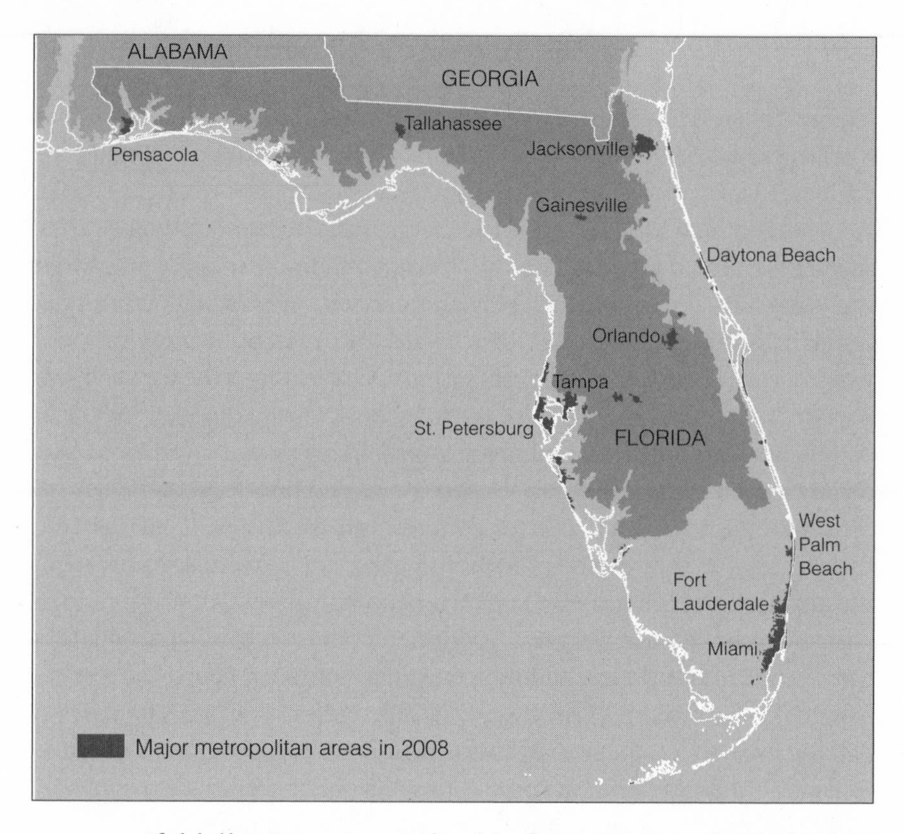

FIGURE 13-2. If global heating continues unabated, in a few centuries several yards of sea level rise will alter the shape of the continents and submerge seashore cities. Here we see the shape of Florida after a six- to seven-yard rise—with Miami forming the base of a new large artificial reef. Adapted from Robert A. Rohde / Global Warming Art.

droughts—are likely to occur with increasing frequency and possibly in areas where they were previously rare or unknown. Picture a future situation in New Orleans: the Gulf Coast is even closer to the city because of land loss caused by human interference with the deposition of silt in the delta, and tides are a foot or so higher as a result of global heating. If a record-size hurricane struck the city in the worst possible location, there would be another dramatic demonstration that vulnerability is not limited to poor countries. Hurricane Katrina actually had weakened to a Category 3 storm by the time it hit—the highest sustained winds were about 130 miles per hour. Even so, in 2005 Katrina was devastating enough and may have owed some of its intensity to global heating.

There will be other unhappy consequences of anthropogenic additions of greenhouse gases to the atmosphere. The global average temperature could rise by almost 10°F during the next ninety years. That would be a dramatic change; in the past when it was 10° *cooler* on average than today, there was a mile of ice over what is now New York City. One obvious effect of global heating will be the movement of tropical diseases such as malaria and dengue into temperate and higher-elevation regions as the mosquito species that carry them expand their ranges. Another will be an increase in deaths and illness from prolonged heat waves, such as those that ravaged Europe in 2003, 2006, and 2007, the first of which killed tens of thousands—although heat waves may be partly counterbalanced by a decline in incidence of frostbite, pneumonia, and death by freezing.

Another consequence of global heating that will affect many people directly will be the melting of the permafrost, which forms the "bedrock" of the Arctic. Much of the infrastructure (roads, buildings, telephone poles, etc.) rests upon it in Alaska, 40 percent of Canada, far northern Europe, and Siberia. Besides exacerbating the warming by releasing large volumes of CH_4 and CO_2, melting of the permafrost will cause extensive and costly damage and disruption. In parts of Alaska, Canada, and Siberia, such damage is already occurring.

As CO_2 is emitted into the atmosphere, roughly a third of it is taken up in the oceans. As mentioned in the last chapter, carbon dioxide forms a mild acid in water whose corrosive effects could disrupt marine food chains by killing zooplankton no longer able to produce shells and reduce the stock of marine animals that people harvest. Acidification of the oceans due to increased CO_2 is also a threat to coral reefs—among the most productive and beautiful of all ecosystems—and ultimately perhaps to all calcifying organisms, which include starfish, clams and oysters, many bryozoans (colonial coral-like animals), and barnacles.

Warming itself also seems already to be causing changes in some ocean ecosystems, in some areas even affecting seabirds that depend for food on fishes and other marine organisms. It is implicated in dramatic declines in coral reefs (about a third of them have been essentially destroyed and another third are threatened), especially causing "bleaching" events in which heat-stressed corals of a reef eject the symbiotic photosynthesizing algae upon which they depend. This causes the corals to fade (the algae provide their color) and die. The IPCC's *Fourth Assessment* notes that more than 80 percent of the heating caused by human-generated greenhouse gases has been taken up by the oceans, and that temperature increases (amounting at present to a fraction of a degree) in the oceans have penetrated as deep as 3,000 meters. Thus the very life of the oceans as we know it may ultimately be at risk.

In our view, the three most serious likely impacts of humanity's greenhouse gas contributions will be changes in regional climates, disruptions of both terrestrial and marine ecosystems, and losses of biodiversity. Rapid change is altering long-established evolutionary trajectories and coevolutionary relationships, with unpredictable but likely deleterious effects on ecosystem services that humanity depends on. Agricultural ecosystems in particular are utterly dependent on climate-moderated services, for both appropriate temperature and sunlight conditions to grow crops, for pollination and pest control, and for water in the form of rain or as supplied by irrigation systems, which in turn are dependent on climate. The popular idea that farmers can avoid difficulties by simply shifting crops to new areas as temperature and rainfall conditions change overlooks the critical issues of suitable soils, day length, novel crop pests, and the frequent need for pollinators. Soils are especially tricky because climate change may also alter their moisture levels and thus salinity, as well as affecting in different ways the soil organisms essential to soil fertility.

Hospitable climates in traditional fertile farming areas may well shift to areas where soil quality is relatively poor. Day length can be critical, since flowering and seed germination of many crops are photoperiod sensitive—the plant gets its cues from the length of the day. New pests may appear in new farming areas, and old pests might well prove to be successful migrants, evading predators and chemical poisons. Pollination is an ecosystem service that can be lost if organisms fail to migrate or to shift the timing of their development rapidly enough for their populations to avoid climate-induced extinction. This will be especially serious where migration paths are blocked by artifacts such as farm fields, superhighways, and cities.

The socioeconomic stresses that will probably accompany attempts to maintain agricultural productivity in the face of rapid climate change are likely to be daunting. They will very likely exacerbate the gap between rich and poor countries and disrupt developing societies by causing increased competition for agricultural water and for food itself. And the people most at risk from crop failures and hunger are among those least responsible for adding greenhouse gases to the atmosphere. For example, wheat production in Pakistan, Nepal, and Bangladesh could be virtually wiped out if heating and drying continue. Wheat, like corn and rice, is very sensitive to heat and is already being grown near the margin of the climatic zone in which it can survive in South Asia. That subcontinent produces roughly a sixth of the world's wheat crop. Rice in Southeast Asia and cereal grains in Africa are also potentially in trouble from global heating and likely changes in precipitation patterns.

Concern about impacts on agriculture is already being expressed, for example, in estimates of the productivity of California's famed San Joaquin Valley. The 2006 heat wave in the valley killed thousands of cattle, and one county alone suffered $85 million in beef, dairy, and poultry losses. A more serious threat is the potential loss of much of the Sierra Nevada snowpack, critical for the irrigation of the valley as well as for provision of water for industrial and domestic use in much of the state. That snowpack is expected to shrink and melt earlier each spring as

FIGURE 13-3. Early spring at an ecosystem-heating experiment at the Rocky Mountain Biological Laboratory being carried out by John Harte. One-fifth of the carbon in the patches warmed by infrared heaters has disappeared and is now carbon dioxide in the atmosphere. In the warmed plots, the once dominant and beautiful meadow flora have been overtaken by rapidly growing sagebrush. Photograph by Dr. Scott Saleska.

global heating proceeds, causing the water to run off quickly rather than feeding streams and reservoirs slowly during the summer dry season. Similar problems can be foreseen for many important agricultural regions around the world and for the large urban areas that depend on mountain glaciers and seasonal snowmelt for water supplies. Such areas include parts of the southwestern United States dependent on Colorado River flows (from the snowpack of the Rocky Mountains), Asia south and east of the Himalayas, South America and the Andes, and Europe and the Alps.

More generally, climate models suggest that the potential exists for events such as the drying of midcontinent grain baskets, leading to drastic cuts in food supply, which might in turn trigger major famines in poor countries as grain shipments from the industrialized world are curtailed to ensure food supplies (or animal feed supplies) at home. Of course, there is also the potential for at least temporary yield increases in agriculture as some areas warm, for example in Russia and parts of Canada. But the regions in which food production might be enhanced by global heating are not as great in extent as those that are threatened with reduced capacity for production caused by higher temperatures, diminished water supplies, or both. Indeed, in early 2007 it was reported that warmer climate had been driving global grain yields below what they otherwise would have been, so the trouble may be here already.[7]

The sort of problems that could occur are foreshadowed by recent events. For instance, in November 2006 we were in northwestern Victoria and southwestern New South Wales, Australia, looking at the effects of the worst drought in that nation's recorded history, one that had entered its sixth year. Everywhere farm fields were barren and trees were dying. We passed a slaughterhouse with several huge trucks loaded with thousands of sheep to be butchered despite low prices because there was no grass left for them to graze on. Populations of many birds were dwindling or going locally extinct. A number of bird species were compensating for the drought by not breeding or by laying smaller clutches of eggs. Such measures could be effective for only a limited time, Australian naturalists told us, and some areas were already being converted from woodland to stony desert.

Soon after our visit, in much of Australia it became difficult to *sell* sheep; the market had collapsed. By December the sheep population of

Australia had been reduced by half, and farm losses mounted to tens of billions of dollars. When we next visited, in November 2007, planning was under way to develop emergency drinking water supplies for towns and cities, and restrictions on water use had been broadly adopted in what was thought to be the worst drought since European settlement. For the first time in world history, climate change became a major issue in a national election, and it clearly played a role in the 2007 defeat of John Howard as prime minister. Australians who a few years ago were claiming that the continent could support 100 million people now realize that it's overpopulated at 21 million. Climatologists fear that this is just the beginning because they think a combination of global heating and depletion of the ozone layer is moving climate bands toward the pole and shifting what was Australian rain into the Southern Ocean.[8] They think that, rather than having "droughts," Australia is moving into a drier "normal" climatic regime. Let's hope they're wrong.

In late 2007 we were in the Pantanal region of southwestern Brazil, where a record drought had led to the burning of large stretches of that traditionally swampy area (figure 13-4). Farther north, in the state of Mato Grosso, remnants of the Cerrado, a now largely destroyed vast savanna ecosystem that once occupied a fifth of Brazil, were also suffering record drought and fire. At the same time, the southeastern United States was enduring an extreme drought that threatened to dry up water supplies of some cities while eastern Texas was experiencing record rains and floods. Since greater extremes of weather are predicted to be among the first effects of global heating, the droughts of Australia, Brazil, and the United States may soon be followed by deluges—perhaps permitting these countries' politicians again to ignore the climate-change threat. But even a deluge will not quickly replace the now-dead trees of Australia, many of which were hundreds of years old. In other subtropical areas besides Australia there probably is simply a long-term climate-change trend toward less precipitation—and soon the term "drought" may no longer be appropriate, any more than it can be reasonably applied to the weather in the Sahara.

Both the degree of disruption of crop production and the success farmers and governments might have in compensating for changes are hard to predict. For example, humanity as a whole has a great potential buffer between climatic reduction of harvests and starvation. Nearly half

FIGURE 13-4. Hungry cattle after fires in the wetlands of the Brazilian Pantanal during a period of record drought, October 2007. Photograph by Rick Stanley.

of the world's grain crops are fed to livestock, as described in chapter 11. Reduction of grain feeding and of the herds or flocks themselves would allow the release of grains strictly for human consumption, thus potentially greatly increasing the available food supply—more than enough if appropriately distributed to guard against famines. It also could slow the flux of greenhouse gases into the atmosphere, since the roughly 400 billion tons of carbon sequestered in Amazonian rain forests are being released as the forests are cleared and burned for such things as inefficient subsistence farming, cattle production, and the growing of soybeans for biodiesel fuel or for export to China as feed for chickens and pigs.

But there are several catches to transferring food from the animals that people eat to people themselves. First, most of the cereals produced as feed for livestock are coarse grains not very suitable for human consumption. Furthermore, grains alone are nutritionally insufficient, except perhaps for temporary relief of a food shortage. Fruits and vegetables, especially protein-rich vegetables such as nuts and legumes, would be required as supplements and partial replacement of animal foods. Nevertheless, even a modest reduction of animal foods (especially beef) in the diets of the world's affluent could significantly relieve the pressure on the food production system and reduce greenhouse gas emissions, as well as improve the health of those people. However, meat-eating human beings so far have shown precious little inclination to reduce their meat consumption and direct extra food toward the hungry.

Destabilization of the climate can undermine biodiversity, exterminating populations and species that cannot adapt or move fast enough to keep up with changing habitats, including species that may play important roles in support of agriculture. Ecosystems may be torn apart as species migrate at different rates and in response to differing changes. Some may survive with changed conditions, and others will disappear. In the Rocky Mountains of the United States around Yellowstone National Park, for instance, pine beetles moving upslope as the climate warms are threatening to exterminate slow-growing white pine trees. That is likely to reduce the snowpack that the trees' shade helps preserve and to threaten populations of grizzly bears that depend on white pine seeds in the fall. And the movement of many species in response to changing conditions will be hindered or blocked by extensive human infrastructure. The extent of these losses—it may be huge—is difficult to predict with much accuracy, and that is an important reason why we find global heating so alarming, especially in light of the irreversibility of many biodiversity losses.

Understanding the inevitable changes arising from climate alteration obviously is essential in order to find ways to ameliorate its effects. Some regions have already begun to assess the impacts and make plans to compensate or adapt. Thus California anticipates increases in floods, droughts, and forest wildfires as a result of climate change. Plans to adjust water use, especially in agriculture (far and away the biggest user), are afoot, as are changes in regulation of coastal development. Some other local and regional governments are also trying to build greater resiliency into their plans, and the number is growing rapidly.

While some degree of adaptation will obviously be required almost everywhere, even more essential will be measures to reduce greenhouse gas emissions drastically so that global heating does not continue to accelerate.

Dealing with Climate Change

What actions should be taken to prevent rapid climate change or to ameliorate its impacts? Which ones are economically or ethically justified? Investigations by the climate science community have led to an assessment that there is roughly a 10 percent chance that most people will escape serious consequences of climate change, and perhaps a 10 percent chance of a worldwide catastrophe. The rest of the probability distribution is spread between these extremes, with possibilities ranging from substantial inconvenience to disaster. In the end, the key question arising from all this is a non-scientific one that can be put in familiar terms: "How much insurance is worth taking out against the more serious possible consequences of climate change?"

There is no straightforward answer to that insurance question, but plenty of debate. One of the most interesting analyses, done by ecologist Stephen Pacala and physicist Robert Socolow, shows that a portfolio of technologies already in hand and more or less practical to deploy could be used to keep the atmospheric CO_2 concentration under 500 ppm (remember, it is now about 383 ppm). That is less than twice the preindustrial level of 280 ppm, a level many scientists believe would avoid many of the most catastrophic possible climatic impacts. In Pacala and Socolow's analysis, the possible technologies are shown as a series of "wedges" imposed on a graph of the future "business as usual" trajectory of CO_2 emissions, which could deflect the curve downward in a direction that would meet the target of less than 500 ppm. The potential wedges include such things as increasing the fuel economy of cars, reducing automobile travel through telecommuting, increasing investment in mass transit and urban redesign, capturing and sequestering the CO_2 released at power plants, increasing use of wind, solar, and nuclear power, making more extensive use of "no-till" cultivation, and reducing rates of deforestation. Sadly, this otherwise fine paper neglected to mention one obvious "wedge" that could contribute—reduction in the population growth rate and (ultimately) population size—and a less obvious one of encouraging a dietary shift away from beef and pork and toward poultry, fish, and vegetarian cuisine.[9] And a needed follow-up study that would evaluate the cost-benefit ratio of each wedge remains to be done.

Other studies have adapted the Pacala-Socolow wedge idea, and some have shown that increasing energy efficiency is the most effective measure that can be taken promptly—the easiest, fastest, and cheapest way to reduce energy-based greenhouse gas emissions. The technology exists today to improve efficiency in myriad ways, and much more could be developed within a few years or decades; what is lacking are appropriate incentives to apply known technologies and develop new ones. And, of course, most efficiency improvements would reduce end-use demand for energy, regardless of the source.

Most of the other proposed wedges would not be easy to implement quickly at the required level, and all would be resisted in some quarters. But society could take many "wedge-making" steps that clearly would have social benefits beyond reducing the chances of climate catastrophe. The United States, for example, could increase taxes on carbon-emitting activities, such as greatly increasing gasoline taxes. In 2007 Congress passed new energy legislation that will boost CAFE (corporate average fleet efficiency) standards for automobiles, light trucks, and SUVs by 2020. Not only will reducing the fuel consumption of U.S. automobiles substantially lower the flux of CO_2 into the atmosphere; it also could somewhat reduce American dependence on imported oil and lessen the medical and social costs of respiratory diseases caused by auto and truck exhausts. Even so, this is only a modest start on tackling a massive problem.

What incentives could be used to encourage introduction of new, low-carbon technologies or to reduce the use of carbon-based fuels? One would be to introduce a carbon tax—a tax on fossil fuels according to their carbon content. Such a tax could be applied at the wellhead or mine mouth, thereby helping to capture the external costs of using coal, natural gas, and petroleum.

Another approach to reducing emissions is to make more extensive use of cap-and-trade systems such as have been used in connection with the Clean Air Act, as discussed earlier in this chapter. Such systems would impose a total "cap" on allowable emissions of greenhouse gases that shrinks over time. The regulating authority would allocate "emissions allowances" to various facilities, with each allowance entitling the holder to a certain amount of emissions. The allowances could be given out for free or, preferably, auctioned to the various facilities and could be traded among allowance holders. The total number of allowances introduced would equal the total emissions cap. Trading of allowances would help ensure that the facilities that could reduce emissions at lowest cost would do so and sell some of their extra allowances to facilities for which pollution reductions are too expensive. The latter entities would instead purchase allowances to cover their shortfalls. The trading of allowances wouldn't affect the total allowances in circulation—it would just change who owns the allowances and the distribution of emissions reductions across facilities. Allowances not purchased would be retired as part of the shrinking cap.

Both carbon taxes and cap-and-trade arrangements are called incentive-based approaches because they raise the cost of emissions, which gives polluters incentives to find ways to cut their emissions. The carbon tax raises the cost of emissions directly through the tax itself, while the cap-and-trade system does it through the market price of emissions allowances. Under either system (or potentially a combination of both), the cost of energy use would inevitably rise, although increased efficiency, which reduces energy use, would compensate at least in part. Nonetheless, increased energy costs would place a burden on many people, especially low-income families. Command-and-control measures such as tightening efficiency standards for appliances and buildings also can increase purchase prices. If gasoline prices rise substantially, many poor people will no longer be able to afford to commute to their jobs in areas with little or no public transportation.

One solution to the dilemma of higher energy costs, as Stanford University economist Larry Goulder has shown, would be "tax shifting." One such shift might be to use part of the increased revenues from the gas or carbon tax to reduce the FICA (payroll) tax, which disproportionately burdens the poor (as economists say, the tax is "regressive"). Some of the revenues from a gas or carbon tax also could be used to upgrade public transportation systems, increase efficiency throughout the energy system, and support research and development of renewable energy sources.

A large array of other measures would be necessary to make any of the above proposals really effective, however, since people are strongly locked into their consumption patterns. Many scientists believe that greenhouse gas emissions must be reduced by at least 80 percent by mid-century if catastrophic climate change is to be averted. If our society were to institute major economic changes by starting a transition away from the use of fossil fuels, many more people than those just mentioned, from coal miners to gas station owners, would unavoidably be hurt. Much could be done to lessen such impacts, however. Coal miners and auto workers could be pensioned off, for instance, as many already have been, while some could be retrained for other jobs. At the same time, transforming the energy system and converting to alternative energy sources present major opportunities for new investments, businesses, and jobs.

A series of geo-engineering "fixes" for the greenhouse gas problem have also been proposed, from increasing the CO_2 absorption in the oceans by fertilizing them with iron to shooting aerosols into the stratosphere or placing giant sunshades in orbit to reduce the amount of solar energy reaching Earth's surface. All such giant technological experiments share some important characteristics—great expense and great uncertainty as to the results. The plan for injecting aerosols, for example, would require a long-term and constant effort beyond any project ever done by humanity, since the cooling aerosols would not remain in the stratosphere anywhere near as long as the heating greenhouse gases will stay in the atmosphere. Furthermore, neither aerosols nor sunshades will do anything to slow the acidification of the oceans. Still, if humanity waits too long and is faced by abrupt and disastrous climate change, such extremely risky and costly measures might need to be taken.

Sequestering CO_2 from power plants by injecting it deep into Earth's crust has more potential; serious experimentation with sequestration is already being carried out by oil companies in connection with secondary oil recovery (techniques for retrieving residual oil in geologic formations after simply pumping it out of the ground is no longer possible).

The carbon tax mentioned above, which could be applied, modified, or removed at the stroke of a pen, seems a much saner starting point for solving what fundamentally is just a more serious, complex, and larger-scale air pollution problem than those we've had considerable success in managing before. Nevertheless, getting oil companies to put additives in gas and getting automobile manufacturers to add catalytic converters to cars were a lot easier than it undoubtedly will be to get either to reduce or even to phase out production.

Even so, the approaching (if not already reached) peak of global oil production, combined with the rapidly intensifying demand from China and other developing nations, plus rising prices, may well soon force curtailment of oil use in the United States and other industrialized countries. And the public clearly has caught on. As gasoline prices soared in 2006 and 2007, demand for SUVs slumped and demand rose for hybrid gas-electric cars. American automakers, rapidly losing money and market share, are belatedly retooling for less excessive gas consumption and expecting, though still fighting, a renewal of efficiency regulation.

The task of addressing climate change is surely daunting, and many economists have for years asserted that the costs would be too high. But a comprehensive study headed by Sir Nicholas Stern, former chief economist of the World Bank, and reported to the British government in 2006 reached a different conclusion. The study estimated that, under worst-case assumptions and in the absence of serious preventive measures taken soon, damages from climate change could eventually cost as much as 20 percent of global gross domestic product annually. Warning that social and economic costs could resemble "those associated with the great wars and the economic depression of the first half of the 20th century,"[10] the Stern report urged that major efforts to avoid such consequences be undertaken by the world community as soon as possible, by investing billions of dollars to reduce greenhouse gas emissions. The clear message was, start paying now or pay much more later.

But perhaps the most serious barriers to satisfactorily dealing with climate change are ethical and political. Who should do what, and what should be the goals? For instance, China is becoming the leading total emitter of greenhouse gases as it passes the previous champion, the United States. How much should the huge past contribution of the United States, much of it still in the atmosphere, be counted against America? Or should responsibility be measured from the time China takes the lead? Should responsibilities be assigned by a nation's total emissions (China would lose) or by per capita emissions (the United States loses)? How much of a right does China have to replicate the Victorian industrial revolution path taken a century ago by the United States and other fully developed nations? Is it ethical for the United States to *increase* its emissions, even a tiny bit? Or should it be required to use only the most modern, energy-efficient, and environmentally benign technologies? These questions give just a small hint of how vexed future international negotiations—and power struggles within some countries—are likely to be.

A Nuclear Atmosphere

Those struggles could be one cause of a large-scale and extremely deleterious anthropogenic global change, one lurking in the background as problems such as acid precipitation and global heating have gradually developed. That would be the result of a medium- or large-scale nuclear war. In 2007 the United States and Russia still had thousands of nuclear warheads on missiles pointed at each other, essentially on hair-trigger alert. Furthermore, the Russian command-and-control system had deteriorated by 1995 to the point that misreading of a preannounced Norwegian research rocket launch as an attack from the United States allowed Russia to get eight minutes into a fifteen-minute countdown to a full-scale retaliatory strike. The exchange of nuclear missiles that would have followed would essentially have spelled the end of the world as we know it.

If such an unthinkable scenario were to unfold, in addition to hundreds of millions of prompt deaths from blast and radiation, millions more would die from other effects. Many would succumb to simple exposure because of the massive destruction of infrastructure and power supply; millions would be killed in vast firestorms ignited all over the Northern Hemisphere. The climatic disruption from those fires in many areas, lofting smoke and particles of debris into the stratosphere, would cripple the ability both to grow food (the chilling and darkening effect is why the phenomenon was christened nuclear winter) and to distribute what food was produced, leading to vast famines among survivors. A recent study suggests that a war involving the detonation of "just" 100 Hiroshima-sized bombs (15 kilotons—equivalent in explosive force to 15,000 tons of TNT) could cause serious climatic effects and damage to the ozone layer.

Even a "medium-scale" nuclear war, involving a dozen or so weapons (as might happen between India and Pakistan), or even one or two 10- to 15-kiloton terrorist detonations in, say, Washington and New York, would trigger significant global change. The biosphere would be essentially permanently altered because of cascading political and economic effects, worsened by the disruption of humanity's capacity to cooperatively tackle the ensuing severe environmental, social, and political problems.

The process of globalization, in all its various guises and with the potential for global catastrophe, has brought humanity face to face with a plethora of very difficult dilemmas. How does one estimate the chances of even a medium-scale nuclear exchange? How does one determine whether a change in some global environmental factor is part of "normal variability" or something new—and especially something caused by human activities? These aren't simple questions, and it is often difficult to find answers. They inevitably bring scientists into the area of statistics, basically because science doesn't ever supply "proof"; indeed, many scientists believe, following the philosopher Karl Popper, that scientific theories can only be disproved.[11]

Scale, Thresholds, Non-linearities, and Lags

Interactions such as those between poverty and vulnerability to storms, along with the underlying factors of population growth, demand for tropical timber and high-priced seafood in rich countries, persisting economic inequities, and other elements of the human predicament, are legion. So too are characteristics of some of those interactions—problems of scale, threshold effects ("tipping points" in the language of sociology), non-linearities, lag times, complexity, and the like—which plague analysis of numerous environmental and social issues. These characteristics also sometimes have surprising consequences and should always be kept in mind when considering a complicated environmental problem.

In the San Francisco Bay Area of the 1860s, the atmosphere could be used as a convenient garbage dump—cooking fires and belching cattle didn't pose much of a problem. By the 1960s, the need for controls on fires, power plants, and automobile exhausts had become all too evident to choking residents—a typical problem of *scale*.

Such problems are nearly ubiquitous in environmental matters, intimately connected with the growing degree of dominance of Earth by our species. Remember, a town of 2,000 people cannot sustain a measles epidemic; a city of 500,000 easily can. The scale of the human population determines whether it will permanently host the disease. If the ancient Greeks ruined Attica, the Greeks suffered. There were perhaps only 200 million people in the whole world in those days, with little or no contact among sizable population centers. But now if more affluent nations keep injecting record amounts of CO_2 and other greenhouse gases into the atmosphere, everyone in the world is likely to suffer. On the more positive scale side, think how much less greenhouse gases the United States would be adding to the atmosphere if its population were less than half as large (as it actually was in 1950), and how much less dependent on foreign oil it would be. If, as is now projected, the U.S. population grows to 400 million in the next four decades, per capita greenhouse gas emissions would need to be reduced by 25 percent just to avoid increasing today's level of emissions into an already overburdened atmosphere.

Problems connected to the scale of the human population or its practices also may contain *threshold* factors—a sudden transition from one environmental state to another, when scale problems go over a brink. Mississippi River water may remain potable as more and more people use the river as a toilet—up to a point. Then suddenly the density of some pathogenic microorganism may cross a threshold so that there are enough organisms in a drink of water to make an average person seriously ill. Similarly, excess fertilizer can flow into the Gulf of Mexico with relatively little effect until a threshold is crossed and a dead zone appears. Such threshold transitions are common in processes of land degradation; soil erosion, for instance, can proceed unnoticed for a long time, and then suddenly the land's productivity may drop disastrously as critical nutrients are exhausted. In a classic ancient case, the vegetation of the Sahara region of North Africa gradually and unevenly declined until somewhat over 5,000 years ago the whole regional ecosystem collapsed into a desert. Systems with such thresholds are special cases of *non-linear* systems; all systems with thresholds are non-linear, but not vice versa.

Climate is an example of a non-linear system, one in which a steady rise in an input does not necessarily result in a steady rate of change but instead may cause a slow response at first, which then accelerates as the input rises. Think of how slow you were when you first tried to learn another language or a new game, and how the rate of learning went up faster than the effort you put into the project as you started to "get the hang of it." Through recent human history, Earth's climate seemed to be in a rather stable equilibrium. For instance, changes in solar input produced in the Northern and Southern hemispheres by the tilt of Earth's axis as it travels around the sun produce what we call seasons. The oscillation does not result in summer spiraling away to a heat death for the planet, or winter getting colder until everything freezes. The perturbation caused by increased solar heating in summer causes not a permanent change but one that returns to the previous state as the solar flux begins to diminish after the summer solstice.

But what is known about the history of Earth's climate suggests that there have been other, quite different patterns, such as the cold climates at the height of glaciations. If we push hard on the system, for example, by continuing to inject greenhouse gases into the atmosphere, a threshold might be passed that would thrust the climate into a very different pattern—one that could be disastrous for society. Development of today's civilization occurred during a long period (roughly 11,000 years) of unusually constant and favorable climate, but there is no guarantee that the climate will remain that way. Remember the Younger Dryas, whose equivalent would be a vast catastrophe if it occurred today. Humanity, in effect, has been gambling through its inaction that we won't run into a non-linearity as Earth is heated by anthropogenic activities. Such a non-linearity, an increasing response to each bit of heating, would have an especially devastating effect on biodiversity and ecosystem services because human modification of Earth's ecosystems has already made them so much more vulnerable and less resilient.

Many of the problems we have described develop with very long *lag times* before consequences materialize. One example obviously is global heating, in which the effects of the greenhouse gases that have been released by human activities into the atmosphere will take decades to become fully evident, even if not another molecule were emitted. Lags are inherent in social trends and systems as well. Ending global population growth by reducing birthrates, for instance, will take at least several more decades to accomplish, even after each couple on average produces not quite enough children to replace themselves, as we saw in chapter 7.

Scale problems, thresholds, non-linearities, and lags all can in various ways complicate and worsen many of the interacting problems we face. Understanding what is going on in the world now and in the future will require much more attention to these seldom recognized features of human-environment interactions. While it is obvious that human dominance has given comfortable lives to a larger minority of human beings in the past century or so, the rise to that position has had many unintended consequences that have imposed or will impose large costs on our species. "Humanity's dominance of Earth means that we cannot escape responsibility for managing the planet," one group of distinguished ecologists has concluded. "Our activities are causing rapid, novel, and substantial changes to Earth's ecosystems. Maintaining populations, species, and ecosystems in the face of those changes, and maintaining the flow of goods and services they provide humanity, will require active management in the foreseeable future."[12] Will we be able to provide that management with respect to the centrally important energy resources, a critical factor in our impact on climate? That is the topic to which we now turn.

CHAPTER 14

Energy: Are We Running Out of It?

"Energy is the lifeblood of all societies."
THOMAS HOMER-DIXON, 2006[1]

ENERGY, the ability to effect change in our physical world, is essential to the human enterprise at every level. The human body, for example, acquires the energy it needs in the form of chemical bonds in the molecules of food and drink. This energy fuels each human being's metabolic activities. It fueled those of our evolving ancestors, and in fact it has fueled the activities of all living beings through all time. Energy also modifies and supports environments; for instance, energy from the sun drives the hydrologic cycle, which among other things erodes landscapes and influences regional climates. Energy is essential for life on Earth and has helped shape the evolution of people and all other organisms. The mobilization of abundant energy, especially in the form of fossil fuels, also makes large-scale human economies possible and powers manufacturing, transportation, commerce, industrial agriculture, and economic growth. It is what made us the dominant animal on Earth. Here we will focus on how humanity mobilizes and uses energy to modify its environments and what alternative paths may exist.

Most—but not all—of the energy mobilized for human use in industrialized countries today comes from chemical fuels such as gasoline, diesel oil, jet fuel, coal, and natural gas. In motor vehicles, fuel is burned in internal combustion engines that provide mechanical energy to the wheels. In many coal-fired plants, pulverized coal is mixed with air and burned in a furnace lined with water-filled tubes. In the tubes, steam is produced under pressure and spins turbines attached to generators that produce electricity. In some natural gas plants, the actual combustion occurs in a turbine (as in a jet engine), causing expansion in the gaseous combustion products; the pressure of that exhaust spins the turbine fans and rotates a shaft attached to a generator. Significant contributions are also made by nuclear power plants that "burn" uranium in a controlled chain reaction and use the heat to make steam and run generators just as in plants powered by fossil fuels. Other contributions come from hydroelectric generation, in which the gravitational energy released by water falling from behind dams is used to spin turbines, and geothermal power, which uses the heat of Earth's interior to generate steam that spins turbines.

In the United States, about 40 percent of the energy used comes from petroleum, 23 percent from coal, 23 percent from natural gas, 8 percent from nuclear power, and 6 percent from hydroelectric and other renewable sources. For the world, the equivalent numbers are somewhat different: 34 percent from petroleum, 25 percent from coal, 21 percent from natural gas, 6.5 percent from nuclear plants, 11 percent from "biomass," 2.2 percent from hydroelectric generation, and 0.4 percent from solar, wind, and geothermal sources.

Most of the world's use of biomass fuels arises from the dependence of a significant portion of humanity—about 2 billion people, mostly in developing countries—on the burning of wood, crop residues, or dung to heat their huts, boil their water, and cook their food. Poor rural populations lack access to—or are unable to afford—"commercial" fuels, so they burn the locally available forms of organic matter. These are generally not counted in energy statistics because they are gathered by households themselves or sold locally.

Altogether, the world's people now use energy at a rate of about 16 terawatts (1 TW = 10^{12} watts = a million million watts). A flow of 16 TW provides in a year (terawatt-year) the rough equivalent in energy content of a billion tons of coal.[2] As with food energy, the flows of commercial energy are unequally distributed among human groups. In 2000, the 800 million people with average incomes of more than $20,000 annually were using 6.3 TW; the 1.1 billion with average annual incomes between $5,000 and $20,000 were using 6.8 TW; and the poorest 4.1 billion people, with average annual incomes of less than $5,000, were using just 2.9 TW.

That means the average person among the richest 800 million

accounted for a commercial energy flow (that expended in manufacturing and using cars and appliances, constructing roads and buildings, bombing other countries, growing food, etc.) of about 8 kilowatts (kW), while an average poor person enjoyed the use of only about 0.7 kW of commercial energy and roughly half that much from biomass. That maldistribution is a fundamental cause (and consequence) of economic inequity. It is one of the main reasons that almost half of humanity lives on less than US$2.00 per day, while Europeans are able to subsidize their cows by an average of US$2.50 per day.

The Uses of Energy

In developed countries nearly all energy is commercial energy. Commercial energy contributes to the planting, fertilizing, irrigating, harvesting, processing, and transporting of our crops, so it makes an essential contribution by helping to feed humanity. It heats and cools our homes and offices, refrigerates and cooks our food, entertains us with music and drama transmitted through the sky, and transports us to work and to the grocery store. There Americans buy the food that energy has helped grow and transport 1,400 miles or so, on average, to the store's shelves. Energy built the car, the store, and the shelves, and energy mined the ores that went into the cans that hold the juices and vegetables, and energy flies us to see the family at holiday dinners. Few things in life that are necessary or pleasurable for readers of this book lack a traceable commercial energy connection (in poor countries, of course, much is traceable to biomass energy).

But commercial energy also runs the bulldozer that is mowing down a patch of tropical forest, all the while spewing carbon dioxide (CO_2) and toxic pollutants into the atmosphere from its diesel engine. Energy is used to power the manufacture of an automobile, build the roads on which it travels, run its engine, and so on—all the while contributing to air pollution, climate change, and the paving over of ecosystems. Energy runs the chemical plant that produces a toxic pesticide and fuels the crop duster that spreads that pesticide, sending some of it into the atmosphere, whereby it ends up contaminating animals all over the world, including the far Arctic. And while contributing to our nourishment, commercial energy keeps artificial ecosystems such as farms going (in place of natural systems, which need no such subsidies from society) while altering biogeochemical cycles and Earth's energy balance, often to our detriment.

Humanity walks a narrow line between energy as a necessary boon and energy as a source of degradation of many aspects of our life-support apparatus. So it pays to keep track of the energy situation. It is particularly important to do so today, as mounting environmental damage, climate change, and tightening competition for dwindling supplies of cheap, easy-to-exploit fossil energy sources are forcing humanity toward a comprehensive overhaul of its system for supplying energy.

Human society as a whole is in no danger of "running out" of fossil fuels, although all nations could face serious problems if alternatives to them, or ways of obtaining and using them without unacceptable environmental disruption, are not developed in a timely manner. Although there is much discussion of exhausting supplies of conventional oil and gas within a few decades, about five to ten times as much energy is available in accessible coal deposits as in oil and natural gas. In turn, the quantities of energy that could be made available from deposits of oil shale, tar sands, and unconventional gas (gas held in reservoirs from which it is relatively difficult to extract) are five to ten times greater than that from coal. The energy available from deposits of the nuclear fuels uranium and thorium (which can be converted to uranium fuel) is greater still. Of course, fossil fuels and nuclear materials are by no means the only potential sources of energy: various forms of solar energy, properly harnessed, could power today's civilization many times over. Sufficient supply in itself is *not* the energy problem, despite the popular view to the contrary.

What society *is* in danger of running out of relative to energy is environment—especially the "away" in which to throw things (as in the expression "throw away"). The environmental drawbacks of fossil fuels, of course, are the damage caused by their extraction, the effects of their effluents on local and regional air quality, and, of increasing importance, their emission of greenhouse gases, especially CO_2, into the atmosphere. The atmospheric sink in which CO_2 and other products of fossil fuel energy use are deposited is a classic open access resource—one nobody owns, and thus one open to overexploitation. Humanity is likely to pay a very high price, beyond the costs already being paid for pollution, for the use of that resource, as we've seen, and the price is not captured in the market price of fossil fuels. There is no tax on gasoline designed to go into a fund to defray the costs likely to be paid as sea level and food prices rise and heat waves become more common, for example. Instead, gasoline taxes are usually spent to build and maintain roads, on which more cars and trucks can travel and burn fuel. In economics, remember, environmental costs that are not captured in market prices, including those deferred to our descendants, are externalities.

A fundamental fact about the human energy situation is that energy use and its hidden costs have increased globally roughly fivefold since 1950, and they seem certain to continue increasing. China alone is installing the rough equivalent of California's total electricity sector every year—and almost all of that is powered by coal. In 2006 and 2007, roughly one new coal power plant per week was opening in China. India's energy use was rising almost as fast as China's, and in the United States it has increased slightly faster than the population since 1990. In Japan and several European countries, however, fossil fuel energy use has begun to decline.

If human society is to become sustainable, people who have studied the energy issue generally agree, we need to reduce sharply and eventually end our dependence on fossil fuels, the principal cause of greenhouse gas emissions. While substantial improvements in efficiency are critical in the short term, a major effort must also be made over the next several decades to develop and deploy alternative energy sources throughout the world.

There are inherent difficulties to keep in mind when considering rapid replacement of a large portion of the fossil fuel–based energy systems, which now provide some 80 percent of the energy used globally. The "sunk costs" (irretrievable investments in, for example, coal-fired power plants) are massive and may lead to political decisions not to replace outmoded power plants even though the environmental costs they will incur will be even greater than the costs of replacement. Meanwhile, some rapidly developing nations are galloping ahead with conventional energy systems, often based on older, inefficient technologies. China's current annual increase in electric capacity of 40 or 50 gigawatts (a gigawatt is a billion watts) can't easily be shifted into wind or solar power. Those industries are not yet large enough to supply that level of demand, to deploy enough wind turbines, solar cells, and so forth. That is especially so since the costs of many forms of such renewable energy are still significantly higher than that of existing conventional coal power plants (ignoring the mounting external costs of pollution from burning coal).

Efficiency as an Energy Source

The quickest, cheapest, and safest new "source" of energy is efficiency.[3] Most nations are enormously wasteful of energy in their practices. In the United States, the abundance of heavy SUVs used for personal transportation in cities and suburbs is a poster-child example. Although it seems a little counterintuitive, we can "harvest" an energy resource from a current inefficient use. If SUVs were replaced with hybrid cars—or, better, trains and buses—the reduced fuel consumption would in effect free up new energy "supply." But American elected officials have generally not supported development of convenient mass transit systems that are secured against crime (crime incidence being a significant reason in some areas for not using them). Public authorities have failed to design cities and suburbs so that people can easily walk or bike to work, and they failed for two decades to insist on improved fuel efficiency in automobiles and trucks, for which the technology already existed.

Efficient use of energy is in part a matter of the choices consumers make in their everyday lives. Energy-efficient behavior means building an adequate, comfortable home rather than an energy-inefficient "dot-com palace" (named after the sprawling mansions bought by many new millionaires created by the Internet boom), buying energy-efficient fluorescent rather than incandescent lightbulbs, heating the house to 68°F rather than 75°F in winter and cooling it to 78°F in summer, recycling, and insisting upon energy-efficient products and packaging.

But wasteful energy use by consumers is only part of the story. Efficiency could also be vastly improved in industrial processes; in the heating and cooling of public and commercial buildings; in lighting of offices, streets, and public places; and in electricity generation and power distribution networks. As energy analyst John Holdren and many others have shown, through efficiency improvements the United States could potentially reduce its energy use to about a *quarter* of what it is now while *increasing* the quality of life. Many of the changes that make this theoretically possible have been promoted by energy guru Amory

Lovins' Rocky Mountain Institute (RMI). For instance, using space-age ultralight, ultrastrong materials (principally carbon-fiber composites) and new propulsion systems, safe, fast, more durable family cars can be produced that, even using gasoline, can get about 100 miles per gallon.

Lovins has long promoted a "soft energy path" focused on efficiency, the use of a diversity of renewable energy resources, cogeneration (e.g., using one fuel to generate electric power and using the "waste" heat given off by the equipment running on that electricity to make steam for industrial processes or to heat buildings), and mobilization of energy near the site of its use. The economic and political barriers to adopting the Holdren-Lovins path to efficiency, however, are great, since they often involve major changes in who does what and who gets paid what for it. Even so, Lovins and his colleagues at RMI have had considerable success in gaining support from many industry and government leaders (the latter more often at the state than the federal level) and initiating significant changes.

Holdren has developed an optimistic (or, as he calls it, "best plausible") scenario for the global population-energy-environment-economic future. It is shown in table 14-1. Why do we call this an optimistic scenario? First of all, it shows world population size peaking around the middle of this century and then slowly declining. This assumes that attitudes and government policies before 2050 will increasingly favor further improve-

TABLE 14-1 **Holdren's Optimistic Scenario**

	2000	*2050*	*2100*
Population (billions)	6.1	9.0	7.5
GDP (trillions, year 2000 $)	45	225	525
GDP/person (year 2000 $)			
Industrial nations	23,000	43,000	70,000
Developing nations	3,000	22,000	70,000
Energy use (TW)	14	27	23
Energy use per capita (kW)			
Industrial nations	6.3	5.1	3.0
Developing nations	1.3	2.6	3.0
Carbon emissions/year (GtC)*	6.4	6.9	2.8

Source: J. P. Holdren, personal communication, 2007.
*GtC = gigatons of carbon.

ment in the education and status of women and make family planning and safe abortion more widely available around the world. It also implies that the efforts in some rich countries to increase their birthrates will largely fail. Even so, this projection is not far from the United Nations' medium population projection to 2050.

Second, the scenario postulates that total human energy use in 2100 could be 23 TW—down from 27 TW in 2050. That's still a lot of energy—more than half again as much as is used now—and it's not clear that Earth's ecosystems can sustain the effects of using more than 20 TW for almost a century, even if the forms in which the energy is obtained and used prove to be generally less damaging than the current fossil fuel–dominated system. In any case, 23 TW is much, much better than the 57 TW projected under a "business as usual" scenario, in which humanity continues more or less to ignore its mounting environmental peril.

Third, the Holdren scenario shows the gap between the rich and the poor disappearing by 2100 as the two groups converge on a per capita energy use of 3 kW. This means that far greater efficiency must be achieved among both rich and poor, with the poor extracting much more quality of life from roughly three times their 2000 energy use, and the rich also improving the quality of their lives but using less than half as much energy as they did a century before. That implies deployment of many technologies already in hand but not in wide use nor as well developed as they might be. Of course, it also implies a great application of political will to change regulations and patterns of subsidy and investment. As the table also illustrates, the carbon efficiency of the global economy (roughly the amount of carbon emitted per unit of gross domestic product in the form of CO_2 from fossil fuel burning) would increase some twenty-five-fold, through switching almost entirely to non-carbon-emitting technologies and sequestering substantial amounts of CO_2.

Alternative Energy Sources: Renewables

How, in addition to increasing efficiency, might Holdren's optimistic scenario be achieved with regard to energy? Without the use of alternative energy sources efficiency alone cannot solve the dilemma, especially given the skyrocketing demand for energy in developing regions.

The most desirable alternative energy sources are a variety of renewable technologies—mostly versions of solar energy, sometimes in the guise of wind, water, or biomass fuels, which are all derived from the energy arriving from the sun. *Wind power* is based on energy from the sun heating surfaces that warm air, cause it to rise, and create air currents in the atmosphere. *Hydropower* (hydroelectric power) is based on solar energy that evaporates surface water and deposits it in highlands, allowing rivers to be continually recharged. *Biomass fuel* (biofuel) is created by plants as they capture, in the course of photosynthesis, a tiny fraction of the sun's energy that falls on them. That energy can then be harnessed by burning the plants, by extracting their oils, or, as pointed out in chapter 12, by converting them through fermentation into ethanol, which can then be used as a portable fuel.

When renewable energy sources are discussed, mention of the sun immediately leaps to the fore, and for good reason. Each square meter of Earth's surface receives on average about 200 watts (averaged over the entire year, 24/7). The surface of the entire Earth receives about 100,000 TW. Remember, 1 TW equals 10^{12} watts, so Earth gets 100,000 million million, or 10^{17}, watts, of which the land receives some 30,000 TW. Given that the world's people now use about 16 TW of power, the potential for solar energy to power our society is obviously large, and in most forms solar power does not add significant amounts of greenhouse gases to the atmosphere once the energy-acquiring installations are built.

The biggest direct use of solar energy at present is simply to heat water, which can be done by an apparatus as simple as a black-painted drum on the roof of a house or a black plastic bag on a camping trip. In a few cases (so far), water or another substance is heated in a large solar thermal power plant in which mirrors focus concentrated sunlight on tubes filled with fluid or a boiler and indirectly or directly produce steam. The steam then spins turbines, just as it would in a power plant fired by fossil fuels. Some of the heat generated during the day in a solar thermal plant can be stored for use in generating power at night, thus overcoming solar power's disadvantage of intermittency and allowing it to become a reliable supplier of electricity. A related technology, solar dishes, are modular units the size of a satellite dish. These dishes do not use steam at all but instead use a Stirling engine, which directly converts heat into motion. The dishes focus heat on an antenna-shaped engine in which two chambers filled with a gas are alternately heated. As each expands, it

FIGURE 14-1. Wind is one of the most attractive options among the alternatives to large-scale dependence on fossil fuels. Wind power, like energy from solar cells and coal-fired plants, is derived ultimately from the sun—although the fossil fuels contain solar energy captured millions of years ago. Photograph courtesy of iStockphoto.

creates a simple piston motion such as you would find in a car, which turns a generator.

Solar energy can be used more directly for space heating and cooling by situating and designing buildings so they can best absorb and hold heat in the winter while reflecting sunlight in summer. This is known as "passive" solar heating and cooling. Passive solar heating deployed on a large scale could rather soon supplant fossil fuel energy as the main source for heating and cooling both homes and large buildings as the movement for "green building" gains momentum. Similarly, solar heating of water for industrial purposes is also growing.

In solar technology, electricity can be produced directly when photons (elementary particles of light) of the right type, as are a substantial portion of the photons arriving from the sun, strike metal or metal-like surfaces (e.g., the metalloid element silicon). Their energy liberates loosely attached electrons, producing an electric current. Research has steadily increased the efficiency of these *solar photovoltaic* (PV) systems in converting solar radiation into electricity, and they can perform even under somewhat cloudy conditions.

Overall, the economic attractiveness of such solar systems is increasing as the costs of associated technology fall and the prices of fossil fuels rise. The process of adopting solar technologies is being accelerated in many areas, such as California, by subsidies at state and local levels. One huge additional advantage to widespread adoption of solar technology would be potentially significant decentralization of the electric power system, reducing society's vulnerability to accidental or terrorist-induced blackouts. Eventually, most homeowners could be connecting their PV systems to the grid and selling excess power back to utilities, helping to solve society's energy problems.

Wind power is also a promising renewable energy source. Perhaps its biggest problem is in siting the enormous number of wind turbines that will be required to make a significant contribution to humanity's energy demands. Large-scale wind installations are most economical when they are close to where the electricity is being used, so that transmission costs are reduced; but unfortunately, many of the best sites for strong, dependable winds (such as the United States' northern plains) are not near cities or major industrial areas. The intermittent nature of winds means that much of the time wind turbines will be operating well below their maximum capacity. This drawback, however, can be compensated for by combining numerous wind towers in a grid; the wind may not be blowing in one location, but it may well be ten, twenty, or fifty miles away.

Some potential locations for wind installations are opposed on environmental grounds because they would spoil scenery or pose risks to bats or migrating bird flocks. These problems can be avoided, though, with careful siting. Large offshore installations have been effectively deployed in northern Europe and could be in the United States and elsewhere. In the midwestern United States, farmers have been happy to host wind installations, which give them a modest, welcome addition to their incomes as well as supplying them and their neighbors with electricity.

Both wind and solar photovoltaics can be very effective under some circumstances when available. Until much more effective electricity storage systems are developed, however, neither will be able to serve as the primary source of base power. Solar thermal systems with heat storage, by contrast, would be able to supply power at night, but they must be located in areas of intense dependable sunlight, such as hot deserts, and can supply power only to areas easily served by transmission lines. Another potential technology for storing solar or wind energy for use when the sun is not shining or the wind is not blowing is compressed air energy storage (CAES). Electricity from solar or wind generators can be used in "off-peak" hours to compress air and pump it into geologic structures. Then, when needed, the pressurized air is released, mixed with fuel, and burned in turbines. The process uses less fuel than would be consumed if it were burned in conventional turbines.

Like solar systems, wind installations offer a flexibility that is lacking in industrial-style power grids. In some remote areas, from which the costs of transmitting power would be too high, wind-generated electricity could be used on-site to decompose water to make hydrogen for use as a portable fuel. This could be part of a transition to a "hydrogen economy" in which the main carrier for energy replacing liquid fuels in vehicles would be hydrogen. Hydrogen can power automobiles directly or by generating electricity in fuel cells (battery-like devices that, instead of storing electricity in a closed system, generate it by chemical reactions in an open system where a flow of reactants must be continuously replenished). But hydrogen as a fuel to be burned would probably find its most important use in airplanes, which for practical reasons can't be run on electricity. They would, however, need to be totally redesigned to use it.

In remote areas (including those off the grid, as is common in rural parts of developing countries), smaller-scale wind installations, backed up with efficient battery storage for windless times, can also provide base power. Both wind and solar PV systems are especially well suited for rural villages in developing regions where no electric grid exists, and many countries, including China and India, are taking advantage of this option. Indeed, environmental writer Bill McKibben has written of seeing tiny solar collectors hung on the yurts of nomads in Tibet, powering a single lightbulb and perhaps a radio.

Biomass fuels could, in some circumstances, help to stretch supplies of fossil fuels. Conversion of corn to ethanol has recently attracted considerable interest, as we saw in chapter 12, and ethanol distilleries are being built all over the midwestern United States. But the intensity of agricultural use of fossil fuels in industrialized countries makes it difficult to turn an energy profit; ethanol from corn yields only about 25 percent more energy than was used in producing it by the most efficient processes. The principal use for corn ethanol today is as a gasoline additive, which promotes more complete combustion and hence less pollution. However, the quantity of ethanol that would be needed to replace *all* the gasoline consumed today in the U.S. transportation system would far exceed the nation's corn harvest, which under the best estimate could be converted into enough biofuel to meet only 12 percent of the demand for gasoline.

So one of the biggest downsides of developing a heavy dependence on biomass fuels derived from major crops is that they would be competing for the production capacity of agricultural land against food, a vital resource that *is* in short supply globally. It seems unwise in the extreme to expand agriculture toward it becoming a major source of portable fuels at a time when the entire global food-producing enterprise is threatened by rapid climate change, and while much of the land now being farmed is being degraded and thus is unlikely to sustain production.

Nonetheless, there may well be a future for biofuels—in some special circumstances and with careful controls, they could contribute something in the long run. Cane sugar is a considerably more energy-rich source than corn, for example. Brazil has very successfully made use of its abundant sugar crop to produce ethanol to fuel its cars and trucks and now is eyeing the possibility of export trade. Cuba, also a major sugar producer, seems poised to enter the world market for ethanol as well.

Another form of biofuel is oil pressed from oilseeds, primarily soybeans. Brazil is also exploiting soy oil, unfortunately to the great detriment of the Amazon forest. Making diesel from soy is more efficient than converting corn to ethanol—biodiesel yields some 90 percent more energy than the quantity of commercial energy that is invested in producing it. Amazonian biodiversity has little chance against numbers like that, and the forest's key climatic functions are also likely to be altered to humanity's disadvantage.

What remains of the rain forests of Malaysia and Borneo, both already well on their way to becoming vast oil palm plantations, could be destroyed within a decade or so by the increased demand for palm oil as a fuel. Along with them will very likely go the last wild populations of orangutan, Sumatran rhino, tiger, and thousands of less charismatic (but arguably more important) organisms. At a saner level, some enterprising Americans have begun fueling vehicles with used frying oil from fast-food outlets. But this, like other such efforts, is only symbolic—equivalent to trying to bail out a flooded basement with a thimble.

Other sources of ethanol for use as fuel are currently being investigated in pilot operations, especially fuels derived from cellulose, a complex, indigestible (by people) carbohydrate present in essentially all plants. Switchgrass, fast-growing trees, and other weedy, cellulose-rich plants could thus be sources of biofuels, and they can be grown in areas unsuitable for most crops, such as on the margins of farmlands and in woodlands, pastures, and "wastelands." Similarly, a portion of food crop residues after harvest (cornstalks, rice stems, etc.) could be used as well, though some residues are essential for maintaining soil fertility and should be left in place.

If a successful fermentation process is developed, with careful management it might be possible to produce rather significant amounts of cellulose biofuel with minimal competition with food-producing acreage and without substantially degrading land or diminishing biodiversity. Furthermore, a change in diet could make room for more biofuel production. Recent work by ecologist David Tilman shows that if total meat consumption could be kept constant, and fish and poultry (which are much more efficient in converting feed grain to meat) could be substituted for various proportions of the beef and pork consumed in the United States, immense amounts of land would be freed for low-input, high-diversity biofuel production and carbon sequestration.[4]

So biofuels may contribute one small "wedge" (chapter 13) to a sensible solution to the climate change–energy dilemma. But the major effort currently under way to switch from fossil fuels to biofuels seems quite misguided. The end result is likely to be higher food prices, fewer jobs, more fertilizer runoff and pesticide poisoning, and diminishing biodiversity. Only a myopic society would, in essence, grind up its most irreplaceable resource and feed it to Hummers.

Hydropower is a traditional energy source that is much less important in the human energy economy than fossil fuels and is likely to remain so. Most of the best energy-producing sites are already in use, especially in developed countries. Also, dams tend to be damaging to downstream ecosystems, and flooding above the dams can be environmentally destructive as well. In particular, dams have caused serious problems by blocking salmon runs in the northwestern United States and elsewhere. Furthermore, recent work by tropical ecologist Philip Fearnside and energy analyst Danny Cullenward suggests that the methane released when tropical ecosystems are flooded to create reservoirs may be a significant addition of greenhouse gases to the atmosphere. Moreover, dams are inherently temporary structures, eventually silting up and becoming waterfalls. In some parts of the United States, dams now are actually being removed to recover the natural landscapes they had altered and to restore natural salmon runs.

Generating electricity from the oceans, whether from thermal, wave, or tidal power, presents daunting technical and economic problems. So does geothermal power—the use of heat from Earth's interior. In some areas, hot groundwater can be pumped directly into homes (as is widely done in Iceland), or underground steam can be tapped to run turbines, but in general the oceans or Earth's hot interior are at best future hopes with seemingly limited potential to contribute to solving the energy-environment dilemma.

A principal drawback of all the potential power sources considered so far, with the partial exception of biofuels, is that they do not directly yield portable fuels, which are needed for long-distance surface transportation and for aviation. Besides converting to hydrogen generation for their production, the *need* for portable fuels could itself be alleviated, as Holdren, Lovins, economist Lester Brown, and many others have pointed out. What would be required is converting current gas-powered automobile fleets to lightweight gas-electric hybrids, built of those new polymer composites. These could be plugged in overnight, when power demand is normally low, and run on electricity alone for short trips. Indeed, plug-in hybrids are already expected to be marketed soon, although they are not yet being made of lightweight composites. There is hope, moreover, that lighter and more efficient batteries will soon enter the market, making plug-in hybrids even more attractive.

While biofuels may help to fill the portable fuel gap, much of the need for them could be reduced by a combination of changes, including substantially improving vehicle efficiency, shifting a significant fraction of commuting to much more efficient mass transit systems such as light rail and bus networks, and redesigning settlement patterns to eliminate the need for most people to travel long distances to work or just to perform daily errands such as grocery shopping. Also, reducing the need for transport of goods (often from distant countries) and increasing the efficiency of distribution and delivery systems could further reduce the need for fuels. In combination, such changes potentially could reduce the per capita demand for gasoline equivalents in the United States by two-thirds or more.

Fossil Fuel Alternatives

Efficiency could also be significantly increased in several types of advanced fossil fuel power systems, although they ultimately have some serious drawbacks. Prime among these are "integrated gasification combined cycle" (IGCC) systems, very complex designs in which coal is turned into a mixture of flammable gases by heating and chemical reactions. That complexity, however, permits a cleaner and more efficient use of coal than is presently employed in the vast majority of power plants. Coal burning produces the most air pollutants and the most CO_2 for a given energy yield of all the fossil fuels, however, and IGCC technology does not in itself alter that relative standing. Recently, IGCC systems have been promoted in combination with technologies for carbon capture and storage (CCS), such as by sequestering the CO_2 generated by the power plant in deep geologic formations—a technology that is far from proven. Energy experts think that the best and cheapest way to utilize CCS will be in tandem with IGCC systems, although it also will work with conventional coal and other carbon-containing feedstocks such as "petroleum coke," a by-product of oil refining. There is also building interest in "polygeneration" technologies, in which electricity, industrial chemicals, liquid fuels, and perhaps even industrial heat and carbon capture are all combined in the same facility using coal.

The catch is that IGCC plants are more expensive to build than conventional coal burners, and CCS adds even more to the cost, along with wasting more of the total energy stored in the coal than do conventional processes. IGCC (with or without CCS) is therefore not so attractive to utility companies, which have proposed ambitious plans to build some 150 conventional coal plants in the United States in the near future. Shifting these plants to IGCC or other advanced coal technologies would be a step forward, but an even better move would be to reduce the need for the new plants by increasing energy efficiency throughout the economy (thereby also reducing pollution) and undertaking large-scale wind and solar PV installations. Indeed, rising public concern about global heating in 2007 led to cancellation or downgrading of many of the proposed coal plant projects, and questions were being raised about many others.

Even less desirable than a boom in inefficient coal use is development of oil shales, which already is causing environmental havoc in large areas of Canada's province of Alberta through strip-mining and wastes from thermal processing. The shales contain enough organic materials to yield oil and burnable gases when they are themselves heated. The energy-intensive process of extracting oil from the shales markedly reduces the net energy yield from the oil while increasing CO_2 emissions. Nonetheless, plans are afoot to greatly expand exploitation of the shales, spurred by the tightening supplies and rising prices of conventionally extracted petroleum.

Nuclear Power

Some energy experts have been promoting renewed deployment of nuclear fission power plants, claiming that nuclear plants don't emit CO_2 and therefore make a relatively slight contribution to global heating. Nuclear fission works by the splitting of heavy nuclei (uranium, plutonium) into lighter ones, which releases massive amounts of energy through the conversion of small amounts of matter. But nuclear power presents several difficulties. First, power plants need to be designed and built that are fail-safe against a catastrophic accident that would lead to a breach of the containment system and a massive release of radioactivity. The sort of accident that almost happened at the Three Mile Island plant in 1979 and did happen at Chernobyl in 1986 potentially could make an area the size of Pennsylvania permanently uninhabitable.

Nonetheless, many experts in the scientific community who have been critical of the use of nuclear power believe that a plant that would be satisfactorily safe from such an accident can be developed. Of course, it would also need to be designed to be terrorist-proof, but that too seems feasible (although the power companies are reluctant to invest in it). Another huge problem that some believe *could* be solved is disposal of millions of tons of high-level radioactive wastes, which must be sequestered for tens of thousands of years. Incorporating them into ceramics and burying them deep may well work. But numerous problems surround the United States' chosen burial site at Yucca Mountain, Nevada, such as potential local seismic activity and leakage into an important aquifer. Moreover, the wastes from existing plants must be transported through some forty states to reach their destination in Nevada. So a major NIMBY (not in my backyard) problem has emerged. Revelations of incompetence and duplicity by the original Atomic Energy Commission and its successor, the Nuclear Regulatory Commission, have left citizens in the United States highly skeptical of government statements that repositories and the transport systems for taking the wastes to them are "fail-safe."

While in operation, nuclear plants themselves emit no CO_2, but the processes of mining, refining, enriching, and transporting the uranium definitely do, while causing considerable environmental damage as well. Moreover, massive amounts of concrete are used to build the power plants, just as they are for dams, and concrete making is itself a significant source of global CO_2 emissions. Furthermore, decommissioning nuclear plants after their approximately forty- to sixty-year life span is expensive and presents problems of disposal of radioactive parts.

At one time it was thought that perhaps the most serious problem with nuclear power was that spreading knowledge of the technology increased the chances of enabling rogue nations and subnational groups to develop nuclear weapons. But since effective measures to reduce that risk have not been vigorously pursued, access to nuclear materials and weapons is now so widespread that the danger of proliferation is, sadly, not quite as powerful an argument against nuclear power as it once was.

Finally, at a time when rapid deployment of energy systems that don't emit greenhouse gases is badly needed, nuclear power plants not only are costly to construct but also require a decade or more from the planning stage to deployment. Because of these considerations and the risks associated with them, investors are not jumping at the chance to build more nuclear plants in the United States, despite the lucrative subsidies for doing so. It seems likely that economic hurdles will forestall the deployment of many (or any) new nuclear power plants in the United States, although many are being planned or built in other countries, including China and India.

Because of the way energy binds together atomic nuclei, a conversion similar to that in fission power plants can be achieved by smashing together lighter nuclei (most commonly, forms of hydrogen), causing them to fuse into a heavier nucleus. This is the process that powers the sun and is the main source of explosive power in a hydrogen bomb, and it has been suggested as eventually a potentially significant source of conventional energy. There are various ways that controlled fusion might be attained without an explosion, but so far all such attempts to do so have required more energy to be put into the process than can be extracted. While the development of nuclear fusion generators seems more attractive in the long term than fission power, the promise of fusion is many decades away from fulfillment.

Meanwhile, industrial-scale wind farms and other renewable power sources, which can be built much more quickly and cheaply than nuclear plants, will soon surpass nuclear power in their share of U.S. energy mobilization. Wind and solar installations are booming in many other countries as well, especially Europe and Japan. Competition from cheaper and more quickly deployed energy sources may well ultimately spell the doom of nuclear fission power.

Toward the Future

A single solution to humanity's "energy problem," a magic bullet to lead us through Holdren's optimistic scenario, is clearly not in the cards. Efficiency can potentially make a huge contribution, but in addition a mix of renewables, modern coal technology with CO_2 sequestration, and eventually perhaps upgraded nuclear fission, fusion, or both will all have to play their parts. To move toward an effective solution to energy supply and related environmental problems, what seems needed is prompt action on efficiency, combined with careful planning, research, and rapid implementation of an array of supply technologies that minimize adverse environmental effects. Such a program is also essential to reduce the vulnerability of society to sudden disruptions of electricity flows or oil supplies through natural disasters or terrorism. Some people have thought that cheap, abundant energy would be a solution to many of humanity's problems. That goal is unlikely to be reached, and it should be remembered that abundant energy is no panacea for the world's ills. A bulldozer run on hydrogen produced from a wind farm can still decimate a tropical rain forest.

The energy problem brings together many crucial issues in the human predicament. The great speed of evolution in the technological dimension of our culture has led to the deployment of a dangerous and unsustainable global energy infrastructure. The slowness of change in the social arena of culture has retarded recognition of this circumstance and its consequences. The consequences of fossil fuel dependence in the context of the needs and demands of a growing human population are clear. They are most readily seen in the dominant species running out of environment (the most obvious symptom being incipient rapid climate change) and in global political instability, which is especially dramatic in the form of wars over petroleum.

CHAPTER 15

Saving Our Natural Capital

"To keep every cog and wheel is the first precaution of intelligent tinkering."

ALDO LEOPOLD[1]

THE SINGLE most important challenge facing humanity today is the set of interrelated yet distinct problems of landscape modification and losses of both biodiversity and ecosystem services. They might be at least as serious in the long run as that other unintended consequence of the human rise to dominance, climate disruption. Both have repercussions that will be irreversible in the conceivable human future, and each has the capacity to worsen the effects of the other. The scientists who are working to preserve our only known living relatives in the universe, along with the crucial services they help to supply, are conservation biologists. Their activities in trying to prevent the worst mass extinction in 65 million years, the result of a dominant animal dramatically and rapidly altering evolutionary trajectories and relationships thousands to millions of years old, are the central topic of this chapter.

Interactions between climate disruption and the loss of biodiversity deepen the difficulty. Land-use change and associated extinctions, especially from deforestation, are significant contributors to global heating, and that in turn is already becoming a significant threat to many species of flora, fauna, and microbes that provide important ecosystem services. An enormous challenge, paralleling that of reducing human-caused greenhouse gas emissions, then, is to devise ways to preserve the other organisms of Earth and the ecosystems within which they function in the face of expanding human activities.

This is a difficult undertaking for several reasons. First of all, the enterprise will require a large transfer of resources to it, and that means a politically thorny reallocation of scarce resources. For example, if the numbers of a species have dropped so low that captive breeding is required, it usually takes decades of work, requiring continuous funding and substantial luck, to accomplish successful reintroductions—as the case of the California condor has illustrated. That bird was nearly extinct when the last few individuals were removed from the wild in 1987 and a long, expensive, and difficult captive breeding and release program was initiated. In 2007 there were still just under 300 condors in existence, almost 150 in captivity, and some 140 wild birds, most from releases in California and Arizona and a few in Baja California. The species is still endangered in nature, especially from lead poisoning, a result of feeding on game carcasses and ingesting the bullets. The program has cost some $40 million, and that's a mere drop in the bucket compared with the billions often needed to preserve large stretches of habitat. Considering that there are millions of species and billions of populations to preserve, even if many share the same habitats, one can get an idea of the scale of the financial problem of saving, by any strategy, a substantial portion of those components of our natural capital.

Second, in many cases, it can be a lot easier to detect declines in biodiversity than to uncover their causes. Honeybee colonies started disappearing in the United States and Europe ("colony collapse disorder") in 2006 and 2007, and yet, although billions of bees were disappearing, the cause remained (as of this writing) mysterious, with suspicion falling on a virus combined with the stress of constantly shifting hive locations.

A third reason why preserving biodiversity is a challenge is that individual species live in and are sustained within their habitats, but we usually lack the detailed ecological information required to restore habitat satisfactorily. Moreover, restoration is almost always far more difficult and costly than the activities that caused the damage and led to a decline in one or many species in the first place. In addition, climate change, which is a moving target and largely unpredictable in the kind of detail needed to ensure preservation or restoration, creates even more uncertainty about prospects for success.

Finally, there is the vexed problem of exactly what is to be saved. To what degree are any populations, species, or ecosystems essential (or disposable)? Should efforts be focused on species of functional groups of recognized importance, such as bees, or pollinators as a group, which are crucial both to agricultural production and to ecosystem preservation in general? How does one estimate the long-term costs of *any* extinction when all organisms are bound up in webs of relationships, and the loss of an apparently "unimportant" species can lead to a cascade of extinctions that removes one or more "important" ones? How do we know that a species that seems unimportant today might not be critically important after rapid climate change?

Why Should We Preserve Biodiversity?

With such daunting challenges ahead, what are the compelling reasons to make the effort? Originally, the main arguments given for preserving the living components of our own life-support system were basically esthetic and ethical. Other organisms are beautiful, intricate, and endlessly interesting, and they have a right to exist. Their beauty and intrinsic interest also produce financial returns through ecotourism, bird-watching, gardening, scuba diving, wildlife art, and so on. Other values have often been cited, such as the potential medicines that may be derived from plant chemicals or the sustenance that fisheries provide. But a half-century of experience shows that these arguments by themselves are inadequate to generate sufficient public and political support for the preservation of biodiversity.

As we've seen, however, those disappearing organisms are critical elements of humanity's natural capital, the ecosystems that provide the steady flow of nature's goods and services on which we all depend. That perspective gives us an argument that may speed progress toward halting the depletion of that capital. It also brings together arguments for maintaining esthetic and ethical (cultural) values and the more traditionally appreciated provisioning, supporting, and regulating services of nature (chapter 10).

When conservation biologists talk about the decline of biodiversity and decay of ecosystem services, they are referring to the loss of significant populations as well as of species. The loss of populations—the individuals of a species in a given area at the same time—is going on at many times the rate of species loss, which is of critical importance because it is populations that deliver ecosystem services. It's the population of blue spruce trees uphill from your Colorado canyon home that keeps your house from being swept away by an avalanche. If the trees are cut, your home will be in immediate peril, even though other blue spruce populations live throughout much of the mountain West. The natural beauty of those spruce trees, and the populations of songbirds that live in them, also deliver cultural (esthetic and recreational) services to you and other local nature lovers. And the population of trout in the local stream provides you with a delicious dinner, eaten at your barbecue underneath the nearest spruce tree.

The importance of the existence of multiple populations of species can be made clear with a thought experiment. Suppose that, miraculously, it were possible to preserve permanently only one small but viable population of each species on the planet. Then by definition there would be no further loss of species diversity, no crisis of species extinctions. But very soon civilization would collapse and all human beings would die, since we couldn't harvest even the small remnant populations of crops or domestic or game animals without destroying each species. And, even if those species that are directly useful as food were all to be exempted from the experiment, disaster would still be upon us, since all crops are dependent on other organisms, such as fungi that transfer nutrients from soils into plants, pollinators that many plant species require to reproduce, predators that protect crops from pests, and so on.

To take a more realistic example, what if all the honeybees died off everywhere in the United States as a result of, say, the disappearance of all colonies in that mysterious colony collapse disorder. Global species diversity would not be diminished because those honeybees would still be present in Europe, Africa, and elsewhere. But pollination services would be drastically reduced for some ninety American crops, at an estimated cost that could be higher than $14 billion per year, depending on the precise mix of crops, the degree of damage, and the prices of produce.[2]

Strategies to Protect Biodiversity

The classic approach to conservation has largely been to identify either species considered at risk or geographic areas that are rich in species but threatened by human development. Then conservation efforts focus on providing protected areas for those "endangered species" and conserving as much habitat area as possible in species-rich "hotspots." Secondarily, important attempts have been made to limit the harvesting of stocks of the organism that are being depleted (restrictions on whale hunting are a classic example) and to restrict the use of toxic substances such as pesticides that threaten other organisms.

International agreements such as the 1973 Convention on International Trade in Endangered Species of Wild Fauna and Flora (CITES) try to stop overexploitation of harvested organisms. CITES established an intergovernmental regulatory structure designed to limit trade and international movement of endangered species. It has had many successes in intercepting at customs barriers shipments of, for instance, endangered parrots or snakes destined for the pet trade or rhinoceros horn fated to be ground into an ostensibly aphrodisiac powder that is highly valued in China and elsewhere. China, indeed, serves as a vast sink for overexploited biodiversity. Besides the giant Chinese market for animals used as folk medicines, many nations, especially in Southeast Asia, ship endangered species there in huge numbers to be sold in markets and restaurants, since they are considered delicacies. Unfortunately, human cultures tend to value things that are rare, such as giant diamonds—thus rich Chinese love to dine on the lips of snappers, endangered reef fishes that might be thought of as diamonds of the sea. The conservation paradox is that the rarer the species, the more economically valuable it often is, and thus the greater is the incentive to elude regulations by smuggling. Smugglers tried to take one entire population (of two known remaining ones) of a cactus, which could have brought a fortune from cactus fanciers, into Germany in fifteen suitcases in 1978—an ultimate example of the high monetary value of extreme rarity.

Many elements of biodiversity are threatened within poor nations because people depend on animals they hunt for a key protein-rich portion of their diets. In some nations the acquisition of this "bush meat" is becoming commercialized to feed workers in mining and timbering projects, greatly increasing the pressure on some populations of animals. Hunting for bush meat is a major extinction threat for primates—including our closest living relatives, the great apes—with bonobos apparently at the highest risk. These activities, unfortunately, are mostly outside the reach of CITES.

Attempts have been made to ameliorate toxic threats in various parts of the world as well, such as the banning of DDT in the United States when it was shown to be wreaking havoc with the reproduction of eagles, peregrine falcons, and other raptors. More recently, India, Pakistan, and Nepal stopped treating cattle with diclofenac, a non-steroidal anti-inflammatory drug (NSAID) like aspirin and ibuprofen. Persisting in the bodies of the animals after their death, the toxic drug had been ravaging vulture populations, for which cattle carcasses are a major food source. The loss of the vultures' decomposition service is a major problem for India, which faces more rabies as a result of growing populations of scavenging dogs, other diseases from exploding rat populations, problems of disposing of cattle carcasses to prevent the spread of disease, and human corpse disposal problems for Parsis, an ethnoreligious group who feed their remains to birds of the air because their religious scruples forbid them from either cremating or burying their dead.

Although programs focused on the adverse effects of species trade and pesticides are important, the efforts to protect biodiversity still mostly involve the establishment of "reserves"—areas set aside with minimal human activity to protect either individual charismatic species or unusual concentrations of species. Centering much effort on reserves was very sensible in the early days of conservation. Then the field was new, and there was little scientific information about threats or conservation strategies, just the obvious fact that human activities were harming other organisms. Isolating nature from people was a reasonable strategy, especially since it still seemed practical to sequester large amounts of land. The reserve approach was further supported by a theoretical framework known as the equilibrium theory of island biogeography.

Island biogeography is the study of the distribution of organisms on islands, and the equilibrium theory was the first real theoretical framework in all of biogeography. It was proposed in a 1963 paper by two leading ecologists, Robert MacArthur and Edward O. Wilson. The equilibrium theory attempted to describe the assembly of a community of organisms on a newly formed island. The basic idea, like many great ideas in science, looks simple in retrospect. It was that the number of species that successfully immigrated to the island would be balanced by the number of species that subsequently went extinct there. Note that the equilibrium here is a *number* of species, not a particular *set* of species. At equilibrium there would still be species turnover: new ones arriving and establishing populations, and populations of old residents going extinct.

The equilibrium number would largely be determined by the distance of the island from the mainland, influencing how easy it would be for an immigrant species to reach the island, and by the size of the island. The smaller the island, the greater the chances of a species going extinct; other things being equal, a smaller island can support fewer species than a larger one can. The theory performed well when tested against data from Krakatoa, a volcanic island in Indonesia that exploded in 1883, wiping out its entire flora and fauna. In this natural experiment, recolonization by birds followed patterns predicted by the theory. The island approached an equilibrium (steady state) in which the number of species did not change, while the turnover of species continued. Each new immigrant species was balanced by an extinction, keeping the total species number constant. A field experiment by Wilson and ecologist Daniel Simberloff in which mangrove islets off the Florida coast were fumigated, after which recolonization by insects, spiders, and mites was recorded, showed a similar close conformity to theoretical predictions.

The key conservation importance of the theory emerged when it proved useful outside the realm of oceanic islands and fit in with the notion that biodiversity should be preserved by setting aside reserve "islands" in which a variety of species could be protected. For example, it was used in early conservation biology studies to estimate how many species could persist in habitat "islands" such as isolated national parks, freshwater lakes, and habitat patches in disturbed landscapes. There the assumption of habitable islands surrounded by an uninhabitable "matrix" was stretched (in the cases of Krakatoa and the mangroves, the matrix was composed of salt water), but, as with most good theories, a little stretching did not destroy its utility. Thus it allowed the first rough calculations of the numbers of species that would go extinct when one large "island" of tropical moist forest was converted into islandlike fragments by clearing. Island biogeographic theory, like Darwin's theory, shows how an understanding of complex systems is advanced by, and indeed requires, a theoretical framework.

In the United States, the species-based strategy of conservation is embodied in the Endangered Species Act of 1973 (ESA). It authorized the determination and listing of endangered species and specified how listed ones were to be protected. Invocations of the act have had some notable successes, such as helping to bring the national bird, the bald eagle, back from the edge of extinction. Under the ESA's requirement to protect the habitat of threatened or endangered organisms, the law has also been used to good effect to shelter large stretches of habitat and preserve the ecosystem services that the habitat provides. For that reason, organisms such as the spotted owl are sometimes referred to as "umbrella species."

In a similar vein but on the habitat-oriented side, the hotspots approach, pioneered by British ecologist Norman Myers, draws recognition to areas with extraordinary species richness, a high proportion of endemic species—those restricted to the area in question—and high levels of threat through habitat destruction. This approach has directed attention and conservation efforts toward such centers of diversity as Brazil's remaining Atlantic rain forests, which hold some 40,000 plant species, 40 percent of which are endemic. California is another hotspot, with much plant endemism, more butterfly species than any other state but one, and high rates of habitat destruction due to urbanization and agriculture. Mostly deforested Madagascar is also a hotspot, with ten taxonomic families of plants, five of birds, and five of primates that live nowhere else. Botanically, the Western Cape region of South Africa is a stunning hotspot. It is home to the unique fynbos, Afrikaans for "fine bush," a flora containing many shrubs with needle-like leaves. The fynbos has 12 endemic families and 160 endemic genera—greater diversity than an equivalent area of tropical rain forest.

The reserve-island approach to preserving biodiversity focuses on the distribution of valued species or on biodiversity richness in general and on finding ways to protect the habitat "islands" they occupy. One intriguing way of attempting to finance establishment of reserves in developing countries, the debt-for-nature swap, was pioneered by ecologist Thomas Lovejoy. Although there are many versions, the basic plan is simple. Many tropical nations are rich in biodiversity but deeply in debt. With a developing nation's agreement, interested parties (sometimes including other governments) buy some of its debt on the open market for a very small portion of its face value, much as if you were to pay $200 for a $1,000 IOU that a bankrupt friend in Mexico owes for a loan from your neighbor. The neighbor sells because she knows she's unlikely to collect the full amount and won't accept Mexican money. In a debt-for-nature swap, the money is then converted into local currency and the poor nation uses that currency (which may be as much as the original debt) to finance agreed-upon conservation activities there. It would be as if your Mexican friend, relieved of a US$1,000 debt, spent an agreed-upon US$200 worth of pesos to plant trees around his house to provide resources for birds and other animals. It's a win-win-win situation—the poor nation reduces its debt cheaply, the lending institution recovers at least some of the money it loaned, and other governments or organizations get desired conservation results, such as saving a hotspot region.

A major effort of conservation biologists has been to use tools such as the Endangered Species Act in the United States and parallel laws and debt-for-nature swaps in other countries to protect the many key areas around the world that should be in reserves. These and other tools are being used in heroic struggles, often by non-governmental organizations (NGOs) funded by contributions from private sources. A typical example is the Calakmul Biosphere Reserve on the Yucatán Peninsula, which is protecting 1.8 million acres of virtually untouched tropical forest. The existence of this priceless collection of natural capital, threatened by logging and other forms of exploitation, depends on the efforts of a small Mexican-American operation called Rainforest2Reef.[3]

Reserves and Corridors

Important as the protection of endangered species and the establishment of reserves have been as focal points, they alone cannot accomplish the job of conserving the broad range of Earth's species diversity. In response to that recognition, conservation biologists are now incorporating them into a more comprehensive approach to maintaining organic diversity and humanity's life-support systems. The new approach is built on three realizations.

First is that the diversity and the geographic distribution of populations are critical to the provision of ecosystem services, those services are essential outside of hotspots and protected areas as well as within them, and hotspots themselves are difficult to define and delimit, even for particular taxonomic groups such as mammals or birds, much less their combination. For instance, hotspots for mammals only partially overlap hotspots for birds. And mammal hotspots defined by the level of species diversity (e.g., those areas holding the most species) are very different from those defined as holding the most threatened species or those containing species with very limited distributions. Thus, valuable as hotspots are in theory, the concept takes us only so far in the practical task of preserving biodiversity.

Second, while reserves are crucial to the success of conservation efforts, they are not—and cannot be—sufficient to preserve either species or population diversity on their own because many are too small, are poorly located, have only temporary boundaries, or all three. Many also are already occupied or are being invaded by local human populations—"paper parks," as they are often called. This is especially true in poor tropical countries, where expanding populations and the need to increase consumption (especially of food) lead to poaching of game and timber in reserves, often to the near destruction of the reserves' capacity to protect biodiversity. Impoverished and growing local communities often need fuelwood, timber, game, and other resources from within reserves for their very survival. In developing countries, temptations are also legion for individuals or groups to exploit poorly protected reserves not from necessity but for profit. On the Indonesian island of Sulawesi in 1996, for example, among the commonest sights we saw were trucks loaded with rattan illegally extracted from reserves and destined for the export market.

A third realization about the limits of reserves for protecting given species and ecosystems is the disruption that climate change is likely to cause in some regions. In the geologic past, when climatic conditions changed, many organisms simply migrated—staying within what for them were satisfactory climatic conditions. Thus many tree species in eastern North America migrated north with the retreat of the glaciers at the end of the last ice age. But today the opportunities for migration are more limited for many species, not only because of the projected speed of climate change but also because of numerous barriers created by expanses of cropland, grazing land, and tree plantations, as well as by roads, fences, cities, housing developments, and other human-built structures coating much of the planet's land area. In many cases organisms will be "stuck" in the areas reserved for them, and their populations will go extinct with the disappearance of the climatic conditions that permitted them to persist.

Because of the problems posed by those anthropogenic barriers, habitat fragmentation, and accelerating climate change, the topic of ecological corridors—slender reserves connecting larger ones—is gaining more attention. Corridors can be used to link fragments into areas big enough to maintain organisms that could not persist in small chunks of habitat. But perhaps more important in a changing world, corridors can provide paths that organisms can use to move rapidly in response to shifting climate; especially important would be corridors with north-south orientations or ones that connect low-elevation areas with cooler higher elevations. Careful research is showing that corridors can be very effective. In a series of large-scale experiments in the southeastern United States, for example, ecologists Nick Haddad, Joshua Tewksbury, Douglas Levey, and their colleagues have shown that habitat fragments connected by corridors conserve more native animal and plant species (and more animal-plant interactions) than do isolated fragments.

Some concerns have been expressed about the creation of corridors. For example, they might also facilitate the spread of diseases, such as the cancer cells injected during mating battles that are now decimating the powerful, fifteen-pound carnivorous Australian marsupials known as Tasmanian devils. Or corridors might speed distribution of the multitude of invasive organisms now threatening biodiversity almost everywhere. Corridors apparently could even create problems for native fauna. For example, in Australia they could allow an extremely aggressive edge-inhabiting bird, the noisy miner, to reach habitat patches in which endangered birds have been thriving in the miners' absence. But many studies do not show any increased spread of invasive species, and there are few data suggesting that the possible negative effects of corridors would outweigh their positive conservation contributions. It also should be noted that under some circumstances invasive species can contribute to evolutionary diversification—by their own populations differentiating in different invaded habitats and by placing new selection pressures on native organisms. It seems unlikely, however, that this could balance the diversity-destroying role so often played by invaders. Where corridors can't adequately do the job of enabling species to move from one habitat to another, there is the possibility of assisted migration—the transplanting of organisms from one place to another—to help them adjust to climate change. This strategy is being hotly debated in the scientific literature. Success in transplant experiments is very difficult to predict, even when the organisms are moved to apparently highly suitable sites, and there is always the danger that transplanted species may act as deleterious invaders.

Issues related to reserves are not restricted to dry land. For example, marine conservation biologists such as Jane Lubchenco have recently been promoting the potentially enormous benefits to be gained from the establishment of marine reserves—areas where fishing and offshore development are excluded. Pilot studies have shown that even reserves of modest sizes can significantly boost production of fish and shellfish populations in surrounding areas. In the reserves, reproduction is increased, often to the extent that migrants from the reserves can replenish adjacent fisheries. If networks of such reserves were established in the world's productive aquatic regions, they could help restock the oceans, which are now being rapidly depleted of commercially harvestable marine life.

Marine reserves could be supplemented with other management schemes such as individual transferable quotas (ITQs), which have been used with some success in New Zealand, Australia, Canada, the United States, and elsewhere. Under one form of an ITQ, a limited "total allowable catch" arrangement, for example, fishers can trade their quotas, which are similar to property rights within the total allowable catch amount. If Joe is entitled to take eight tons of salmon but time and equipment constraints mean he can catch only four, he can sell his right to the other four tons to another fisher. Fishing can be both limited and made more efficient because owners of ITQs don't need to overinvest in elaborate fishing gear to compete with other vessels, as they would if the resource were open access. Government agencies that regulate the fisheries can adjust the total allowable catch in response to the health of the fish stocks.

There are some problems associated with ITQs, though, such as establishment of a basis for the initial allocation of rights, fraud in reporting, poaching, and a practice known as "high-grading"—discarding less valuable fishes when there are substantial age or sex differences in value. For instance, when a fisher is allowed to keep 100 fishes over a foot long but catches 200 eligible fishes, the fisher discards the smallest 100 of those, most of which die from the handling. Nonetheless, ITQs seem a step in the right direction. We should note, however, that none of the schemes to manage fisheries addresses other important threats to oceanic productivity, such as destruction of coastal wetlands, acidification and heating of oceans, marine pollution, and damage to ocean-floor habitats caused by bottom trawling.

Beyond Reserves and Corridors

But theory moves on, and it has become clear that the reserve approach, for which island biogeography provided such useful background, even with the development of corridors is inadequate to the task of protecting most biodiversity and ecosystem services on a broad scale.

A number of conservation biologists have begun to look again at land outside reserves. In the course of research in Costa Rica, Stanford University ecologist Gretchen Daily, for example, an innovative thinker in the field, realized that many species of tropical "forest" birds were also found in the agricultural countryside of Costa Rica, in an area that forty years before had been essentially continuous forest. That observation led her to develop what she called "countryside biogeography." That subdiscipline of conservation biology attempts to discover which elements of biodiversity can survive outside of reserves, which of them are most important in helping to supply ecosystem services to humanity, and what might be done to make already disturbed areas more hospitable to crucial elements of biodiversity.

Such studies thus broaden the conservation focus to include, besides reserves, agricultural landscapes and other areas already substantially modified by human activities. Daily's research on many groups of organisms in Costa Rica, for example, has demonstrated that if agriculture were not too intensive, farmlands could have considerable conservation value. They could support populations of many organisms and help maintain crucial ecosystem services such as pollination, pest control, and soil preservation. Work led by Taylor Ricketts, now director of conservation science for the World Wildlife Fund, showed that preserving small forest patches adjacent to coffee farms could enhance the value of the coffee crops by tens of thousands of dollars by providing necessary habitat for pollinators, whose activities increased the size and quality of the crop. Research has begun to show that in some circumstances relatively small changes in the configuration of countrysides, at little or no economic cost, or even with economic benefit, could substantially enhance their conservation value and at the same time enhance agricultural productivity and sustainability.

Daily has also made pioneering efforts in other areas with huge potential for improving human sustainability, most importantly in promoting the idea of nature's services, the goods and services supplied to humanity by natural ecosystems. This revolution in conservation, involving a fusion of ecology and economics, traces in large measure to Daily's research and public outreach. Her 1997 book *Nature's Services* brought home to many within the scientific and environmental communities the importance of ecosystem services and helped to make preserving ecosystem services a second major goal of conservation along with con-

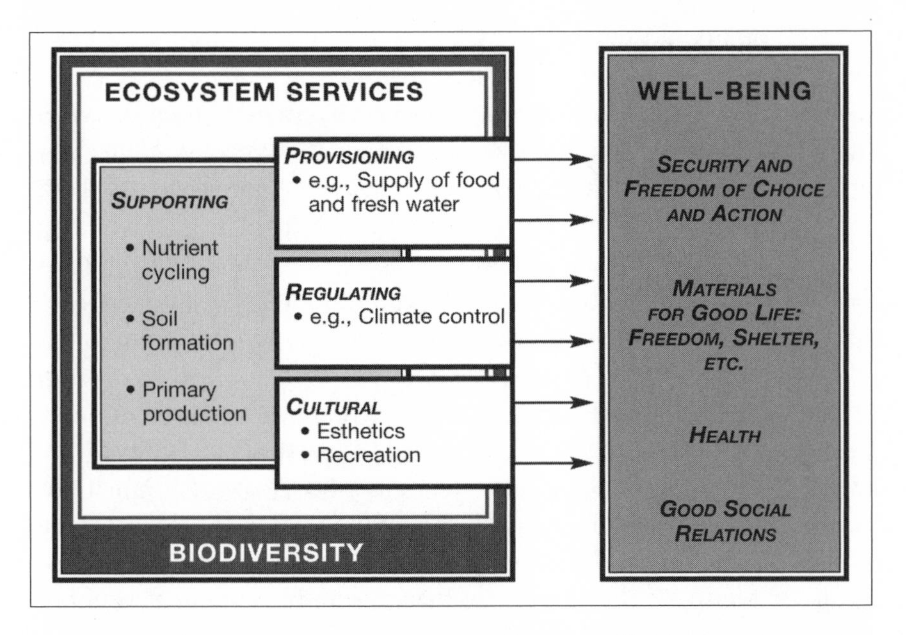

FIGURE 15-1. Links between ecosystem services and human well-being. Adapted from Millennium Ecosystem Assessment, *Ecosystems and Human Well-Being: Synthesis* (Island Press, Washington, DC, 2005), p. 50.

serving biodiversity, while broadening the latter from conserving species diversity alone to conserving population diversity.

A very promising way to preserve ecosystem services, according to the thinking of Daily and her colleagues, would be to align this conservation goal with financial incentives. A well-known approach is spending money to protect the watersheds of cities and their biodiversity through incentives for conservation measures among landowners in the watershed area, a strategy that can turn out to be much cheaper than building water purification plants, as in the New York City case discussed in chapter 10. Similarly, financial incentives can be aligned with conservation by the use of integrated management of crop pests. As discussed in chapter 2, integrated pest management (IPM) features the encouragement of a pest's natural enemies, the planting of mixed crops rather than monocultures, destruction of pest hideaways, and only occasional use of pesticides (as scalpels rather than axes). IPM is not only more ecosystem friendly but also, especially in the medium to long term, often much more cost-effective than the broadcast use of large quantities of pesticides. One outgrowth of the rethinking of conservation has been an ambitious new collaborative effort between universities and environmental NGOs.

The Natural Capital Project was launched by an agreement between Stanford University, The Nature Conservancy, the World Wildlife Fund, and members of the financial community, but the hope is that it will ultimately bring in many other institutions and groups. The goal is to align economic forces with conservation—to develop new scientific methods, new financial instruments, and new corporate and government policies to make preservation of natural capital as conventional as preservation of human-made and human capital is now.

The Natural Capital Project is focusing its initial efforts on four model systems: the Upper Yangtze River basin of China; the Eastern Arc Mountains of Tanzania, in the Afromontane region of eastern Africa that stretches roughly from Ethiopia to Mozambique; Hawai'i, especially the island of Hawai'i (the "Big Island") itself; and the Sierra Nevada of California. In each case, efforts will center on developing ways to incorporate ecosystem service values into decisions regarding land use. Elements of natural capital and services flowing from them will be mapped to motivate the public and policy makers to recognize their value and stimulate people to press for their preservation. Various approaches, including the development of markets for ecosystem services, will be explored to finance the necessary operations. The model systems, and subsidiary ones, will be used as test beds to see whether and how natural capital and ecosystem services can be made a standard feature of community land-use and investment decisions.

Restoration Ecology

A complementary approach to countryside biogeography and the Natural Capital Project involves efforts to restore stretches of degraded lands to relatively natural habitat that can serve as refuges for biodiversity. Instead of simply enhancing the hospitality of intensively used areas, the goal of restoration ecology in essence is to restore habitats that have been seriously degraded or totally destroyed. Candidates for such efforts include areas that have been strip-mined, heavily logged, or carelessly overcultivated or overgrazed and subjected to nutrient depletion, severe soil erosion, or both.

In Guanacaste Province in northwestern Costa Rica, for example, ecologist Daniel Janzen has taken pioneering steps to restore biodiversity in a large area set aside for conservation. For example, he allowed a local orange juice plant to use part of the conservation area as a dumping ground for the huge quantities of orange peel generated. Spread around, the orange peels made a wonderful mulch that sped the return of native plants. Another important project is an attempt to restore part of Brazil's Atlantic rain forest, explicitly forests growing along rivers and streams. That biodiversity-rich forest once stretched over some 400,000 square miles of southeastern Brazil, but only 7 percent of it remains, largely fragmented. Part of the stimulus for the project came from the disappearance, related to deforestation, of once abundant groundwater needed by the area's farmers. Plans include the planting of some 2 billion seedlings of carefully selected rain forest trees and reintroduction of other biodiversity elements to the replanted forests. It's a delicate and complex job, and only time will tell if it can succeed. In late 2007 we visited the Atlantic rain forest, just twenty years after our first trip. Much of its former expanse had been replaced by overgrazed wasteland (figure 15-2), but the avifauna of one of the remaining upland patches, Itatiaia National Park, seemed not to have declined, suggesting that many of Aldo Leopold's "cogs and wheels" of the higher portions of the Atlantic rain forest still exist to be used in restoration.

FIGURE 15-2. Grazed hillside that once was occupied by the biodiversity-rich Brazilian Atlantic rain forest. Photograph by Rick Stanley.

Perhaps the brightest spot in the restoration picture is that considerable progress can be made by agriculturalists with little or no professional help. For instance, in arid and desperately poor Niger on the fringe of the Sahara, farmers themselves seem to have turned the tide of desertification in many areas. Rather than removing saplings from their fields, farmers have protected and nurtured them—and have begun to reap benefits from the greening of the countryside.[4]

The reintroduction and management of endangered animal species is another facet of restoration ecology. It can be very controversial, and such was the case when attempts were first made to reestablish wolf populations in Yellowstone National Park and other areas where cattle graze. The absence of wolves in Yellowstone had severe but at first unrecognized impacts on the local ecosystem. Without the wolves, elk populations boomed and decimated stands of cottonwoods, aspens, and willows. Lacking aspens to make lodges and dams, and willows to eat, the beavers disappeared, and with them went entire wetland ecosystems that the beavers had created. Reintroduction of the wolves a decade ago has allowed the beavers to return, and restoration of the previous ecosystem dynamics has begun. The move was not originally popular with local ranchers (who feared loss of cattle) and hunters (who feared competition), but careful management has lessened the opposition.

As the wolf reintroduction showed, the scientific, social, economic, and political issues involved in virtually all restoration efforts can be daunting. One of the most intellectually challenging issues is determining what the conservation goal should be. Suppose, for example, Americans wished to "restore" the entire Yellowstone ecosystem. Should the goal be to return it to its state in the early twentieth century? Or to its state when Europeans first saw it? Or would returning wolves to the area be enough of a change? And in conjunction with public decisions on what the goal should be, how would restoration ecologists determine what those early states were like?

This is obviously a place where scientific and value judgments overlap. Our own answer would be to restore areas to a state that supplied habitat for as much biodiversity as practically possible, with an eye to maximizing ecosystem services for nearby human settlements. Decisions would necessarily be ad hoc. Others may consider this bar to be too low, and indeed it may be, especially in such places as Yellowstone, which have status as great heritage sites.

The most ambitious move in the direction of restoration has been led by the population biologist and founder of the discipline of conservation biology Michael Soulé and by environmentalist Dave Foreman. They and their colleagues established the Wildlands Project, which can be thought of as a continent-scale corridors project. The goal of the Wildlands Project is "rewilding" much of western North America—creating huge connected areas where wolves, bears, and other large animals can range freely and help maintain ecosystems in undeveloped areas that people value for wilderness and recreation. The Wildlands Project's leaders plan to do this by creating what they call "MegaLinkages." One technique for that is to create wildlife bridges and underpasses ("eco-ducts") across and beneath highways to keep the highways from splitting areas into habitat fragments too small to maintain large animals. Connectivity of areas is the key Wildlands principle—which, of course, could lead to the same sorts of problems and benefits as do corridors between smaller habitat patches.

Notwithstanding those issues, the present lack of connectivity can produce quite general problems for conservation efforts. For example, in the San Diego area of California, patches of chaparral that are too isolated for coyotes to reach have fewer birds than do those with coyotes. The reason is that coyotes suppress foxes, raccoons, and especially cats, which are deadly predators of birds. In such cases, species such as wolves and coyotes can be considered "keystone species," ones whose influence on their communities or ecosystems is disproportionately large relative to their abundance. For that reason, creating conditions that support large keystone species is an important element of the Wildlands campaign.

The Wildlands plan proposes more dramatic and large-scale change than does any other approach to conservation. In the western United States, creation of many of the needed linkages would require that lands be managed differently: establishing conservation easements on private lands, purchasing some private lands for reserves, and so on. It is therefore not surprising that four-wheel-drive and off-road recreationists; "wise use" groups, who have favored opening all government lands to logging, grazing, and mining; and other development-oriented groups have been violently opposed to recreating vast stretches of habitat in parts of the United States. Ironically, some of those same groups now find themselves making common cause with environmentalists in opposing the boom in oil and gas exploration and drilling launched by the George W. Bush administration all over the West. Even so, the Wildlands Project often gets a more welcoming reception in other countries. For such projects really to flourish, we suspect, a social transformation will be required in the way population growth, consumption, and natural capital are viewed by most Americans—indeed by most human beings.

Conserving Natural Capital

The ultimate issue in conserving biodiversity and ecosystem services is dealing with the drivers that make success increasingly difficult to achieve, mainly human population size and growth, overconsumption by the affluent, and the use of environmentally faulty technologies (e.g., commuting in SUVs) and sociopolitical-economic arrangements (city design that encourages automotive travel). No matter how many principles of conservation are developed by environmental scientists, their efforts will be fruitless in the end if those critical drivers are not addressed effectively, rapidly, and with much attention paid to the needs of poor people who depend directly and often heavily on many of the landscapes and fisheries of concern.

Ecological economists have been making progress in finding market-based solutions to some of the problems of protecting living natural capital, which could help these efforts. Among these are the use of tradable fish-harvesting quotas such as those discussed earlier, tradable water allocations in water-short agricultural areas, and cap-and-trade pollution abatement schemes (such as those described in chapter 13), which have been helpful in abating acid rain and thus protecting biodiversity.

Of course, there are many areas in which market solutions are unlikely to be helpful and direct regulation will work better. For instance, it is generally impractical to use markets to control the release of ecosystem-disrupting synthetic chemicals such as polychlorinated biphenyls (PCBs). And frequently, attempts to limit total amounts of pollutants released will be aided by regulatory requirements to employ the best available control technology. That means that a firm can't spew massive amounts of pollution into the environment, regardless of whether it can afford to by purchasing permits to pollute. Furthermore, even market-based systems usually involve command-and-control elements. For instance, governments often directly regulate the length of the harvesting season of a fishery in which quotas are tradable, and protect the piscine capital stock by setting the total size of allowed catch within which the quota market operates, as well as setting up the rules for permit trades in the first place. Similarly, the "cap" part of a cap-and-trade emissions control regime is the government requirement to keep all emissions within the shrinking "cap," even though emitters are free to rearrange their shares of the permitted pollution.

Growing understanding of the value of living natural capital and the ecosystem services that flow from it, and the need to conserve it, culminated in 2005 with the release of the report of the Millennium Ecosystem Assessment, *Ecosystems and Human Well-Being*. The report noted that, since 1950, "humans have changed ecosystems more rapidly and extensively than in any comparable period of time in human history, largely to meet rapidly growing demands for food, fresh water, timber, fiber, and fuel. This has resulted in a substantial and largely irreversible loss in the diversity of life on Earth."[5]

The changes, as we've described in previous chapters, have occurred on an unprecedented scale: nearly a quarter of the world's land surface is now devoted to crops, shifting cultivation, concentrated animal-feeding operations, or freshwater aquaculture; more land has been put under cultivation since 1945 than in the eighteenth and nineteenth centuries combined; some 40 percent of the world's coral reefs have been either destroyed or degraded; and it is estimated that 23 percent of mammals, 12 percent of birds, 25 percent of conifers, and 32 percent of amphibians are threatened with extinction in the coming decades. Since industrial fishing began, the world's fish stocks have been reduced by 90 percent.

The Millennium Ecosystem Assessment received little more attention from the press than did the warning statements from the scientific community a decade earlier, unfortunately. But scientists share some of the blame for the lack of action in protecting biodiversity. Conservation biologists need to find ways to overcome their own conservatism, and so do researchers in more traditional (but key) disciplines such as ecology. Public interest in and support for ecology have increased since people became aware of environmental problems. But unfortunately many ecologists continue to pursue scientifically trivial problems with the excuse that it is "curiosity-driven research," and they also decline to get involved in "applied" problems.

This is a hangover from an outdated attitude that science could be divided into "pure" research (with no immediate application to human problems) and "applied" research (that with obvious application). In the old days, the best science was thought to be pure, and there are indeed innumerable examples of pure science discoveries that later yielded practical applications—nuclear physics is an excellent example, although the value of some of those applications is questionable.

Yet problems of trying to analyze and then avert or moderate the interrelated consequences of human overpopulation, overconsumption by the rich minority, biodiversity loss, global heating, and so on are at least as basic and challenging as solving most apparently "pure" scientific problems.

Walter Reid, leader of the Millennium Ecosystem Assessment's core authors, summarized the current situation well:

> The bottom line . . . is that we are spending Earth's natural capital, putting such strain on the natural functions of Earth that the ability of the planet's ecosystems to sustain future generations can no longer be taken for granted. . . . Clearly, the dual trends of continuing degradation of most ecosystem services and continuing growth in demand for these same services cannot continue. . . . But the assessment shows that over the next 50 years, the risk is not of some global environmental collapse, but rather a risk of many local and regional collapses in particular ecosystem services.[6]

On a more hopeful note, he continued, "We can reverse the degradation of many ecosystem services over the next 50 years, but the changes in policy and practice required are substantial and not currently under way."[7]

CHAPTER 16

Governance: Tackling Unanticipated Consequences

"Behold, my son, with how little wisdom the world is governed."

AXEL GUSTAFSSON OXENSTIERNA,
SWEDISH STATESMAN, 1648[1]

OUR HUNTER-GATHERER ancestors had relatively simple and direct ways of ensuring that their societies would remain viable. Anyone who didn't like the way the group was run could leave and join (or found) another group. Disruptive individuals often were simply executed with informal agreement—a possibility that was still discussed among the Inuit as late as the 1950s. And the idea of moving to find better governance (a society with more tolerance of divergent views and less threat of punishment) was a major feature of European migrations to the Western Hemisphere in centuries past, just as it was in westward migration in North America and, indeed, in some refugee movements in the twentieth and twenty-first centuries.

But the human rise to dominance has been so spectacular and far-reaching that affairs today are far more complex, and civilization's problems can't be solved just by dealing with individuals or migrating as groups in search of opportunities for better governance. Many of the most serious environmental risks menacing our species are already global, from climate change and loss of biodiversity to planetary toxification and the emergence of new diseases.

Americans can't avoid these problems by moving away, any more than can Bangladeshis, though the ways in which these problems affect them may be quite different. Despite the threats these issues pose—or because of them—few among the citizens of the only superpower are acting cooperatively to help solve them, and there's not much more cooperation along these lines elsewhere either. Like most people everywhere, we have yet to realize that our small-group world has faded away as globalization proceeds along unregulated economic pathways. We are global citizens whether we like it or not, and occurrences in the farthest corners of Earth can affect us for good or ill. Within the world community there is, however, no real plan and little organized capacity as yet to deal with the economic, social, and political issues generated by a global society—to say nothing of the severe environmental problems it has engendered. And the problems of formulating policies to make this complex system sustainable have yet to be recognized even by many decision makers.[2] In short, despite our dominance, we may have evolved ourselves into quite a mess.

The situation is made more challenging because a formidable obstacle in the way of actions national governments need to take is globalization itself. Action by a single nation alone usually can have only a small effect on a global dilemma such as oceanic fisheries depletion, so it is difficult for politicians or the general public within each nation to justify sacrifice or restraint. Exercising such restraint unilaterally can lead to what economists call a "free rider" problem—it enables a person or other entity to enjoy the benefits of another's action without paying a fair share. If some nations curtail their marine harvests, there will be more fishes for the free riders—at least in the short term. If the free riders still will not exercise the necessary restraint, allowing natural reproduction to restore declining stocks to everyone's benefit, then overexploitation will destroy the fisheries, which are common property resources.

The tendentious issue of what is a fair share often delays action while parties argue over who should do what—something exemplified by the haggling over whether the United States should ratify the 1997 Kyoto Protocol on reducing greenhouse gas emissions. The main bone of contention was whether the United States would be paying more than its "fair" share, compared with some high-polluting developing nations that had not been responsible for much past greenhouse gas release but would soon become major emitters. That those developing countries were also signatories to the agreement, under which they were committed to start reducing their own greenhouse gas emissions under an anticipated new arrangement after the Kyoto Protocol expires in 2012, was apparently not understood by the American public and many members of Congress.

Globalization

What exactly is globalization? It can be thought of as increasing economic, social, informational, and environmental integration transcending national boundaries. Economists see it as a process several centuries old, which accelerated greatly after World War II and especially since 1990 and the end of the cold war.

Today the process of international integration is knitting the major economic features of human society into a single global system, superseding the millennia-old centering of economies in state and national units. Although the volume of trade relative to the scale of economies is not that much greater now than it was a century ago, globalization today features nearly instantaneous movement of capital and information without regard for national boundaries.

Much of this activity is carried out by giant transnational corporations, often independently of any government oversight. When an American buys a car, it may well have been manufactured in Japan using parts made in several other countries; in winter, people in Europe and North America eat fresh fruit grown in the Southern Hemisphere; someone in Argentina seeking help with a crashed computer might have her phone call directed to a technical representative in Bangalore, India.

The benefits to consumers of much of this activity are obvious. But the results may include largely unforeseen disadvantages to workers and farmers because globalization also promotes the international migration of resources, jobs, and labor and often results in the undercutting of prices of indigenous manufactured and agricultural products as well. Moreover, globalization tends to accelerate the process of submerging small, often less technologically advanced cultures under large, dominant cultures—American, European, Chinese, and so forth. Wherever one goes in the world, Coke is available, and so in most areas are Big Macs, tacos, and pot stickers. Contaminated pet food and lead paint on toys imported from China kills cats and threatens children in North

America; Latin American parents give their kids names they learned on U.S. television; and American blue jeans are popular around the world.

A major component of globalization is greatly enhanced communication, as exemplified by the World Wide Web, without which global trade would be severely hindered; the rapidity of information and funds transfers, though, makes control over trade very difficult. At the same time it has become increasingly hard, though not impossible, for any society to remain very isolated from the world or for a dictator to control a population by limiting what people can learn about the outside world or their own government.

Some economists see globalization as an unalloyed good because international markets increase efficiency by encouraging each nation to do what it can do best, and that helps the gross global product to grow. Indeed, global trade has become more and more necessary as the human population expanded beyond a very few billion and as societies have industrialized, simply because no country possessed all the resources or agricultural productivity needed to support very large populations, especially with a high standard of living. The United States came closer than most, but in the past several decades it has become increasingly dependent on foreign sources for important resources, notably petroleum.

Ecologists generally are not so enthusiastic about economic globalization because it promotes globalization of environmental problems as human dominance imposes profound and far-reaching changes on Earth's land surface, atmosphere, and oceans. Actions carried out within national boundaries thus are now often of global concern. For example, pollution from the hundreds of coal-fired power plants that China is building will augment global warming, and that in turn may contribute to the problems of citizens of New Orleans, Florida, and Central America by intensifying hurricanes and speeding sea level rise. Furthermore, globalization, with its transport of food and other products around the world, considerably increases the energy costs of transportation and, of course, the flow of greenhouse gases into the environment and the movement of invasive species.

The World Trade Organization (WTO), established in 1994 to oversee trade agreements, enforce trade rules, and settle disputes, has the power to override national laws that may impinge on trade, including those designed to curb greenhouse gas emissions and provide other environmental safeguards, which can hinder solutions. For example, Japan's attempt to adopt restrictive fuel efficiency standards in its own country was opposed within the WTO by the United States, which saw this as a move to restrict imports of American gas guzzlers. Agricultural subsidies to farmers in the United States and some other rich countries that export foods have also been a bone of contention in the WTO.

Many social scientists and other knowledgeable observers are also unenthusiastic about globalization because they think it creates more problems for the world's poor. The WTO's ineffectiveness at reducing agricultural subsidies in rich countries is a case in point. The subsidies cause overproduction and export dumping—and reducing the prices of exported grains has the effect of undercutting the price of the foodstuffs produced by poor farmers in developing countries.

Such connections are already beginning to affect the everyday lives of millions of people. For instance, there's the cashmere-cough connection. Cashmere sweaters are now very cheap in the United States, thanks to a flood of inexpensive ones from China. One element of China's economic boom has been the construction of many cashmere factories and an accompanying expansion of goat herds to supply the key natural resource. That resource is special hair that must be combed from under a goat's coarse "guard hairs," pliable fibers that are spun into a wonderfully soft yarn that is made into sweaters.

But gigantic herds of goats, in the service of the cashmere industry, are now desertifying the grasslands of the harsh, dry Alashan Plateau on the Mongolian border of China. A consequence is the generation of huge dust storms, with plumes of dust big enough to reach the United States. One Chinese storm reduced visibility over Colorado enough to make the national news. Other Chinese-generated pollutants have been swept across the Pacific Ocean in sufficient amounts to exceed the U.S. limits for air to be healthy to breathe.[3] The impacts of dust storms within China itself and in Korea and Japan are, as you might expect, much more severe. The cashmere-cough connection traces to Mao Tse-tung's push for migrants to open up China's western frontier, including the Alashan Plateau, to provide resources for its burgeoning population. Not enough consideration was given by Mao to the predictable effects on the region's marginal environment, so deforestation and overgrazing have led to today's growing calamity.

Mao badly needed education in environmental sciences, not least in agricultural ecology, as do most of today's political leaders. With a basic understanding of how agriculture, environment, and human culture are intertwined, more politicians might even learn to care about the fates of people's traditional ways of life—in this case, pastoral nomads on the Alashan who had grazed sheep, cattle, goats, and camels sustainably since the days of Genghis Khan. The nomads were forced to settle in villages and are now living in an area that is being desertified, with the grass largely gone and the lakes and streams dried up.

So, on balance, with serious problems for some people in China and some adverse environmental effects far beyond its borders, while providing benefits for others in China and cheaper, more abundant cashmere globally, is the growth of the cashmere industry a good thing? Like most issues in globalization, there is no clear answer. The actual distribution of economic benefits and the impacts of the movement of jobs in globalization present a complex picture in which people in some areas have done well and others badly. Overall, though, the gap between the incomes of the richest and the poorest countries has widened in recent decades, and some poor nations, especially those in sub-Saharan Africa, have become more impoverished, as even the pro-trade International Monetary Fund recognizes.[4]

The environmental effects are complicated too. In Latin America, as in China, pollution—air pollution, declining water quality, toxic wastes, and so on—has generally increased with development and globalization. This has occurred largely because of a failure of government oversight. Such failure is usually seen as a lack of "institutional capacity," a grab-bag term for governance problems in poor nations—an inability to manage economic policy, establish and enforce environmental regulations, maintain law and order, provide public services, resist the investment intrusions of aggressive international corporations, and so on. In China, the institutional capacity is no doubt greater than it is in sub-Saharan Africa or much of Latin America, but so is the magnitude of the problems.

Very poor nations in a globalizing world generally have a comparative advantage in industries that specialize in extracting and exporting natural resources or in mass-producing goods. These industries include timber, mining, and certain types of manufacturing, all of which can have serious, even devastating, environmental consequences. When these industries concentrate in developing countries, the industrialized world has essentially "outsourced" its pollution and environmental damage by purchasing the products of resource extraction and manufacturing from poor countries. Globalization undeniably has increased the economic standards of living in much of the developing world in the past few decades, even if the degree of inequality between rich and poor has increased. While the *number* of people living in poverty (defined as those with incomes of US$2.00 per day or less) remains extremely high, as we've seen, the *proportion* of the very poor has significantly fallen, and the numbers of people who have attained some level of comparative affluence—the "new consumers"—are growing rapidly, especially in Asia. That "new consumers" soon will be driving more than 200 million of the world's some 800 million cars, and that they already represent the fastest-growing market for cars, shows the handwriting on the wall.

Part of the reason that new consumers are aspiring to Western lifestyles can be seen in the power of the global electronic media, whether American CNN, British ITV or BBC, Arab Al Jazeera, or the Internet. These institutions seem to have made the force of ideologies and aspirations ever more far-reaching, whether for good or ill. The American all-consuming lifestyle as presented in living color around the world has certainly, and deliberately, increased dangerous consumption,

FIGURE 16-1. U.S. combat troops in Iraq, engaged in an attempt to secure long-term access to that nation's massive petroleum reserves. In an age of increasing demand for fossil fuels and less easily found supplies, resource wars are likely to become more common. Photograph courtesy of iStockphoto.

while offending the religious sensibilities of some devout Muslims, among others. And opposition to the Vietnam and Iraq wars in the United States and elsewhere was certainly galvanized by exposure to bloody on-the-spot television coverage.

With its growth in consumption and population, humanity is continuing to outstrip its resources and destroy its natural capital on a world scale, if anything at faster rates than ever. Why has so little been done around the world thus far to address these issues and the inequalities related to them? There are, of course, many reasons, but among the central ones, in addition to simple ignorance, are discounting by distance or time (not taking seriously the consequences of events that are perceived as far away or far in the future), outdated institutions, belief that there are panaceas that will solve problems (rather than the "wedges" approach discussed in chapter 13), and maldistributed power within societies. Moreover, established ideologies—the prevailing strong cultural beliefs of individuals, groups, and government officials about what is good, true, and important—frequently conflict with potential solutions to environmental problems. For most of history the impact of ideologies was largely local or continent-wide, but now ideologies, especially those guiding major powers, have great global significance, as many of them have been widely adopted in other nations.

Problems in Governance

Let us explore a few aspects of those stumbling blocks to action on environmental issues, with a primary focus on problems of governance in nation-states, exemplified by the United States, still the dominant power in the world today. A failure of the American government to adapt to modern conditions has, in our view, helped make it ineffectual in dealing with the most serious problems facing the nation and the world, the most important of which are environmental problems broadly defined. Unfortunately, the government's structure is highly vulnerable to influence by monied and other powerful special interests. It is the responsibility of leadership in the executive branch and Congress (in addition, of course, to an informed citizenry) to recognize and respond to the serious threats such interests represent. This has been done in the past, for instance in the Jacksonian democracy movement, which expanded participation in government before the Civil War; the

Populist Revolt at the end of the nineteenth century, which strengthened farm interests politically against industry; and Franklin Roosevelt's New Deal in the Great Depression years before World War II. But no strong counterforce to the entrenched power of money seems on the horizon today.

Monied interests have clearly played a role in the problems that beset government in the United States, especially in the ignoring or downplaying of environmental issues in favor of immediate profit. On a broad scale in the United States a well-financed, fundamentally right-wing, campaign has in recent years spread disinformation on global heating. It has been very successful at asserting either that climate change is not under way or, if it is, that it is by and large a "natural" phenomenon and does not present serious problems. This ideological effort, amplified by the media,[5] not only has influenced the center of the political spectrum but has even misled environmentally naïve commentators on the far left[6] and had some impact outside the United States. At the state level, for example, in 2006 the Chevron Corporation and other oil companies spent more than $95 million to fund a misinformation campaign and defeat a California initiative (Proposition 87) for a state tax on oil production revenues that would have been used to support renewable energy programs. The advertisements included untrue claims that the initiative would raise gasoline prices and increase dependence on foreign oil, although the proposed tax would in fact simply have been similar to those already in place in all the other oil-producing states.

Powerful coal and oil interests have made substantial contributions to the election campaigns of key congressional and presidential candidates, as well as providing other inducements, to produce energy legislation to benefit their industries rather than society as a whole. A major factor giving industries leverage in these (and other) areas is the need of politicians, especially in the U.S. House of Representatives (because of the frequency of their elections), to raise huge sums of money to pay for election campaigns.

Millions of dollars spent by fossil fuel industries, along with automakers, for nearly two decades blocked attempts to reduce the use of fossil fuels, such as by raising the CAFE (corporate average fleet efficiency) mileage standards for cars and light trucks (including SUVs) or by increasing funding for research and development of alternative energy sources. Auto companies in the United States profited from these and other campaigns during the 1990s, but more recently the car-buying public began turning to more efficient cars made by companies overseas, pulling the financial rug out from under Detroit.

How might our system of governance be improved to enhance the ability of the United States to deal with the issues the nation and the world now face? One obvious measure would be to make legislators less susceptible to self-interested pressures by removing most or all private funding from elections. It might require an amendment to the U.S. Constitution to allow Congress to limit campaign spending, since the Supreme Court has declared that what in essence amounts to buying votes is a form of political free speech and therefore constitutionally protected. Campaign finance reform and greater use of the Internet—far less costly than television advertising—could shift freedom of political speech away from dollars and back toward people. It will be difficult to accomplish, as the persistence and success of the industry-sponsored campaign of disinformation about climate change has shown. But with determined public support, some measure of balance could be restored.

To get environmental action or to pass legislation, it isn't necessary to convince every citizen in a democracy to support a particular program. In the United States and other democratic nations, to enact a policy change it is essential to generate a sizable "issue public," a critical mass of people who care deeply about an area of public policy and are willing to take action (donate money, lobby, vote, protest, etc.) to support their

TABLE 16-1 **Issue Publics in the United States**

Issue	Percent
Abortion	31%
Government social service programs	21%
Global warming	18%
Gun control	17%
U.S. military spending	16%
Capital punishment	14%
Women's rights	12%
Race relations	10%
Unemployment	9%

Sources: National election studies and general social surveys. Numbers are from various years since 1980 (surveys are not done regularly). Global warming number is from a survey conducted jointly by Stanford University, *ABC News*, and the *Washington Post* in 2007.

preferences. One composite list of leading issue publics in the United States as of 2006 and 2007 is shown in table 16-1; sadly, though climate change registers quite high on the list, many of the most pressing environmental problems (loss of biodiversity, toxification of the planet, and the like) do not now have substantial "publics."

Psychologist Jon Krosnick, who gave us the data in the table, believes that the key elements of getting action on an issue are convincing people that (1) a problem exists, (2) it will be bad for people if it isn't addressed, (3) there is certainty that people caused it, and (4) the problem can be solved. By 2007 with regard to climate change, most Americans were already quite convinced of the first three of these elements, but some confusion remained about the fourth—not surprisingly, since a multitude of solutions will be needed over the next several decades to reduce greenhouse gas emissions substantially.

Public concern and the efforts of scientists and non-governmental organizations have produced countervailing forces to corporate dollars, and an explosion of action on climate change has already begun at state and local levels. For example, California, long the most active state in environmental protection, has enacted several measures designed to reduce carbon dioxide (CO_2) emissions, including the Pavley Act, which mandates a substantial reduction in those emissions by automobiles and light trucks in the state beginning in the 2009 model year, escalating to a 30 percent reduction by 2016. By the end of 2007 some twelve other states had adopted California's standards, and four others were poised to join as well. Similarly, California governor Arnold Schwarzenegger signed an executive order calling for the state to achieve an 80 percent reduction in emissions below 1990 levels by 2050.

In December, after Congress passed a somewhat weaker federal CAFE standards law, the U.S. Environmental Protection Agency rejected California's petition to enforce its stricter standards. California and the other states, along with several environmental organizations, promptly filed suit against the EPA; but even if the suit fails, a new administration in 2009 might well support California's case.

A parallel trend to state-level action has been adoption of the U.S. Conference of Mayors Climate Protection Agreement, in which mayors of more than 700 U.S. cities, large and small, have pledged to meet the Kyoto Protocol target for U.S. reductions of CO_2 emissions, 7 percent below the 1990 level, by 2012. Given that U.S. emissions by 2007 were some 16 percent higher than in 1990, the task will be far from easy. Nonetheless, following the mayors' lead, in 2007 dozens of county governments undertook a similar pledge to reduce emissions. How successful these actions will be remains to be seen, but it is a serious start. It also illustrates a process that has succeeded historically many times: state and local governments functioning as laboratories for innovative lawmaking. Eventually, however, new needs and ideas must find a hearing in Washington for effective action on a larger scale, and not just on emissions reduction but on a whole range of issues, from agricultural reform to pollution control and protection of living natural capital.

At the national level, in recent decades power and influence in United States government have shifted from the legislative to the executive branch. Thus the government has become a less democratic or, more technically, a less republican (small "r") representative government. The separation of powers—the system of independent legislative, judicial, and executive branches of government—so precious and so important to the framers of the Constitution, has been weakened. It has headed the United States toward a greater concentration of power in the hands of the few: possibly the same sort of course that occurred when Julius Caesar crossed the Rubicon and the Roman Empire moved from governance by a republic to dictatorship.[7] Some might claim that it would be easier for oligarchies, dictatorships, or other forms of government with concentrated power to deal with the environmental crisis, but it would be unusual (to say the very least) for concentrated power in a gigantic population to work for the public good no matter what its rhetoric—something dramatically demonstrated by the lack of citizens' freedom and the environmental degradation that were hallmarks of the Soviet Union.

Is there a way that the United States can avoid such increasing concentration of power and allow more environmental issue publics to be heard in the (at least temporarily) dominant society of the dominant animal? In addition to whatever mechanisms might be invented to effect serious campaign finance reform for all national elections, other critical structural or procedural reforms that in our view are needed to restore congressional power and broaden representation on environmentally related issues include the following:

- Stop gerrymandering—the redrawing of electoral districts into weird shapes to favor one party or the other.

- Require more disclosure of congressional activities.
- Delegate more congressional authority to isolate some key environmental activities from immediate reelection concerns.
- Strengthen federal support for scientific research and education.
- Strengthen the independence of some agencies, such as the U.S. Food and Drug Administration and the Environmental Protection Agency.

Congress already has budgetary and other powers vested in it by the Constitution and with a little courage could press the executive branch to act on some key issues, such as developing an agenda for achieving environmental sustainability in the years ahead. To allow time for that and to increase global security, Congress should insist in the short term that the executive branch work with Russia on what may be the most crucial environmental problem of all—the threat of a humanly and ecologically catastrophic nuclear war. There is a pressing need to reduce the number of nuclear missiles on both sides and to take both Russia's weapons and ours off short-reaction-time alert. At present, having thousands of warheads at the ready in a situation in which an error could unleash the end of civilization is a constant but preposterously unnecessary danger that could end human dominance.

Corporations and Wealth

Corporations have played a large and growing role in a globalizing and environmentally deteriorating world. Many companies outstrip entire nations in sheer size—roughly half of Earth's top economies are not nations but corporations. The General Motors Corporation is roughly as big in terms of its income as New Zealand, Hungary, and Ireland combined. Royal Dutch Shell's revenues exceed Venezuela's gross domestic product, and the revenues of Wal-Mart Stores, Inc. are more than those of Indonesia. The overwhelming majority of major corporations are based in rich countries and of course use their power in various ways to advance what they see as their own interests.

Corporations as individuals are "legal fictions," discrete legal entities separate from the stockholders who own them. The owners can change while the corporation continues to exist, and the liability of the owners for the actions of the corporation is limited. Corporations have been given or claim many rights that accrue to actual individuals, and the law regards them similarly in many respects. Corporations can own property, go into debt, enter contracts, sue and be sued—legal endowments that help formalize and organize sophisticated economic activities found throughout the modern world.

Sometimes, however, corporations assert rights granted to citizens, such as freedom of speech, that many on the political left think should never be given to corporations. Others, ideologically committed to the notion that unrestrained capitalism is the best and fairest way to run the world, claim that all the privileges of free speech (such as distorting what's known about climate change to promote a corporation's self-interest) are perfectly acceptable. We're on the restraint side: we think giant companies, many of which operate in dozens of nations, should have much restricted rights in relation to the political system. In particular, election reform should deny them the right to "vote" with corporate dollars funneled one way or another to candidates, and severe penalties should be placed on the use of those dollars to buy votes in Congress or offer favors to government officials, such as travel in corporate jets or lodging in fancy resorts.

The power modern corporations exercise in shaping legislation and administrative regulation (environmental and other sorts), as well as influencing international trade agreements, is legendary. Most large corporations are generally law-abiding. That is what one might expect, since in countries such as the United States, corporations as a group more or less make the laws through their lobbying and financial support of legislators. But as they operate more and more outside of industrialized nations and in poor countries, they can evade taxes and environmental, employment, and fair trade regulations, as well as other legal requirements that they can't block or evade in their home countries. They are thus "individuals" that can simultaneously enjoy the benefits of a secure base and commit acts abroad that would be punished at home, acts that are either legal or not prosecuted in the foreign nation.

Corporations can't really have ethics, no matter how many codes they write; they have no theory of mind and no possibility of being empathic. Their primary goal (and responsibility, executives assert) is to make profits. Corporations can, however, have social responsibilities, not just to stockholders but also to their national and international employees and the communities and environments they affect.

Charging corporations with responsibilities and developing ways to enforce their compliance can be legislated. One of the most successful ethics laws ever passed in the United States was the Foreign Corrupt Practices Act, which outlaws bribery in foreign transactions. Many in Congress at the time argued that, if passed, the FCPA would disadvantage U.S. corporations, but the corporations themselves quietly supported passage of the law, which would shield them from bribery competition. Congress did pass it, and later defended it against an effort to have it repealed. More than a success, the FCPA has become something of an ethical gold standard globally, and similar legislation has been slowly adopted by many trading partners of the United States. Of course, there are many ways in which legislation can help align corporate interests with environmental ones, through regulations and incentives that level playing fields so that corporations can profit (or at least not lose) by being good environmental citizens.

Large corporations also have played a major role in increasing income inequity in the United States. In 1960 the average salary of the top 100 corporate executives was 30 times the income of the average worker; today it is 1,000 times larger. This huge disparity exists almost exclusively in the United States; European and Japanese chief executive officers have much less flamboyant salaries. Recent research has indicated a global relationship between income inequity and environmental deterioration.[8] The cause is unknown, but one possibility that might be examined is whether individuals who have accumulated money (and power) can insulate themselves from many of the more immediate noxious effects of environmental deterioration. That being so, they are less likely to promote or even support measures that clean up the environment for most folks. In addition, the ability of people to take corrective action can simultaneously be compromised by wealth distribution heavily skewed to the rich. The rich can buy protection of their operations politically, no matter how environmentally destructive they may be. As former Supreme Court justice Louis Brandeis said, "You can have wealth concentrated in the hands of a few, or democracy, but you cannot have both."[9]

Discount Rates and Concern about the Future

The value one gives to the future as opposed to the present helps to explain a common feature attending government inaction and corporate and individual decisions. Since sustainability, even for a society of dominant animals, is a matter of considering the future when taking actions in the present, we need to ask, How should the wealth (or environmental well-being) of future generations be considered in human actions today?

How we view the future is a question that has been of substantial interest to some economists, who try to consider future generations in decisions made today about the costs and benefits of projects. Predicting the future is more than an interest of environmental scientists who fear for the safety of our life-support systems; it is an obsession. Will the human population be far smaller than today's level in 2030 because of decimation by a novel plague or nuclear war? Will there be enough pollinators left to service crops adequately in 2050? Will southern Florida be submerged beneath the sea in 2100? More generally, what will happen if population after population of organisms involved in supplying ecosystem services and goods go extinct?

Human beings, some other animals, and financial markets (or at least the people involved in them) tend to discount the future—to care less about what will happen later than what happens now. For example, most people would value $1,000 collected today as worth much more than $1,000 (adjusted for inflation) to be collected ten years from now. This makes sense from the standpoint of an individual: you might get hit by a truck (or win the lottery) in that decade, or the opportunity to get any money might dissolve before then. Most important, the $1,000 invested today might yield $1,700 in ten years. Moreover, most of us are impatient: you probably would rather have some new clothes to wear to the party tonight than to a party next year. This "time value of money" helps to explain interest rates, which are supposed to compensate you as the lender for the chance that the promised money will not materialize and for your not being able to use that capital in the interim. Junk bonds pay high interest rates because the risk of losing your money is relatively high; government bonds pay little because they're very low-risk.

Interest rates can thus be seen as a measure of how lenders (be they loan sharks, pawnshop owners, banks, or purchasers of bonds) are discounting future consumption—how much they want to be paid to compensate for giving up the use of their money in the face of an uncertain future instead of spending it in the present. High discount rates (a form of interest rate) indicate uncertainty and a high perceived risk. In a sense, they indicate a lower value placed on the future, assuming it to be very different from the present (e.g., society on the whole will be much richer or poorer, technology will be able to solve problems that today seem insoluble, you will be dead rather than alive). Eight hundred years ago, Europeans assumed that whatever the vicissitudes of everyday life, the world was essentially unchanging. They were willing to invest time and money in building cathedrals that would not be finished in their children's lifetimes. In this respect, they did not discount the future much. Imagine the interest rate a lender would need to offer today for a payoff a century in the future! We live in what might be thought of as a high-discount-rate world.

By tradition, economists tend to think of individual discounting (by real individuals or corporations) as being done at positive rates. The discount rate applied to a potential investment is the market rate of return the individual could have earned by alternatively putting the money in an investment offering a safe return—for example, putting the funds in a money market account. In making investment decisions, an investor compares the rate of return on an investment she is considering with the rate of return she could have earned from keeping the money in the money market (or the bank). This is mathematically equivalent to determining whether the *discounted* value of the stream of returns expected from the investment exceeds the initial amount she was thinking of investing. Thus a landowner with a forest growing in a watershed, for example, will compare the income she could get from putting money into maintaining the forest and allowing it to continue to grow, with the return from clearing the forest now and putting the money in a safe investment.

There would be beneficial externalities associated with the intact forest. They might include reduction of downstream flooding, less erosion and silting of the river, maintenance of fishers' catches and wildlife habitat, sequestration of atmospheric CO_2, amelioration of climate change for future generations, and so on. Ignoring these externalities from forest preservation leads the landowner to underestimate the social return from the forest and cut it prematurely (from the standpoint of society as a whole). There are two problems in this example: first, wrong timing decisions attributable to not considering the externalities (but not the discount rate), and second, inappropriate discounting—attributable to the fact that the standard discount rate (the market rate) doesn't give sufficient weight to future generations. That is, the individual discount rate, which underlies the market rate, is too high from an ethical point of view—people are not sufficiently concerned about future generations.

Philosophers and economists distinguish between the market interest rate and a social discount rate. The latter incorporates ethical issues not captured by the market rate. It is generally agreed that the market rate is higher than the social discount rate. This implies that, in making these investments, individuals will discount the future too heavily by common ethical standards, giving too little weight to future generations. The market rate is often taken to be the rate of interest on government bonds. This rate might range from 3 to 5 percent. Suppose a rate of 4 percent is used to determine how much society should invest today to avoid a dollar's worth of costs from climate change a hundred years from now. How much should it invest today? The answer is under 3 cents—paying more would be throwing money away because 3 cents invested at 4 percent would yield more than a dollar in a hundred years. Employing such a high discount rate in an analysis sends the message that it would be sensible not to incur the costs of transitioning to smaller, energy-efficient automobiles to reduce the flux of greenhouse gases, and to continue to enjoy the use of SUVs in city driving. In contrast, at a social discount rate of, say, 1 percent, it would be worth spending some 35 cents to avert a dollar's worth of damage a century hence—a very different situation.

High discount rates tend to be used by politicians in democracies because the time horizon of their tenure is usually less than a decade. They seldom undertake social actions, such as adopting policies to discourage people from having large families, whose benefits won't be evident before the next election. Politicians can thus be said to discount the future at a high rate—after all, what did posterity ever do for them?

The urge to postpone needed action is often seen, and much foolishness sometimes attends discussions of discount rates. For example, Bill Gates has argued that we shouldn't invest heavily in curbing greenhouse gases today because in thirty years much greater technological prowess will be available and everyone will be much richer and able to pay for the technology. But how is the technology to be developed without investing in it, and how likely are technological fixes to be cure-alls? What about the choice between preventing now or dealing later with costly problems such as aquifers invaded by seawater, loss of agricultural productivity, and rising epidemics?

Low social discount rates, in contrast, favor actions that are paid for now and provide benefits to future generations; steps to reduce the flux of greenhouse gases into the atmosphere and support of family planning programs generally fit that description. But low discount rates are not uniformly good for the environment—it depends on how environmentally sensitive the prospective investment is.

High social discount rates, in our view, derive in part from unwarranted assumptions about future wealth—ones that even eminent economists are susceptible to. If future generations will be very wealthy compared with people today, then a high discount rate makes ethical sense: push the costs off on those who can easily afford to pay them, and take care of people in need now. But what if, as seems more likely, future generations will be no wealthier than today, and perhaps even poorer? In those circumstances, low or even negative discount rates would be ethically justified from an economic viewpoint, as we would be paying today to reduce the burden on people more in need tomorrow. Whether effective social mechanisms can be created to encourage such behavior is an open question.

But it would seem sensible to try very hard in the case of climate disruption. First, huge costs are likely to be associated with global heating, dwarfing those we have already begun incurring: replacement of infrastructure inundated on coastlines; rearrangement of aqueducts and irrigation systems to track changes in precipitation patterns; increased costs of food after floods and droughts; disruption of various ecosystem services; and so on. These costs could increasingly require a diversion of resources from consumption and lead to lower standards of living. And that's if we're lucky and climate change doesn't induce widespread famines or epidemics or trigger disastrous wars over water and food. So the prospect of disruptive climate change itself casts grave doubt on the assumption that future generations will be better off than we are. So do many other environmental trends.

The economist Partha Dasgupta likes to point out that negative discount rates are too little considered, especially since their use would imply that society was taking out insurance to buffer future generations from nasty consequences. Recommending such negative rates is an example of applying (in economic jargon) the so-called precautionary principle. That principle is: in the absence of strong evidence that serious damage will not ensue from a given proposed action that scientists say entails heavy risks, the burden of demonstrating its safety rests on the action's promoters.

Foresight Capability

Real solutions to the dilemmas humanity faces demand that we place substantial focus on the world to be occupied by our grandchildren. Oddly, the United States government currently has almost no institutionalized foresight capability that would allow it to detect slowly developing multidimensional problems and address them in any way—other than by the use of threats or of military force—before they get out of control.

This was not always so; an attempt at developing such foresight was launched in 1977 by the Jimmy Carter administration. In 1980 the resulting *Global 2000 Report to the President* was published jointly by the Council on Environmental Quality (CEQ) and the U.S. Department of State. The study combined projections from a dozen agencies in the U.S. administration, from the Census Bureau and the Environmental Protection Agency (EPA) to the Agriculture, Interior, Defense, and Energy departments—groups that normally had little or no contact with one another—to depict the likely state of the world in 2000. The world in 2000, the report predicted, would be "more crowded, more polluted, less stable ecologically, and more vulnerable to disruption than the world we live in now"—an outlook that has proven generally accurate, although a few trends deviated in some respects from the projections. Among key findings was that the independent assumptions and expectations of the various departments often led them to optimistic conclusions, whereas the combined projections showed that sufficient resources to fulfill the expectations of all the agencies—capital, energy, water, and land—might well fall short.

The CEQ at the time indicated its intention to develop an ongoing foresight capability, but after President Carter lost his reelection bid in 1980, incoming president Ronald Reagan had all remaining copies of the three-volume report destroyed. Administrations since 1980 have progressively downgraded the role of the CEQ, have eliminated its once extremely valuable annual *Environmental Quality* report to Congress, and have shown little or no interest in developing a foresight capability. President Bill Clinton did little to revive the CEQ, but he did appoint the President's Council on Sustainable Development, which sought input from a wide variety of experts outside the government and worked toward a consensus until 1997, when political concerns led to its being disbanded.

Foresight capability in the United States has been further reduced by a failure of government leaders, especially in the George W. Bush–Dick Cheney years, to pay much if any attention to the advice available from the scientific community. Between 2000 and 2006, both the Republican-dominated Congress and the Bush administration increasingly rejected scientific findings as guidance in setting policy on a multiplicity of issues. These ranged from stem cell research and drug safety to management of endangered species and the problems of oil dependence and global heating. Indeed, even scientists in federal agencies have frequently been discouraged by the Bush administration from sharing their views with Congress or the public, while Congress has often ignored what expert advice it has sought when that advice didn't fit its prejudices. Years ago it abolished the Office of Technology Assessment, whose crisp reports were a source of unbiased information and analysis that benefited not only members of Congress but also the public at large.[10]

Maintenance of any form of democracy in the United States or other nations (as well as the proper functioning of markets) depends upon citizen access to as much information as possible. Transparency may, indeed, be the main hope for our small-group species to govern itself successfully. President James Madison put it well in 1822: "A popular government without popular information, or the means of capturing it, is but a prologue to a farce or a tragedy, or perhaps both. Knowledge will forever govern ignorance, and a people who mean to be their own governors must arm themselves with the power which knowledge gives."[11]

Foresight capability is limited at the international level as well. Cultural evolution in a few thousand years has transformed *Homo sapiens* from an animal living in groups of a few hundred individuals at most, greatly subject to environmental vagaries, to a global community of billions dominating Earth's biosphere. But the glacial pace of cultural change in issues of human relations has yielded only a single truly global governance body, and one that is none too powerful at that. The United Nations and related international entities have no commanding capacity for dealing directly with world-scale environmental problems. The UN's environmental agency, the United Nations Environment Programme (UNEP), was established in 1972 in Nairobi, Kenya, with the expectation that, being in such an out-of-the-way place, it couldn't be very effective—an expectation that has been largely fulfilled.

Nevertheless, the UN does have the ability to raise global awareness and help focus concern about major environmental and social issues through its periodic world conferences and publications on the environment, population, development, and women's rights. Examples include the work of the World Health Organization (WHO) to raise awareness of the threats posed by malaria, a flu pandemic, and HIV / AIDS. Particularly notable, of course, have been the efforts of the Nobel Peace Prize–winning Intergovernmental Panel on Climate Change (IPCC), under the auspices of the World Meteorological Organization, which raised public consciousness about climate change and brought the international community to the realization that something must be done about global heating. Another achievement is the Millennium Ecosystem Assessment, called for by UN secretary-general Kofi Annan in 2000 and coordinated by UNEP, which produced a massive evaluation of the condition of the biosphere and brought it to public attention in 2005.

Conflict over Resources

Wide international discussion of the sorts of issues considered in these UN-related efforts is badly needed, since they, of course, have already begun to cause tensions in various areas of the world. An increasing problem in international relations on our hugely populated and resource-stressed planet is likely to be an acceleration in conflicts over resources. Most people naturally think of oil as a prime source of conflict, and so it is. Oil played a central role in World Wars I and II, and interest in having a client state in the oil-rich Middle East has been a key underpinning of United States support for Israel. Oil, and possible threats to its continued flow through the Strait of Hormuz, the only waterway to the open sea for large parts of the oil-exporting Persian Gulf states, was the prime underlying cause of the United States' invasion of Iraq and the Bush administration's plan to establish permanent military bases in the region. As former Federal Reserve chairman Alan Greenspan, a Republican, put it in 2007: "I am saddened that it is politically inconvenient to acknowledge what everyone knows: the Iraq war is largely about oil."[12]

As the world passes the global peak of conventional oil production—especially if little effort is made to increase efficiency and develop alternative energy sources—and as developing nations escalate their use of oil, competition and tension over shrinking reserves will inevitably also intensify. Increasing pressure on the resource was indicated as the world oil price soared to more than $100 a barrel early in 2008. Less appreciated so far is that reserves of natural gas are also becoming seriously depleted, especially in the United States, which is tearing up its western states to find more deposits while increasingly importing its supplies. In the not too distant future, international competition for natural gas deposits may well parallel that for oil, where declining supplies are predicted, along with other energy resource shortages, to increase unrest and fuel wars in the near future.[13]

Energy demand is far from the only potential source of resource conflict. With rapid industrialization in countries such as China, India, Brazil, and Indonesia, various industrial materials may well become bones of contention. Water has been an important flash point of conflict in the past, and it will surely be even more important in the future. The combination of continuing population growth, intensifying agriculture, and industrial development guarantees that demand and needs for water will rise substantially, often in areas where water is already in short supply. The conflict over the Jordan River, which runs between Israel and Jordan, is a case in point.

By world standards the Jordan is tiny, its flow being roughly 2 percent of the Nile's. But it is a crucial resource for the populations of both Israel and Jordan. Diplomatic efforts by the United States led to an agreement between Israeli and Jordanian engineers on water sharing in 1955, which almost resulted in a formal treaty. But politics intervened, since Arab politicians did not want any treaty signed that would imply that Israel had a right to exist. Tensions naturally grew as the Israelis sped the construction of projects to divert the Jordan's water to coastal cities and the Negev Desert. Then the League of Arab States in 1960 came up with a plan to block the headwaters of the Jordan that in effect would remove much of the water flowing into the Israeli canal system. Israel considered this a grave threat, and when in 1964 Syria began work on the project, an escalating series of military clashes, including aerial dogfights, between Israeli and Syrian forces ensued. Israeli aircraft bombed dam construction sites deep inside Syria in April 1967, Egypt expelled the UN force on the Sinai Peninsula, removing the buffer between Israeli and Egyptian forces, and war broke out on June 5, 1967. While there were many other issues between the Arabs and Israelis, the conflict over water was primary because both sides viewed it as a grave national security issue.

After Israel's overwhelming victory in what became known as the Six-Day War, its occupation of the Golan Heights and West Bank removed any Arab threat to the headwaters of the Jordan and gained for Israel control of the lower Jordan and valuable aquifers, but it did not end tensions over water. Indeed, they worsened as Israeli allocation of West Bank water gave Jewish settlers five to eight times the amount of water granted the Palestinians. No comprehensive settlement has yet been reached between Israel, Jordan, and Syria. Meanwhile, the combined population of the three countries and the Palestinians is growing rapidly and expected to be almost twice as large in 2050 as it was in 2007 (64.7 million, up from 36.9 million), greatly increasing demand for water.

The potential for serious conflict over water exists for many river basins around the world, from Egypt's Nile to the Tigris and Euphrates rivers, shared by Turkey, Iraq, and Syria, and the Indus River basin, drawn on by two nuclear powers with rising populations, Pakistan and India. Indeed, water may in the long run prove more of a spark for resource wars than petroleum has been. Water simply has no substitute, and rising average global temperatures are likely to exacerbate tensions by changing the distribution of water and making much water-handling infrastructure obsolete. Climate change may already be affecting precipitation patterns and clearly is melting mountain glaciers, which are major sources of water in many regions.

Not all wars over water are likely to be between nations; droughts are also highly correlated with *intra*national violence, which can have international ramifications. It is estimated that some 1.5 billion people around the world already are under severe water stress, and with rapid climate change the numbers of thirsty people are likely to increase.[14] But, as with petroleum, innumerable opportunities for more efficient use of water exist, and, unlike petroleum, water can be reused and recycled.

The Environment and International Governance

Many issues that could potentially escalate to resource wars are already on the international agenda, from dealing with greenhouse gases and sustainable fisheries management to control of weapons of mass destruction and the legitimacy of preemptive strikes against terrorism. There is even discussion of the activities of what Hannah Arendt called "desk murderers"—those who order modern remote killing by dropping bombs and launching cruise missiles with little concern for the lives of non-combatants.[15] Also on the agenda are managing international conflicts, protecting human rights and health, and maintaining some level of international law, and they all fall into the lap of the United Nations.

Although the UN has never been given much political power, its dozens of programs and agencies nonetheless have assisted people around the world. To name just a few: The Food and Agriculture Organization of the United Nations (FAO) and related agencies have done much to increase food production and provide emergency famine relief for poor countries. The International Atomic Energy Agency (IAEA) monitors the development of nuclear power plants and nuclear weapons making (but has little power to prevent nuclear arms buildups). The International Court of Justice (World Court) settles any number of disagreements between nations but is better known for holding trials of war criminals. The United Nations Children's Fund (UNICEF) focuses on children through numerous community-based programs in developing countries.

Of particular interest to us is the previously mentioned United Nations Environment Programme (UNEP), which monitors international agreements in such important areas as international trade in endangered species and in hazardous materials, and keeps important environmental issues alive. It also has sponsored and monitors multinational regional environmental cleanups and regulation in many parts of the world.

In addition, the United Nations Conference on Environment and Development (UNCED, also known as the Earth Summit) of 1992 and similar meetings in past decades have laid much of the groundwork for managing human interactions with Earth's natural systems, notably the Rio Declaration on Environment and Development and Agenda 21, both adopted at UNCED. The United Nations Framework Convention on Climate Change (UNFCCC) of 1992, though little known in the United States, was a remarkable example of vision and leadership. Unlike the situation with the Kyoto Protocol, which the United States has never ratified and which was intended to begin implementing the UNFCCC, the

United States did ratify the climate treaty and is legally bound by it. Some progress has been made in meeting the goals of those very ambitious agreements, which have been formally accepted by many nations, although often neglected or only partially met. Furthermore, a large number of international environmental treaties are already in force. Some, like the Montreal Protocol on Substances That Deplete the Ozone Layer (chapter 13), have been quite successful; others, much less so. Unlike environmental legislation within nations that can be enacted and enforced by governments, there is no world government to enforce environmental treaties—the UN basically can only use friendly persuasion, and what is essentially peer pressure among nations often must suffice. The same is largely true for international agreements on managing nuclear materials and arms control, although the United Nations Security Council can impose economic sanctions against nations that act against perceived global interests.

Clearly, the needs and ideals that inspired the United Nations' founding are more compelling today than they have ever been. Accordingly, there is considerable interest now in making all those treaties more effective, but the question is, how? The best strategy appears to be, as in con-

FIGURE 16-2. A meeting of the United Nations General Assembly. The world must find ways to expand its governance of the interlinked environment, which supports all nations. UN Photo/Eric Kanalstein.

servation finance, to find ways of aligning participants' (in this case, nations') self-interests with the goals of the treaties. In essence, that would make the treaties self-enforcing. A treaty can generally be written so that it is to no nation's advantage to withdraw from it; negotiations should make both threats and promises agreed to by the participants credible; and the treaty should be perceived by participants as fair. Accomplishing all this is a substantial challenge, but fortunately there have been successes in the past, especially with regard to nuclear arms control and some environmental conventions.

Responding to rising concerns about global problems in the 1990s, and building on the earlier agreements on environment, development, and population, the United Nations established an impressive list of Millennium Development Goals for the first fifteen years of the twenty-first century:

1. Eradicate extreme poverty and hunger.
2. Achieve universal primary education.
3. Promote gender equality and empower women.
4. Reduce child mortality.
5. Improve maternal health.
6. Combat HIV / AIDS, malaria, and other diseases.
7. Create a global partnership for development.
8. Ensure environmental sustainability.

These are surely worthy goals, and *in theory* they are attainable, but funding for the program has been sadly lacking, in no small part because of reluctance of the United States government to contribute financially to the UN or to lead the way in providing significant aid to poor nations in general. Progress has thus been at best minimal, and the goals are very unlikely to be met by 2015.

Succeeding with such ambitious goals will require a massive change in human priorities; token gestures will not suffice. While virtually everyone subscribes to these goals, success requires a much greater dedication of funds and effort than has so far been undertaken. And if goal number 8, ensuring sustainability, isn't seriously addressed soon, achievements among the others are likely to be short-lived. It is a circular dilemma.

In our view, moving much further toward solving the equity problem in the world will be essential for achieving environmental sustainability. Unless given a fair shake, the poor are unlikely to be able to take the needed actions in their own situations, nor are they likely to be enthusiastic about cooperating globally. Achieving the eight goals above, realizing John Holdren's optimistic scenario for limiting greenhouse gas emissions (chapter 14), and closing the rich-poor gap would all seem to be key objectives in the struggle to preserve a livable planet, and thus a sustainable civilization.

In assessing progress toward these goals, it is important to appreciate and amplify small steps in the right direction. For example, integrated pest management and ecologically sound agriculture are in place in some areas, although huge monocultures and high inputs of pesticides and fertilizer are still the rule in most places. Wide application of alternative approaches that can match the productivity of chemical-based, energy-intensive industrial agriculture will, in economic terms, require internalizing the externalities associated with industrial agriculture (soil degradation, toxification, losses of biodiversity and genetic variation in crops), something big agriculture would fiercely oppose. Nevertheless, public awareness of the health consequences of industrial agriculture is growing rapidly, in rich countries particularly, and along with it demand for "organic" foods is rising, which are some hopeful first steps.

Similarly, hybrid automobiles are less harmful for the environment than standard autos, let alone SUVs, but by themselves they cannot be a general solution to the problems associated with gigantic automobile cultures. If every car and truck produced in the world over the next few decades were a hybrid, it would have a significant but minor impact on the environmental damage resulting from designing the world around automobiles. Many other measures, such as redesigning cities and transportation systems to move both people and goods efficiently, are also essential to prevent further environmental deterioration. Preventing suburban sprawl also helps preserve productive farmland (too often undervalued by city folks) and natural areas that provide ecosystem services.

Recycling has its own importance, though at present it doesn't make much of a dent in the growth of the throwaway society. But it does give individuals a sense of participation in environmental protection; and, again, if it were more widely and fully adopted and buttressed by programs that curbed excess packaging, it could make a real contribution.

Giving "microloans" to poor people, as pioneered by the Grameen Bank in Bangladesh, has helped millions, but *billions* are still in need.

Debt-for-nature swaps also have preserved many natural ecosystems to the benefit of neighboring human groups. Innumerable steps such as these are helping to move humanity slowly down the track toward sustainability; they involve many people in key actions, and together they can add up to real progress. But now, in the opinion of many, we have reached a turning point where we need express trains moving us down that road before some critical bridges collapse.

The United Nations system at large addresses many important problems, from providing emergency food supplies to peacekeeping in many areas around the world. But, through influence on their government delegations, vested interests often swing decisions to favor themselves or to impede action. They hold even more sway at the World Bank, which controls investments in developing nations. In both the energy and environmental sectors, many of their decisions reflect the beliefs of traditional economists and businessmen and fail to recognize the environmental imperatives of the human predicament.

The United Nations itself suffers from an outdated structure (especially that of the Security Council—the UN organ charged with maintaining peace and security), a far-flung set of agencies with too little coordination among them, and a lack of concrete and consistent support from some member nations. Reforming the Security Council to reflect a contemporary set of important powers is a crucial step. Perhaps the reform should include some mechanism for revising the set of five or six permanent members at regular intervals, say every twenty years. Unhappily, though, amendments to the Charter of the United Nations must, among other things, be approved by all the permanent members of the Security Council.

The stakes of political error have become far greater for everyone since World War II. If the United Nations should fail to intervene in a dispute between India and Pakistan, or between the United States and China or Russia, hundreds of millions, or even billions, of people could die if nuclear weapons were unleashed, heading civilization toward a breakdown. If responsible nations don't cooperate to reduce the flux of greenhouse gases into the atmosphere, the negative consequences might be almost as devastating, if not as sudden. As ecosystems (with their non-linearities and lag times) come under ever-increasing assault, the past becomes an ever less reliable guide to the future.

The difficulty and cultural risks of bringing Masai and Tibetan herdsmen, Mexican, Nigerian, and Chinese villagers, New Guinean and Amazonian forest dwellers, Caribbean and Malagasy fisherfolk, Palestinian physicians and Israeli farmers, and many others into a global discussion of what should be done will be truly daunting, even if the rich nations turn out to be truly cognizant of the seriousness of environmental deterioration, eager to halt it, and determined to keep the discussion democratic rather than to dominate it.

One thing in humanity's favor in trying to get more cooperative progress on the environmental front, according to sociologist Riley Dunlap, is the falsity of a widely held belief that environmental quality is not an issue for the poor. Many think that concern about the environment is a luxury enjoyed by the relatively affluent in rich countries and not considered important by the poor, especially in developing nations. But surveys in both rich and poor countries that Dunlap and others have conducted since the early 1990s contradict this conventional wisdom. There is a difference in that affluent people may put more value on esthetic aspects such as scenery and wildlife, whereas the poor are more concerned about cultural values (e.g., sacred burial grounds) and the environmental underpinnings of their livelihoods. Even so, in rich countries environmental justice groups have increasingly tapped into concerns about toxic hazards and severe pollution in poor neighborhoods.

Environmental concerns, of course, are not the only ones shared by people around the world; others include common aspirations, perceptions of fairness, and hopes for the future and for future generations. Recognizing the shared attitudes and values of both rich and poor people is essential, since humanity has no choice but to develop a discourse among diverse peoples in various economic situations and to struggle collectively to improve human systems of governance.

But how do we get the global paradigm shift needed for humanity to recognize and repair its predicament? The real challenge is to close the gap between the sorts of behavior recommended implicitly and explicitly in the 1992 *World Scientists' Warning to Humanity* and the *Joint Statement of the World's Scientific Academies*, in 1993, and the actual conduct of societies.

A few years ago we suggested that the nations of the world, through the United Nations, might be persuaded to inaugurate a Millennium Assessment of Human Behavior (MAHB)—so named to emphasize that it is human *behavior*, toward one another and toward the planet that sustains us all, that requires assessment and modification in the short term. The idea was that an MAHB might become a basic mechanism to expose society to the full range of interconnected issues involving population, environment, resources, ethics, and power and thus be a major tool for conscious evolution—for deliberately and openly dealing with the problems besetting a small-group animal striving to live in a gigantic global civilization. In other words, an MAHB would strive to generate ethical strategies for dealing with supercircumscription—an extension of Robert Carneiro's ideas of the political consequences of an inability to move away from undesirable neighbors, a recognition that we're all stuck together on this small planet.[16]

All of this may seem extremely impractical, especially when the forces moving civilization toward environmental (and thus economic) disaster seem so very strong. But societies can change extremely rapidly when the time is ripe. Dramatic improvement in race relations in the United States in the past fifty years, catalyzed by the civil rights movement and sustained by farsighted leadership, is a good example. More recently, the sudden collapse of the Soviet Union and the retreat of communism show that unexpected and extremely rapid transformations on a global scale are possible. And the current surge of global awareness and concern about climate change may prove to be another case. The early twenty-first century provides humanity with its best chance to start making the political and social reforms required, nationally and globally, to achieve sustainability while changing for the better the ways we treat one another and our environments.

In the face of unusual uncertainty about the future, perhaps the greatest intellectual challenges confronting the human community are the interrelated problems of the scale of the human enterprise, the complexity of the interlinked global social-biosphere system, how long growth in the physical economy can (or should) continue, and the maldistribution of wealth. It's the classic issue of "limits to growth," complexified by an additional issue of the "limits to inequity."

Many ecologists and ecological economists are convinced that, with continuation of anything like current behavior, humanity is near or above Earth's long-term carrying capacity for human beings. But much of the general public, and many businesspeople, government officials, and economists appear to believe that population and per capita consumption can continue to grow indefinitely, and that eventually all economic inequities can be eliminated by growth itself.

Long ago, the great leap forward changed *Homo sapiens* from an evolutionary to a revolutionary animal by laying the groundwork for an accelerating flowering of technology. That change now may have put all of humanity on a course resembling the fate of ancient civilizations that collapsed. The various revolutions of our past—cultural, agricultural, writing, scientific, industrial, and (perhaps) computer and information—have placed humanity in a totally unprecedented position. Our species is overshooting the capacity of its planetary home to support it in the long run; our margin of error has shrunk dramatically, and powerful nations are playing the game as if it were still the nineteenth century (or, indeed, the first century!). The penalties for continued ignorance, malfeasance, and folly among opinion makers and the leaders of society—indeed, all of us—have escalated enormously, and now those penalties may have global rather than merely local or regional consequences. We have utterly changed our world; now we'll have to see if we can change our ways.

Epilogue

"We still can't see beyond the white wall of fog in front of us, but now we have knowledge we can use like a compass—to help us, together, choose our path through a future full of surprises, danger, and opportunity."

THOMAS HOMER-DIXON, 2006[1]

WE HAVE BEEN privileged as few human beings are today, or previously have been, to see many of the traces of long-vanished civilizations that litter our planet. In an early evening in the great Roman amphitheater at Vienne, on the Rhone River in France, we could imagine a mob of theatergoers filing into the seats, as might happen in almost any big city today. The theater was built less than a century after Julius Caesar conquered Gaul and crossed the Rubicon to begin the end of the Roman republic. Then, in what may be an eerie preview of our times, the transition to empire was disguised by preservation of the forms of the republic as despotism replaced it. Will modern nominal republics follow the same course as the environment-resource situation deteriorates?

While enjoying the magnificent causeways, plazas, and towers of the Mayan city of Tikal, we tried to imagine what it was like when fully populated with those amazing people, before that ancient civilization's collapse, in part, research suggests, because of climate change. We've wandered through the deserted ruins of Angkor Wat in Cambodia, almost expecting a Khmer to come around a corner and confront us. Angkor, once the capital of the Khmer state, has survived more than eight centuries of sackings, wars, and jungle encroachment, standing in such contrast to the situation of the Khmers in modern times that the first Western explorers refused to believe that Khmers were Angkor's architects and builders. Will future hunters and gatherers, as writers are beginning to imagine, wonder over the origins of remains of skyscrapers, railroads, and freeways, and invent religions to explain them? Or will future intergalactic visitors think it impossible that the depauperate tribal people they encounter are actually descendants of the builders?

Each ruin has reminded us of questions and answers pertinent to the career of the dominant animal today. At Canyon de Chelly in Arizona, how much did climate change contribute to the fate of the Anasazi? The magnificent Inca ruin at Machu Picchu symbolizes the fate of one powerful empire at the hands of another that seems to have brought pathogenic microorganisms as allies. The Great Zimbabwe ruins show that magnificent civilizations have sprouted from the minds of people with black skins as well as paler shades. The Pyramids, Karnak, Carthage, Petra, and Persepolis remind us that North Africa and the Middle East were not always ecologically desolate, agriculturally backward, short of water, and overpopulated. But overall our visits have impressed on us the impermanence of past civilizations—and prompt the larger question, one that we suggested at the end of the previous chapter: Will the first truly global civilization collapse just as past regional ones have?

The world in general seems to be gradually awakening to a realization that our long evolutionary story is, through our actions but not our intentions, coming to a turning point. A product of evolution ourselves, shaped by the environments of our past, we have attained dominance by increasing our numbers, diverting resources, and reshaping the world's environments to sustain our huge, still growing population. That dominance has now led to a progressive destabilization of the global systems that sustain us.

Since the great leap forward some 50,000 years ago and the dispersal of modern human groups out of Africa, the human story has been characterized by ever more rapid cultural evolution and the proliferation of distinct cultures and languages. Each culture came to respond in its own way to the particular environmental setting in which it evolved and the problems of making a living. A restless animal, subject to climatic and environmental changes and impingement from other cultures, conquering or being conquered, or exploring new territory, *Homo sapiens* gradually occupied, and began to modify, all of Earth's habitable land.

Over the past millennium, as some groups were still finding and settling in the last remote, uninhabited regions, others had begun to establish much larger societies and to expand in a new burst of exploration and conquest. Both movements were facilitated by technological improvements in long-distance transportation. In the past century or so, the process of cultural differentiation may have reversed as the dominant cultures originating in Europe and, to a lesser degree, in China and Japan have overwhelmed and marginalized many small tribal cultures around the world. We may never know what has been lost to humanity with the disappearance of hundreds of traditional cultures and languages. Each represented a unique worldview and a history of environmental experience that potentially had something to offer our global civilization in contending with today's situation.

The world community is now confronted by multiple dilemmas, all of which demand concerted attention. In 1993, the *World Scientists' Warning to Humanity* was quite explicit about the sort of reconfiguration of world priorities that will be required in order to achieve a sustainable world for future generations. In brief, it said, as did the near-simultaneous statement of the world's scientific academies, that humanity must promptly deal with the problems of population, resources, poverty, equity, and environment that we've discussed in this book.

Many people will look to technology to somehow accomplish these tasks, but new technologies ordinarily produce not only benefits but also costs. Indeed, blind faith in technology as a panacea often seems most intense among people with the least understanding of science, people not trained to consider systematically the uncertainties that always attend proposed solutions. Technological advances will be essential to solving many of our looming dilemmas, but the record of past claims for ambitious possible technological fixes, such as feeding billions of people via nuclear agro-industrial complexes, is not encouraging.[2] Technological advances alone can't save us. And they seldom address important quality-of-life or distribution of power issues. Science and technology might eventually permit 12 billion people to live sustainably on Earth, but most likely in the style of factory ("battery") chickens.[3] Is that a desirable goal?

Most people, given such a prospect, obviously would say no. But other possible futures are even less attractive: mounting conflict over diminishing resources, and an increasingly inhospitable world as various natural and agricultural systems collapse from a combination of overexploitation, decline in ecosystem services and productivity, and changing climate. Not far behind: possibly severe economic disruption, famine, and massive migrations of environmental refugees.

Clearly more necessary than technological boondoggles is social change that can produce the will to begin tackling the problems, which can be done largely with technologies already in hand. A crucial question is whether the world's rich minority will cooperate in a movement that will put us on a more viable course. Achieving a sustainable society—one not susceptible to collapse from destroying its environmental underpinnings and resource base, from undergoing extreme social upheaval, or both—will require cooperation on a scale that is likely to come only through addressing global equity problems that have worsened over many decades.

Ironically, two advances that have contributed to the plethora of problems now facing the world—global trade and revolutions in modern communications—may prove to be keys to their solution. People in most nations today are connected to one another in a vast network of trade and economic ties. Indeed, trade in food and essential resources has fostered a significant interdependence. The idea that any country can be independent and self-sufficient while maintaining its lifestyle—even one as large and richly endowed with natural resources as the United States—is in this day and age a myth.

The current imbalance in wealth deepens the human predicament, but a more equitable and economically sensible arrangement of these relationships could potentially take us a long way toward sustainability. One example, already being discussed, is to rationalize the food system to encourage consumption of foods grown locally and thereby reduce the energy-consuming process of long-distance and international transport of foods. At the same time, adjustments in global food pricing and agricultural subsidies in industrialized nations could encourage increased food production—and consumption—in developing regions.

With the rise of China, the unification of Europe, and the growing power of a few oil-rich nations (on which others are increasingly dependent), the world is rapidly becoming a multipolar system, replacing that of the United States as a lone superpower. This change clearly carries risks if it means rising competition and conflict over scarce resources. But it also could make international cooperation and sharing of sustainable technologies (especially those not dependent on fossil energy) easier to manage.

Similarly, and in parallel with world trading networks, communications have become a global phenomenon, from worldwide television networks to the Internet, all facilitated by satellites and the dominance of English (and secondarily Chinese and perhaps Arabic) as a lingua franca. A few years ago, the United States and the United Kingdom essentially held a monopoly on satellite-transmitted news and informational television broadcasting. But now half a dozen other groups, beginning with Al Jazeera in the Middle East (in English and Arabic) and recently joined by broadcasts from Russia, China, and Venezuela, are competing with the BBC, CNN, and the like for attention. Although many of these broadcasts are filled with trivia and self-serving propaganda, the needed information on important matters often seems to be getting through. At the same time, the Internet is beginning to replace television, radio, and newspapers as a source of information. Despite its capacity for mis- and disinformation, the growth of direct communication that the Internet fosters among individuals around the world is already breaking down barriers between nations and cultures and providing information and knowledge to people and groups previously denied it. Through a variety of channels, apparently, the increasingly alarming reports of the Intergovernmental Panel on Climate Change and other scientific reports on climate change have helped to focus the world's attention on the global environment and the need for collective action on its behalf.

The growing universality of information channels, along with the diversity of viewpoints offered, presents an unprecedented opportunity to channel cultural evolution toward resolving the world community's multiple dilemmas. As a series of United Nations conferences on population, development, women's issues, and the environment have demonstrated over nearly four decades, there has been growing agreement in the international community—people who are active and aware on an international level—on both what the problems are and, to a large extent, how they should be managed.

A great deal is known about a range of solutions for many of our problems. But educational, economic, and bureaucratic barriers, as well as vested interests, too often stand in the way of giving those potential solutions the attention they need and promoting public discussion of them.

Here again, the influences of globalization and global communications can be either a help or a hindrance.

Humanity's globalizing civilization must take this enhanced opportunity to explore conscious evolution and try new ways of organizing societies to cooperate to solve its burgeoning global problems. The world must soon strive to find new answers to questions that economists such as Karl Marx, John Stuart Mill, Joseph Schumpeter, and John Maynard Keynes all examined: how society is to evolve, and whether and when the economic problems of scarcity and equity can be solved within a biophysical steady state—one in which the physical economy no longer grows. We need to set practical goals of how to live and determine how to organize ourselves to reach those goals. And humanity must do this even without assurance that the steps taken will be successful. Dealing with such profound questions along with the consequences of overpopulation, economic inequity, and the erosion of environmental resilience surely will not be easy. But each day that we do nothing forecloses options for creating a better future for ourselves and our fellow inhabitants of Earth. The qualities that made it possible for us to become the dominant animal could now be put to use in creating a sustainable future for ourselves and the rest of the living world.

Notes

Prologue

1. E. M. W. Tillyard, *The Elizabethan World Picture* (Random House, New York, 1941), p. 26.

Chapter 1: Darwin's Legacy and Mendel's Mechanism

1. T. Dobzhansky, Nothing in biology makes sense except in the light of evolution, *American Biology Teacher* 35 (March 1973): 125–29.
2. C. Darwin, *Autobiography* (1876; repr., Collins, London, 1958), pp. 119–20.
3. There are many versions of natural selection on characteristics that are inherited genetically, including genic selection, individual selection, group selection, kin selection, and so on—but individual selection is the most basic modern concept.
4. How the DNA instructions are copied and move from cell to cell and parent to offspring is a complex process essential to evolution. When a cell divides to produce two daughter cells, how do genes get into the new cells? This question applies to the process of producing new single-celled organisms (such as bacteria and amoebae), as well as when a fertilized egg cell starts to divide to make a daisy, a dog, a person, or any other organism made up of very many cells (your body has trillions of cells).

   And why, in each generation, when a sperm fuses with an egg, doesn't the number of DNA strands (and the chromosomes that carry them) double? An intricate chemical mechanism copies the strands of DNA when cells divide in the process of making new cells for growth, repair, or reproduction. In most cell division, daughter cells receive full copies of the DNA in the parent cell—two complete sets of chromosomes, one of each pair inherited from the organism's father and one from the mother.

   A somewhat different divisional process, which halves the chromosome number and the amount of DNA, occurs in the special cell line that produces gametes (eggs and sperm in animals). In this process, each gamete gets half the number of chromosomes as are present in other cells. Gametes, when they unite, combine the genetic contributions of the two parents, restore the full complement of chromosomes, and produce a zygote. In human beings, the zygote is the stage in the continuous life cycle that many feel represents the start of an individual, but, of course, one could also define an egg or a sperm as an individual as well.

In this halving process, the DNA molecules undergo complex mixings, called recombination, in the cells that are going to become gametes. Those mixings add to the genetic variation in populations of sexually reproducing organisms, and thus to the precious store of genetic variation. Without the variation produced by recombination, natural selection would have less "raw material" to sort, and evolution would proceed much more slowly. After recombination and the union of gametes, the zygote then will divide into daughter cells, which in turn will produce many more generations of daughter cells, and eventually the mass of cells will mature into an individual of the next generation.

In both kinds of divisional processes, DNA copying occurs, and this process of duplication (replication) is normally quite accurate. Although sometimes scientists describe DNA as a "self-replicating" molecule, you should always remember that a bottle of DNA left in your closet will not explode from DNA perpetually copying itself. DNA can copy itself only with the help of an extremely complex cellular apparatus. In that sense, it is no more self-replicating than this page, which can be copied, but only with the help of a copying machine and a complex system that supplies energy to that machine.

5. J. T. Bridgham, S. M. Carroll, and J. W. Thornton, Evolution of hormone-receptor complexity by molecular exploitation, *Science* 312 (2006): 97–101.
6. E. Culotta and E. Pennisi, Breakthrough of the year: Evolution in action, *Science* 310 (2005): 1878–79.
7. C. Darwin, *From So Simple a Beginning: The Four Great Books of Charles Darwin. The Voyage of the Beagle. On the Origin of Species. The Descent of Man. The Expression of the Emotions in Man and Animals*, edited and with introductions by E. O. Wilson (W. W. Norton, New York, 2006), pp. 489–90.

Chapter 2: The Entangled Bank

1. S. Freeman and J. C. Herron, *Evolutionary Analysis* (Prentice Hall, Upper Saddle River, NJ, 2001), p. 403.
2. J. N. Thompson, *The Geographic Mosaic of Coevolution* (University of Chicago Press, Chicago, 2005), p. 6.
3. J. A. McMurtry, C. B. Huffaker, and M. van de Vrie, Ecology of tetranychid mites and their natural enemies: A review. I: Tetranychid enemies: Their biological characters and the impact of spray practices. *Hilgardia* 40 (1970): 331–90.

4. J. Burger et al., Absence of the lactase-persistence-associated allele in early Neolithic Europeans, *Proceedings of the National Academy of Sciences USA* 104 (2007): 3736–41.

Chapter 3: Our Distant Past

1. C. Darwin, *The Descent of Man and Selection in Relation to Sex* (John Murray, London, 1871), p. 1239 in C. Darwin, *From So Simple a Beginning: The Four Great Books of Charles Darwin. The Voyage of the Beagle. On the Origin of Species. The Descent of Man. The Expression of the Emotions in Man and Animals*, edited and with introductions by E. O. Wilson (W. W. Norton, New York, 2006).
2. http://emporium.turnpike.net/C/cs/evid1.htm.
3. This myth was thoroughly investigated by scientists, whose results were reported in J. R. Cole and L. R. Godfrey, eds., The Paluxy River footprint mystery—solved, *Creation/Evolution* 5, no. 1 (Winter 1985). The conclusion: that the putative human "footprints" could have been formed in numerous ways other than by actual walking people.
4. Darwin, *From So Simple a Beginning*, p. 889.
5. Ibid., p. 890.
6. Hominins do not include chimps and gorillas, which, like people, are all members of the family Hominidae.

Chapter 4: Of Genes and Culture

1. J. C. Venter, A dramatic map that will change the world, *London Daily Telegraph*, 14 February 2001.
2. E. B. Tylor, *Primitive Culture: Researches into the Development of Mythology, Philosophy, Religion, Art, and Custom* (John Murray, London, 1871), p. 1.
3. R. Linton, *The Tree of Culture* (Alfred A. Knopf, New York, 1955), p. 29.
4. For an overview of the human ability, see A. M. Leslie, Pretense and representation: The origins of "theory of mind," *Psychological Review* 94 (1987): 412–26. For the problems of determining whether chimps have a theory of mind, see D. J. Povinelli and J. Vonk, Chimpanzee minds: Suspiciously human? *Trends in Cognitive Sciences* 7 (2003): 157–60.
5. P. R. Ehrlich, *Human Natures: Genes, Cultures, and the Human Prospect* (Island Press, Shearwater Books, Washington, DC, 2000), p. 110.
6. M. S. Gazzaniga, *The Ethical Brain: The Science of Our Moral Dilemmas* (HarperPerennial, New York, 2006), p. 99.
7. N. Wade, Ideas and trends: The story of us; the other secrets of the genome, *New York Times*, 18 February 2001.
8. J. C. Crabbe, D. Wahlsten, and B. C. Dudek, Genetics of mouse behavior: Interactions with laboratory environment, *Science* 284 (1999): 1670–72.

Chapter 5: Cultural Evolution: How We Relate to One Another

1. A. Furst, *The Foreign Correspondent* (Random House, New York, 2007), p. 94.
2. G. C. Leavitt, The frequency of warfare: An evolutionary perspective, *Sociological Inquiry* 47 (1977): 49–58.
3. B. Ehrenreich, *Blood Rites: Origins and History of the Passions of War* (Henry Holt, New York, 1997), p. 10.
4. E.g., T. Ashworth, *Trench Warfare 1914–1918: The Live and Let Live System* (Pan Books, London, 1980); D. Grossman, *On Killing: The Psychological Cost of Learning to Kill in War and Society* (Little, Brown, Boston, 1995).
5. It is instructive that science, which is culturally constrained to assume there is a real world "out there" and constantly tested against that environment, changes its basic conclusions only very slowly, and it carefully retains aspects that keep passing the tests (consider classical physics, still employed throughout most of technology despite the quantum revolution). See D. Lindley, *Uncertainty: Einstein, Bohr, and the Struggle for the Soul of Science* (Doubleday, New York, 2007). It is a very readable book about the development of quantum mechanics, arguably the greatest jump in the cultural evolution of science since Darwin. It will give you a feel for how counterintuitive science can be when dealing with the extremely small and the extremely large. That could well be a result of natural selection shaping human intuition to relate to things at scales we readily observe. While you're reading about quantum weirdness, think of its possible meanings for the concept of free will.

Chapter 6: Perception, Evolution, and Beliefs

1. I. Rock, *Perception* (W. H. Freeman, New York, 1984), p. 3.
2. R. B. Edgerton, *Sick Societies: Challenging the Myth of Primitive Harmony* (Free Press, New York, 1992), p. 1.
3. See, e.g., R. Adolphs et al., A mechanism for impaired fear recognition after amygdala damage, *Nature* 433 (2005): 68–72.
4. R. Ornstein and P. Ehrlich, *New World/New Mind: Moving toward Conscious Evolution* (Doubleday, New York, 1989), p. 12.
5. W. B. F. Ryan and W. C. Pitman, *Noah's Flood* (Simon and Schuster, New York, 1998). The issue remains contentious.

Chapter 7: The Ups and Downs of Populations

1. A. Coghlan, Pro-choice? Pro-life? No choice, *New Scientist* (21 October 2007).

Chapter 8: History as Cultural Evolution

1. W. D. O. Gore, in *Christian Science Monitor*, 25 October 1960.
2. Baron de La Brède et de Montesquieu, *The Spirit of Laws*, bk. VI, chap. 2 (J. Nourse and P. Vailant, London, 1752) (see http://www.constitution.org/cm/sol_06.htm#002).

3. D. Porch, *The Path to Victory: The Mediterranean Theatre in World War II* (Farrar, Straus and Giroux, New York, 2004), p. 35.
4. Ibid., p. 24.
5. C. Barnett, *The Desert Generals*, 2nd ed. (Cassell and Company, London, 1983), p. 103.
6. A. M. Schlesinger Jr., Folly's antidote, *New York Times*, 1 January 2007.

Chapter 9: Cycles of Life (and Death)

1. W. H. Schlesinger, *Biogeochemistry: An Analysis of Global Change* (Academic Press, San Francisco, 1997), p. 3.
2. A few get theirs from radiation or chemical reactions at deep-sea vents, or similar situations.
3. This term is actually a misnomer. The warmth in greenhouses is there largely because the glass blocks breezes that would carry away the heat absorbed from the sun.

Chapter 10: Ecosystems and Human Domination of Earth

1. A. Tansley, The use and abuse of vegetational concepts and terms, *Ecology* 16, no. 3 (1935): 284.
2. See A. J. Beattie and P. R. Ehrlich, *Wild Solutions: How Biodiversity Is Money in the Bank* (Yale University Press, New Haven, CT, 2001).
3. Millennium Ecosystem Assessment, *Ecosystems and Human Well-Being: Current State and Trends* (Island Press, Washington, DC, 2005), p. 6.
4. Union of Concerned Scientists, *World Scientists' Warning to Humanity* (Union of Concerned Scientists, Cambridge, MA, 1992).
5. National Academy of Sciences USA, *A Joint Statement by Fifty-eight of the World's Scientific Academies: Population Summit of the World's Scientific Academies* (National Academy Press, New Delhi, India, 1993).
6. E.g., W. Vogt, *Road to Survival* (W. Sloane Associates, New York, 1948).

Chapter 11: Consumption and Its Costs

1. Martin Luther King Jr., excerpt of acceptance speech for the Planned Parenthood Federation of America's Margaret Sanger Award, May 5, 1966.
2. See United Nations Environment Programme (UNEP), *GEO4: Environment for Development* (UNEP, Nairobi, Kenya, 2007).
3. On the basis of such calculations, increased consumption and unnecessarily inefficient ("faulty") technologies contributed about half as much as did population growth (20 divided by 6.5 gives 3.1, so the comparison is between 3.1 and 6.5) to environmental impact over the past century and a half.
4. For simplicity of exposition, we ignore social capital, the institutions and rules that govern relationships among various members of society.
5. Measured by purchasing power parity (PPP) per capita, a measure designed

to correct for differences in exchange rates. For example, a Big Mac might cost $5 in the United States and 25 drachmas in a fictitious country, Anonostan. If the same ratio held over a "basket" of goods and services, an income of $100,000 per year in the United States would have the purchasing power parity of 500,000 drachmas in Anonostan, and an Anonostani would have to have an income of 10 million drachmas to be as rich as an American who made $2 million annually.

6. W. Vogt, *Road to Survival* (W. Sloane Associates, New York, 1948), p. 284.
7. You do pay the private costs of the firms responsible for bringing you the fuel, but you don't pay the total (social) cost of the product and its use. When social costs exceed private costs, negative externalities are present. There are also positive externalities. When your neighbor paints her house and makes the neighborhood look better, the value of your house rises. You, not having paid for the painting, are the beneficiary of a positive externality.
8. *New York Times*, Antibiotic runoff (editorial), 18 September 2007.

Chapter 12: A New Imperative

1. See W. A. Steffen et al., *Global Change and the Earth System: A Planet under Pressure* (Springer, Berlin, 2004), p. 1.
2. J. Lovelock, quoted in *Africa Geographic* 15, no. 7 (August 2007): 112.
3. MSNBC News Services, U.N.: Ocean "dead zones" increasing fast: Experts estimate 200 worldwide, up from 149 just two years ago, 23 October 2006; K. R. Weiss, A primeval tide of toxins, *Los Angeles Times*, 30 July 2006; J. Jackson, pers. comm., 8 December 2007, Sackler Symposium, National Academy of Sciences, Irvine, CA.
4. K. Cassman, Perspective: Climate change, biofuels, and global food security, *Environmental Research Letters* 2, no. 1 (2007): 011002.
5. J. Vidal, Global food crisis looms as climate change and fuel shortages bite, *The Guardian*, 3 November 2007.
6. E.g., J. C. Orr et al., Anthropogenic ocean acidification over the twenty-first century and its impact on calcifying organisms, *Nature* 437 (2005): 681–86.
7. See B. Worm et al., Impacts of biodiversity loss on ocean ecosystem services, *Science* 314 (2006): 787–90.
8. R. Black, "Only 50 years left" for sea fish, *BBC News*, 2 November 2006.
9. M. Wackernagel et al., Tracking the ecological overshoot of the human economy, *Proceedings of the National Academy of Sciences USA* 99 (2002): 9266.
10. W. E. Rees, Ecological footprint, concept of, p. 230 in *Encyclopedia of Biodiversity*, vol. 2, edited by S. A. Levin (Academic Press, San Diego, 2001).
11. G. D. Stone, Fallacies in the genetic-modification wars, implications for developing countries, and anthropological perspectives, *Current Anthropology* 43 (2002): 611–30.

Chapter 13: Altering the Global Atmosphere

1. Holdren was then president of the American Association for the Advancement of Science. Reported in K. Human, Experts warming to climate tinkering, *Denver Post*, 15 September 2006.
2. J. Farman, Unfinished business of ozone protection, *BBC News*, 17 September 2007.
3. Intergovernmental Panel on Climate Change, *Climate Change 2007: The Physical Science Basis; Summary for Policymakers*, contribution of Working Group I to the Fourth Assessment Report of the Intergovernmental Panel on Climate Change (IPCC Secretariat, Geneva, Switzerland, 5 February 2007).
4. That means that nearly all of the radioactive form of carbon (carbon 14, ^{14}C), which is continually created in the upper atmosphere and integrated into the global carbon cycle, has decayed away to non-radioactive carbon 12 in the long-buried coal, petroleum, and natural gas. Carbon emitted into the atmosphere from other sources such as forest fires and decomposition of dead animals and plants contains a standard ratio of the two forms of carbon (^{14}C and ^{12}C). So the anthropogenic emissions from fossil fuel burning are calculated by measuring the increase in the ratio of ^{12}C to ^{14}C in the atmosphere. In this case, the correlation did indicate causation, but it took independent confirmation to be quite sure. Remember, "quite sure" is the best we can do—science never "proves" anything.
5. M. S. Torn and J. Harte, Missing feedbacks, asymmetric uncertainties, and the underestimation of future warming, *Geophysical Research Letters* 33 (2006): L10703, doi:10.1029/2005GL025540.
6. M. Byrnes, Antarctic melting may be speeding up, Reuters, 23 March 2007; J. E. Hansen, Scientific reticence and sea-level rise, *Environmental Research Letters* 2 (2007): 024002.
7. E.g., D. B. Lobell and C. B. Field, Global scale climate–crop yield relationships and the impacts of recent warming, *Environmental Research Letters* 2 (2007): 014002, doi:10.1088/1748-9326/2/1/014002.
8. M. Byrnes, Scientists see Antarctic vortex as drought maker, Reuters, 23 September 2003.
9. A substantial saving in greenhouse gas emissions could be achieved—see, e.g., G. Eshel and P. A. Martin, Diet, energy, and global warming, *Earth Interactions* 10 (2006): 1–17.
10. N. Stern, *Stern Review: The Economics of Climate Change* (Her Majesty's Treasury, London, 2006), p. vi.
11. K. Popper, *The Logic of Scientific Discovery* (Routledge, London, 1935; Hutchinson, London, 1968). This was Popper's most famous book. His basic logic went like this—if you had a theory that all swans are white, then no matter how many swans you looked at that proved to be white, the next one could

always be black. So you could never "prove" the theory with more observations. But the discovery of a single black swan would disprove, or "falsify," the theory that all swans are white (there are, by the way, black swans). Popper's views are much debated, and they don't blend well with a probabilistic approach to science (for example, just how unlikely should a result be before it is judged to have falsified a hypothesis?). The philosophy of science is a complex and contentious area, and there even is no agreed-upon definition of "science," although in most cases "we know it when we see it."

12. P. M. Vitousek et al., Human domination of Earth's ecosystems, *Science* 277 (1997): 494–99.

Chapter 14: Energy: Are We Running Out of It?

1. T. Homer-Dixon, *The Upside of Down: Catastrophe, Creativity, and the Renewal of Civilization* (Island Press, Shearwater Books, Washington, DC, 2006), p. 36.
2. A terawatt can be viewed as just a gigantic rate of energy flow. But it's helpful to know something about how energy flows and amounts are normally measured. A terawatt is 10^9 kilowatts. A kilowatt—1,000 watts—is a unit used to measure power—that is, the *flow* of energy, the rate at which energy is processed or, to put it another way, movement of units of energy divided by time. Technically a kilowatt is 1,000 joules per second (a joule is about the energy needed to lift a pound of material an inch and three-quarters). A kilowatt is slightly over 1.34 horsepower. The confusing thing is that time is already in the unit—ten 100-watt lightbulbs turned on have a kilowatt of electricity (1,000 joules *per second*) flowing through them. In an hour, they use one kilowatt-hour. Kilowatt-hours are *quantities*, not *flows*. The 16 terawatt-years of energy humanity uses annually is the equivalent of the energy in 17 billion tons of coal, so that each person uses each year the energy equivalent of 17 billion divided by 6.6 billion, or 2.6 tons of coal.

   Kilowatts you can think of as the *power* (flow of energy) needed to light ten 100-watt lightbulbs. Another way to envision it is that your body uses energy at a rate of about 100 watts, so ten of you would heat a swimming pool about as much as submerging those ten lightbulbs in it. At the other extreme, a terawatt is about one-sixteenth of all the power humanity now uses: 16 times 10^{12}, or 16 trillion watts.
3. This is sometimes referred to as "conservation," but to many this smacks too much of sacrifice, so we use "efficiency" here. The two actually can be differentiated, with efficiency referring to less use of energy to achieve the same results—e.g., achieving more miles per gallon by having a superior engine in a car of the same capacity and performance as the old one, and conserving energy by driving the same car less. Both are often lumped together as "conservation" or "efficiency," as we sometimes do.

4. On November 30, 2007, Professor Tilman wrote to us: "The US annually produces about 42 million tons of livestock and fish, of which cattle are 28%, pigs are 23%, chickens are 37% and fish are 13%. Because about 7, 4, 3 and 1.3 kg of dry grain are needed to produce 1 kg of beef, pork, poultry and fish, respectively, dietary shifts in the proportions of these four protein sources can have large impacts. For instance, if US total meat and fish consumption were kept at 42 million tons, but the proportion of fish and poultry were increased to balance off a 50% decrease in beef and pork consumption, the decreased demand for feed grain would free up 17 million hectares (42 million acres) of fertile agricultural land.

"The lands freed from production of annual grain crops could be planted with high-diversity mixtures of native prairie perennials that received very low inputs of agrichemicals. The annually harvested biomass could be used to produce sufficient synthetic gasoline and diesel fuels . . . to displace about 10% of US gasoline consumption [with] . . . a 50% shift away from beef and pork . . . high-diversity plantations of native perennials provide excellent habitat for birds, mammals, insects and other organisms, recharge aquifers with high-quality water, reduce erosion and flooding, restore soil fertility and provide other ecosystem services."

Chapter 15: Saving Our Natural Capital

1. A. Leopold, *Round River: From the Journals of Aldo Leopold* (Oxford University Press, New York, 1953, 1993), pp. 145–46.
2. M. Berenbaum, Losing their buzz, *New York Times*, 2 March 2007.
3. See http://www.rainforest2reef.org.
4. L. Polgreen, In Niger, trees and crops turn back the desert, *New York Times*, 11 February 2007.
5. Millennium Ecosystem Assessment, *Ecosystems and Human Well-Being: Synthesis* (Island Press, Washington, DC, 2005), p. 1.
6. Millennium Ecosystem Assessment, *Living beyond Our Means: Natural Assets and Human Well-Being* (Island Press, Washington, DC, 2005), p. 5.
7. Millennium Ecosystem Assessment, *Ecosystems and Human Well-Being: Synthesis*, p. 2.

Chapter 16: Governance: Tackling Unanticipated Consequences

1. A. G. Oxenstierna, letter to his offspring, 1648.
2. Technically we now have a tightly coupled complex adaptive system—see S. Levin, *Fragile Dominion* (Perseus Books, Reading, MA, 1999). A complex adaptive system (CAS) is characterized by being the self-organized result of diverse individual components that have their own agendas, interact locally, and produce emergent properties, which can be global. The emergent prop-

erties often feed back on the behavior of the individual elements. Such systems often feature non-linearities (e.g., positive feedbacks and thresholds) that make predicting their behavior problematic. Levin provides an accessible discussion of the characteristics of such systems.

3. E. Osnos, Your cheap sweater's real cost, *Chicago Tribune*, 16 December 2006.
4. International Monetary Fund, Globalization: Threat or opportunity? An IMF brief, http://www.imf.org/external/np/exr/ib/2000/041200to.htm#III.
5. S. Begley, Global-warming deniers: A well-funded machine, *Newsweek*, 13 August 2007.
6. E.g., A. Cockburn, The greenhousers strike back and strike out, *The Nation*, 11 June 2007.
7. There are currently many discussions of the domestic consequences of U.S. imperial dreams; see, e.g., C. Johnson, *Nemesis: The Last Days of the American Republic* (Metropolitan Books, New York, 2007), and C. Hedges, America in the time of empire, 26 November 2007, http://www.truthdig.com/report/item/20071126_america_in_the_time_of_empire/.
8. G. M. Mikkelson, A. Gonzalez, and G. D. Peterson, Economic inequality predicts biodiversity loss, *PLoS ONE* 2 (2007): e444.
9. Quoted in B. Moyers, Restoring the public trust, 24 February 2006, http://www.tompaine.com/articles/2006/02/24/restoring_the_public_trust.php.
10. There is, ironically, *some* foresight capability in the U.S. government, and it does have an environmental element. That capability rests within the military structure in the three-decades-old Defense Advanced Research Projects Agency, the Pentagon's foresight agency. DARPA has concluded, for instance, that urban misery in poor countries will increase and that guerilla wars of the future will mostly be fought in urban slum environments, as has been the case in Iraq, rather than rain forests.
11. J. Madison, letter to W. T. Barry, 4 August 1822, in *The Writings of James Madison*, edited by G. Hunt, vol. 9 (G. P. Putnam's Sons, New York, 1900–10), p. 103.
12. A. Greenspan, *The Age of Turbulence: Adventures in a New World* (Penguin, New York, 2007).
13. A. Seager, Steep decline in oil production brings risk of war and unrest, says new study, *The Guardian*, 22 October 2007.
14. S. Leahy, Thirstier world likely to see more violence, Inter Press Service News Agency, 17 March 2007.
15. See C. Johnson, *Nemesis: The Last Days of the American Republic* (Metropolitan Books, New York, 2007), p. 21. The term fits beautifully the leaders of the United States who arranged the deaths of hundreds of thousands of women

and children in their struggle to control Iraq's oil. Arendt originally used the term to describe Adolf Eichmann, who organized the deportation of millions of Jews to their deaths.

16. An MAHB could try to sketch out what an environmentally sustainable, socially and economically equitable global society might look like. On the technical side, the MAHB could promote examination of factors that could add resilience to societies in cases of disaster: for example, adjusting power grids to be less susceptible to propagating outages; stocking antiviral drugs and planning quarantine measures against the spread of novel epidemics; increasing the ability of societies to be at least temporarily self-sufficient in food and fuels. Central to developing enough citizen interest to permit widespread discussion of such issues is the universal need for better education about the environment. People are unlikely to be convinced of the importance of placing strictures on wealth as long as they believe that growth in aggregate wealth has no biophysical limits. Persuading people to change their reproductive and consumptive behavior is difficult enough when they are aware of the stakes; it seems well nigh impossible when they think environmental problems are no more serious than most other difficulties facing society. Since the MAHB is envisioned as an ongoing effort, not all the goals would need to be reached immediately, although consensus would be needed on challenges to be tackled first. We need to discover whether 7 billion or more partly irrational creatures can organize a sustainable global society. But if the scientific diagnosis of humanity's approaching collision with the natural world is accurate, what alternative is there to trying?

Epilogue

1. T. Homer-Dixon, *The Upside of Down: Catastrophe, Creativity, and the Renewal of Civilization* (Island Press, Shearwater Books, Washington, DC, 2006), p. 308. He calls the stresses building up beneath our society "tectonic stresses"—analogizing them to the stresses that build up in a seismic fault and are eventually relieved by a cataclysmic earthquake. Look especially at the parts of this book on building resilience in the complex systems humanity depends upon.
2. For another example, see N. W. Pirie, Leaf protein as human food, *Science* 152 (1966): 1701–05. An example of another "technofix" on the food front that never panned out.
3. The term refers to the egg production system, in which each hen is debeaked to avoid pecking injury to its neighbors, loaded with antibiotics, and given the absolute minimum in space to live and exactly the monotonous food required for it to maintain its egg production while being gradually deprived of the calcium it needs. Typically each hen is crammed inside a tiny wire cage

with several other individuals, amid tiers of identical cages in gloomy sheds holding 50,000 to 125,000 other terrified birds. At the end of their egg-productive lives, the battered and ulcerated birds are cruelly slaughtered so that their flesh can be used in meat pies, soups, and other products that conceal their degraded state. We owe the analogy to our colleague Gretchen Daily.

Selected Bibliography

General Works

Daily, G. C. 1997. *Nature's Services: Societal Dependence on Natural Ecosystems*. Island Press, Washington, DC. The foundational volume on a new way to look at human-ecosystem relationships.

Daly, H. E., and J. Farley. 2004. *Ecological Economics: Principles and Applications*. Island Press, Washington, DC. Discusses many issues of relevance to the topics in this book.

Darwin, C. 2006. *From So Simple a Beginning: The Four Great Books of Charles Darwin. The Voyage of the Beagle. On the Origin of Species. The Descent of Man. The Expression of the Emotions in Man and Animals*. Edited and with introductions by E. O. Wilson. W. W. Norton, New York. This is the best and most accessible current source for Darwin's most crucial works, assembled by a leading evolutionist.

Dasgupta, P. 2007. *Economics: A Very Short Introduction*. Oxford University Press, Oxford. Brief but authoritative book by one of the world's finest economists. The best summary we know of the basics of how humanity functions from an economic viewpoint, dealing with such often overlooked factors as *trust* and the significance of discount rates. Highly recommended.

Diamond, J. M. 1997. *Guns, Germs, and Steel: The Fates of Human Societies*. W. W. Norton, New York. Modern classic of cultural macroevolution.

Diamond, J. M. 2005. *Collapse: How Societies Choose to Fail or Succeed*. Viking, New York. Detailed descriptions of a series of social disintegrations. See especially his classic analysis of the ecological collapse on Easter Island.

Ehrlich, P. R. 2000. *Human Natures: Genes, Cultures, and the Human Prospect*. Island Press, Shearwater Books, Washington, DC. Contains more detail on evolution and human evolution than is given here.

Ehrlich, P. R., and A. H. Ehrlich. 2005. *One with Nineveh: Politics, Consumption, and the Human Future*. Repr. with new afterword. Island Press, Shearwater Books, Washington, DC. A recent overview of the human predicament.

Ehrlich, P. R., A. H. Ehrlich, and J. P. Holdren. 1977. *Ecoscience: Population,*

Resources, Environment. W. H. Freeman, San Francisco. Getting old, but contains detailed discussions of many population and environment issues.

Ehrlich, P., and M. Feldman. 2007. Genes, environments, and behaviors. *Daedalus* 136 (Spring): 5–12. Part of chapter 4 is based on this discussion.

Futuyma, D. J. 2005. *Evolution*. Sinauer Associates, Sunderland, MA. An excellent text on modern evolutionary theory.

Godfrey-Smith, P. 2003. *An Introduction to the Philosophy of Science*. University of Chicago Press, Chicago. An excellent brief introduction by one of today's leaders in the field. If you want the deeper side of the "how-science-works" issues that we introduce briefly, this is the place to start.

Homer-Dixon, T. 2006. *The Upside of Down: Catastrophe, Creativity, and the Renewal of Civilization*. Island Press, Shearwater Books, Washington, DC. A broad overview by a political scientist who understands the crucial role of population size and growth in environmental problems.

Levin, S. 1999. *Fragile Dominion*. Perseus Books, Reading, MA. Focuses conservation attention on fundamental issues, viewing the biosphere as a complex adaptive system and providing an accessible discussion of the characteristics of such systems.

Millennium Ecosystem Assessment. 2005. *Ecosystems and Human Well-Being: Current State and Trends, Synthesis*. Island Press, Washington, DC. A comprehensive look at ecosystems and ecosystem services. Full treatment can be found in the Millennium Ecosystem Assessment Series of publications available at http://www.islandpress.org; for an overview, see http://www.islandpress.org/matoolkit/MAToolkit.pdf.

Myers, N. 1979. *The Sinking Ark*. Pergamon Press, New York. Pioneering work on the loss of biodiversity.

National Academy of Sciences USA. 1993. *A Joint Statement by Fifty-eight of the World's Scientific Academies: Population Summit of the World's Scientific Academies*. National Academy Press, New Delhi, India. One of two key warnings issued by the world's top scientists, and ignored by the media (see also Union of Concerned Scientists 1992).

Raven, P. H., and L. R. Berg. 2006. *Environment*. 5th ed. John Wiley and Sons, Hoboken, NJ. A modern environmental sciences text, and a good place to check for more on ecosystems, biogeochemical cycles, and the like.

Raven, P. H., G. B. Johnson, J. B. Losos, and S. R. Singer. 2005. *Biology*. 7th ed. McGraw-Hill, New York. A brilliant modern text that gives more detail on many genetic and evolutionary topics.

Sapolsky, R. *Biology and Human Behavior: The Neurological Origins of Individuality*. 2nd ed. A wonderful taped course from The Teaching Company (http://www.TEACH12.com) presented by one of Stanford's great lecturers.

Steffen, W., A. Sanderson, P. D. Tyson, J. Jäger, P. A. Matson, B. Moore III,

F. Oldfield, et al. 2004. *Global Change and the Earth System: A Planet under Pressure*. Springer, Berlin. A fine overview from the viewpoint of earth scientists.

Union of Concerned Scientists. 1992. *World Scientists' Warning to Humanity*. Union of Concerned Scientists, Cambridge, MA. One of two key warnings issued by the world's top scientists, and ignored by the media (see also National Academy of Sciences USA 1993).

Prologue

Kareiva, P., S. Watts, R. McDonald, and T. Boucher. 2007. Domesticated nature: Shaping landscapes and ecosystems for human welfare. *Science* 316:1866–69. Excellent analysis of human dominance.

Tillyard, E. M. W. 1941. *The Elizabethan World Picture*. Random House, New York. Details on the great chain of being.

Chapter 1: Darwin's Legacy and Mendel's Mechanism

Bradshaw, W. E., and C. M. Holzapfel. 2001. Genetic shift in photoperiodic response correlated with global warming. *Proceedings of the National Academy of Sciences USA* 98:14509–11. More details on pitcher-plant mosquitoes.

Bridgham, J. T., S. M. Carroll, and J. W. Thornton. 2006. Evolution of hormone-receptor complexity by molecular exploitation. *Science* 312:97–101. Important work showing that "intelligent design," the idea that complex molecular interactions had to be designed because they could not evolve, is actually nonsense.

Grant, P. R. 1999. *Ecology and Evolution of Darwin's Finches*. Repr. with new afterword. Princeton University Press, Princeton, NJ. Overview of research on Darwin's finches.

Grant, P. R., and B. R. Grant. 2006. Evolution of character displacement in Darwin's finches. *Science* 313:224–26. More recent results from the Grants' research.

Lenski, R. E., C. Ofria, R. T. Pennock, and C. Adami. 2003. The evolutionary origin of complex features. *Nature* 423:139–44. A theoretical demonstration that mutation and natural selection can produce all the evolutionary novelties we observe.

Lenski, R. E., and M. Travisano. 1994. Dynamics of adaptation and diversification: A 10,000 generation experiment with bacterial populations. *Proceedings of the National Academy of Sciences USA* 91:6808–14. A now-classic example of using microorganisms to study long-term evolutionary processes.

Losos, J. B., T. W. Schoener, R. B. Langerhans, and D. A. Spiller. 2006. Rapid temporal reversal in predator-driven natural selection. *Science* 314:111. Extremely rapid natural selection demonstrated in nature.

Majerus, M. E. N. 1998. *Melanism: Evolution in Action*. Oxford University Press, Oxford. A thorough examination of the peppered moth story.

Voight, B. F., S. Kudaravalli, X. Wen, and J. K. Pritchard. 2006. A map of recent positive selection in the human genome. *PLoS Biology* 4:e72. Humanity still evolving.

Chapter 2: The Entangled Bank

Becerra, J. X. 2003. Synchronous coadaptation in an ancient case of herbivory. *Proceedings of the National Academy of Sciences USA* 100:12804–7. Coevolution in the past.

Berenbaum, M. R. 1983. Coumarins and caterpillars: A case for coevolution. *Evolution* 37:163–69. More on butterfly-plant coevolution by one of the leaders in the field.

Coyne, J., and H. Orr. 2004. *Speciation*. Sinauer Associates, Sunderland, MA. A good overview of recent work on speciation.

Ehrlich, P. R. 2005. Twenty-first century systematics and the human predicament. In *Biodiversity: Past, Present and Future*, Vol. Ser. 4, 55 (Suppl. III), *Proceedings of the California Academy of Sciences*, edited by N. G. Jablonski, 130–48. San Francisco, CA. An example of our views where they are heterodox—not part of the broad taxonomic consensus—see especially the species question.

Ehrlich, P. R., and I. Hanski, eds. 2004. *On the Wings of Checkerspots: A Model System for Population Biology*. Oxford University Press, Oxford. Several chapters bearing on key questions in plant-herbivore coevolution.

Ehrlich, P. R., and P. H. Raven. 1969. Differentiation of populations. *Science* 65:1228–32. A view of ours that is still somewhat heterodox.

Mavarez, J., C. A. Salazar, E. Bermingham, C. D. Jiggins, and M. Linares. 2006. Speciation by hybridization in *Heliconius* butterflies. *Nature* 411:302–5. Details of the longwing case.

Moran, N. A., P. Tran, and N. M. Gerardo. 2005. Symbiosis and insect diversification: An ancient symbiont of sap-feeding insects from the bacterial phylum *Bacteroidetes*. *Applied and Environmental Microbiology* 71:8802–10. An early stage of becoming an organelle?

Pennisi, E. 2007. Variable evolution. *Science* 316:686–87. Overview of current excitement in coevolutionary biology.

Price, T. 2007. *Speciation in Birds*. Roberts and Company, Greenwood Village, CO. This is the book to read if you're a bird-watcher fascinated by speciation. If your fascination is limited, Coyne and Orr 2004 is more general.

Rice, L. B. 2001. Emergence of vancomycin-resistant enterococci. *Emerging Infectious Diseases* 7:183–87. Resistance showing up in a "last-ditch" antibiotic.

Rieseberg, L. H., and J. M. Burke. 2001. A genic view of species integration.

Journal of Evolutionary Biology 14:883–86. A possible solution to the "what binds species into units" dilemma raised by Ehrlich and Raven (1969). This issue of *Evolutionary Biology* is a series of papers that will give you a feel for the modern ferment in speciation theory.

Ritland, D. B., and L. P. Brower. 1991. The viceroy butterfly is not a Batesian mimic. *Nature* 350:497–98. As science advances, classic examples are sometimes reassessed—in this case, that the poisonous monarch is mimicked by the tasty viceroy is called into serious question, since viceroys in at least some populations are unpalatable and monarchs in some populations are tasty.

Schwarzbach, A. E., and L. H. Rieseberg. 2002. Likely multiple origins of a diploid hybrid sunflower. *Molecular Ecology* 11:1703–15. The same species evolving multiple times.

Singer, M. C. 2003. Spatial and temporal patterns of checkerspot butterfly–host plant association: The diverse roles of oviposition preference. In *Butterflies: Ecology and Evolution Taking Flight*, edited by C. L. Boggs, W. B. Watt, and P. R. Ehrlich, 207–28. University of Chicago Press, Chicago. An introduction to some of the complexity of plant-herbivore relationships; also shows another aspect of the role plant-butterfly relationships play as a model system for scientists interested in coevolution.

Thompson, J. N. 2006. Mutualistic webs of species. *Science* 312:372–73.

Zimmerman, E. C. 1970. Adaptive radiation in Hawai'i with special reference to insects. *Biotropica* 2:32–38. Amazing diversification in a limited area—speciation on the fast track.

Chapter 3: Our Distant Past

Daeschler, E. B., N. H. Shubin, and F. A. Jenkins Jr. 2006. A Devonian tetrapod-like fish and the evolution of the tetrapod body plan. *Nature* 440:757–63. Description of *Tiktaalik*, the best fish-to-tetrapod missing-link fossil.

Hazen, R. M. 2005. *Genesis: The Scientific Quest for Life's Origin*. Joseph Henry Press, Washington, DC. An excellent overview.

Klein, R. G. 2009. *The Human Career: Human Biological and Cultural Origins*. 3rd ed. University of Chicago Press, Chicago. This is the premier source for the physical evolution of human beings—and it will give you a good idea of the detailed studies that are required to sort out our history. Forthcoming.

Stringer, C. B., and P. Andrews. 2005. *Complete World of Human Evolution*. Thames and Hudson, London. Excellent overview for the non-scientist.

Stringer, C., and W. Davies. 2001. Those elusive Neanderthals. *Nature* 413:791–92. Look at this to get a feel for some of the complexity of the "fate of the Neanderthals" question.

Thorpe, S. K. S., R. L. Holder, and R. H. Crompton. 2007. Origin of human bipedalism as an adaptation for locomotion on flexible branches. *Science* 316:1328–31. New theory of origins of upright posture.

Wolpoff, M. H., J. Hawks, B. Senut, M. Pickford, and J. Ahern. 2006. An ape or *the* ape: Is the Toumaï cranium TM 266 a hominid? *PaleoAnthropology* 2006:36–50. Discussion of the status of *Sahelanthropus*, a fossil find that may represent either the oldest known hominin or the last shared ancestor of hominins and apes.

Chapter 4: Of Genes and Culture

Allman, J. M. 1999. *Evolving Brains*. Scientific American Library, New York. Well-illustrated comparative overview.

Berton, P. 1977. *The Dionne Years: A Thirties Melodrama*. W. W. Norton, New York. Shows how different the life trajectories of genetically identical human individuals can be, even if their early years are passed in nearly identical environments.

Deutscher, G. 2005. *The Unfolding of Language: An Evolutionary Tour of Mankind's Greatest Invention*. Henry Holt, New York. A brilliant book—highly recommended.

Edelman, G. M. 2006. The embodiment of mind. *Daedalus* 135 (Summer): 23–32. The author estimates the number of synapses in the cortex as a quadrillion. This issue of *Daedalus* has a series of interesting articles on the relationship of mind to body.

Ehrlich, P. R., and M. W. Feldman. 2003. Genes and cultures: What creates our behavioral phenome? *Current Anthropology* 44:87–107. Detailed discussion of the role genes play in behavior.

Goren-Inbar, N., N. Alperson, M. E. Kislev, O. Simchoni, Y. Melamed, A. Ben-Nun, and E. Werker. 2004. Evidence of hominin control of fire at Gesher Benot Ya'aqov, Israel. *Science* 304:725–27. The latest on when our ancestors took up one of their most critical tools.

Horner, V., A. Whiten, E. Flynn, and F. B. M. de Waal. 2006. Faithful replication of foraging techniques along cultural transmission chains by chimpanzees and children. *Proceedings of the National Academy of Sciences USA* 103:13878–83. The experiment demonstrating cultural transmission over generations.

Humphrey, N. K. 1992. *A History of the Mind: Evolution and the Birth of Consciousness*. Chatto and Windus, London. Interesting speculation on the evolution of consciousness.

Klein, R. G. 2009. *The Human Career: Human Biological and Cultural Origins*. 3rd ed. University of Chicago Press, Chicago. Besides being the best source on the bones of our ancestors, this is the best on the stone tool kits they invented. Forthcoming.

Povinelli, D. J., K. E. Nelson, and S. T. Boysen. 1992. Comprehension of role reversal in chimpanzees: Evidence of empathy? *Animal Behaviour* 43:633–40. Important early work on theory of mind.

Scarre, E., ed. 2005. *The Human Past: World Prehistory and the Development of Human Societies*. Thames and Hudson, London. See especially recent material on origins of agriculture.

Waal, F. B. M. de, ed. 2001. *Tree of Origin: What Primate Behavior Can Tell Us about Human Social Evolution*. Harvard University Press, Cambridge, MA. A relatively recent and useful source.

Watters, E. 2006. DNA is not destiny: The new science of epigenetics rewrites the rules of disease, heredity, and identity. *Discover* 22, no. 11 (November). A brief introduction to epigenetics, the study of heritable changes in gene function that cannot be explained by changes in DNA sequence.

Whiten, A., V. Horner, and F. B. M. de Waal. 2005. Conformity to cultural norms of tool use in chimpanzees. *Nature* 437:737–40. Lab experiment showing divergent cultural evolution in chimpanzees.

Chapter 5: Cultural Evolution: How We Relate to One Another

Axelrod, R. 2006. *The Evolution of Cooperation*. Rev. ed. Basic Books, New York. Game theory showing how cooperation can emerge from competition.

Boyd, R., and P. J. Richerson. 2005. *The Origin and Evolution of Cultures*. Oxford University Press, Oxford. A recent overview of this topic by two of the leaders in the field; advanced.

Carneiro, R. L. 1988. The circumscription theory: Challenge and response. *American Behavioral Scientist* 31:497–511. Final article of a series in the same issue of the journal on the theory. The articles are generally positive, and this response was written by the author of the classic 1970 paper that proposed the theory.

Cavalli-Sforza, L. L., and M. W. Feldman. 1981. *Cultural Transmission and Evolution: A Quantitative Approach*. Princeton University Press, Princeton, NJ. Foundational book on the topic. Technical but worth the effort if one is interested in the scientific approach to how cultures change.

Diamond, J. M. 1991. *The Rise and Fall of the Third Chimpanzee*. Radius, London. One of Diamond's great books; on how we treat one another, see the material on genocide, pp. 250–78.

Ehrlich, P. R., and S. A. Levin. 2005. The evolution of norms. *PLoS Biology* 3:943–48. Much of the discussion of norms in the chapter is based on this article.

Goodall, J. 1991. Unusual violence in the overthrow of an alpha male chimpanzee at Gombe. In *Topics in Primatology*, vol. 1 of *Human Origins*, edited by T. Nishida, W. C. McGrew, P. Marler, M. Pickford, and F. B. M. de Waal, 131–42. University of Tokyo Press, Tokyo.

Hawkes, K., J. F. O'Connel, N. G. Blurton Jones, H. Alvarez, and E. L. Charnov. 1998. Grandmothering, menopause, and the evolution of human life histories. *Proceedings of the National Academy of Sciences USA* 95:1336–39. Details on the "grandmother hypothesis."

Hua, C. 2001. *A Society without Fathers or Husbands*. Zone Books, New York. A book that dramatically reveals the complexity of human relationships by detailing what may be the most unusual extant society from the viewpoint of its male-female relationships.

Johnson, A. W., and T. Earle. 1987. *The Evolution of Human Societies: From Foraging Group to Agrarian State*. Stanford University Press, Stanford, CA. A classic on the increasing complexity of human societies.

Keeley, L. H. 1996. *War before Civilization: The Myth of the Peaceful Savage*. Oxford University Press, New York. Details on what is known about warfare among non-industrial societies.

Kelly, R. C. 2000. *Warless Societies and the Origin of War*. University of Michigan Press, Ann Arbor. Very interesting discussion illustrating the complexity of the issue.

Martin, D. L., and D. W. Frayer, eds. 1997. *Troubled Times: Violence and Warfare in the Past*. Gordon and Breach, Amsterdam. Interesting discussion of archaeological evidence for prehistoric violence.

Mithen, S. 2001. The evolution of imagination: An archaeological perspective. *SubStance* 2001:28–54.

Rogers, D. S., and P. R. Ehrlich. 2008. Natural selection and cultural rates of change. *Proceedings of the National Academy of Sciences USA* 105:3416–3420. Describes research comparing evolution in functional and decorative patterns of Polynesian canoes.

Sapolsky, R. M. 2006. A natural history of peace. *Foreign Affairs* 85 (January–February): 104–20.

Sapolsky, R. M., and L. J. Share. 2004. A pacific culture among wild baboons: Its emergence and tradition. *PLoS Biology* 2:0534–41. Cultural evolution in aggression.

Waal, F. B. M. de. 1997. *Bonobo: The Forgotten Ape*. University of California Press, Berkeley. If you want to become familiar with bonobos, this is the book for you.

Chapter 6: Perception, Evolution, and Beliefs

Barbujani, G. 2005. Human races: Classifying people vs. understanding diversity. *Current Genomics* 6:215–26. An overview of recent genetic evidence on its subject, which comes to similar conclusions reached by biologists after the publication of Wilson and Brown's pioneering paper in 1953.

Boyer, P. 1994. *The Naturalness of Religious Ideas: A Cognitive Theory of Religion*.

University of California Press, Berkeley. Provocative and demonstrates how tough explaining religion can be.

Dennett, D. C. 2006. *Breaking the Spell: Religion as a Natural Phenomenon*. Viking, New York. A superb example of the religion of scientism.

Edgerton, R. B. 1992. *Sick Societies: Challenging the Myth of Primitive Harmony*. Free Press, New York. An interesting view on whether we can make value judgments about cultures.

Gazzaniga, M. S. 2005. *The Ethical Brain*. Dana Press, New York. See this for mental disorders and religion.

Gopnik, A., D. M. Sobel, L. E. Schulz, and C. Glymour. 2001. Causal learning mechanisms in very young children: Two-, three-, and four-year-olds infer causal relations from patterns of variation and covariation. *Developmental Psychology* 37:620–29. Cause and effect in children's thinking.

Jablonski, N., and G. Chaplin. 2002. Skin deep. *Scientific American* (October): 74–81. A modern discussion of the evolutionary significance of skin color.

Jebens, H. 2004. *Cargo, Cult, and Culture Critique*. University of Hawai'i Press, Honolulu. A recent work on cargo cults, perhaps the best-known revitalization movements—essentially religions in the process of being born.

Laqueur, W., ed. 2001. *The Holocaust Encyclopedia*. Yale University Press, New Haven, CT. A readable standard reference on the Nazi holocaust, which contains about all the examples of human inhumanity toward defined "others" as one can stand.

Ornstein, R., and P. Ehrlich. 1989. *New World/New Mind: Moving toward Conscious Evolution*. Doubleday, New York. A main source on perception and environmental problems.

Rock, I. 1984. *Perception*. Scientific American Library, W. H. Freeman, New York. A classic source, written by one of the great experimenters in perceptual psychology. Well illustrated.

Rosenberg, N. A., J. K. Pritchard, J. L. Weber, H. M. Cann, K. K. Kidd, L. A. Zhivotovsky, and M. W. Feldman. 2002. Genetic structure of human populations. *Science* 298:2381–85. On how careful genetic analysis can come close to pinpointing an individual's continent of ancestry—a sign of the geographic variation in the human genome.

Sacks, O. 1993. To see or not to see. *New Yorker* (10 May): 1–21. Stories of blind people whose sight has been restored.

Sapolsky, R. M. 2005. *Monkeyluv: And Other Essays on Our Lives as Animals*. Scribner, New York. See pp. 161–79, "The Cultural Desert," on the way desert and rain forest environments can influence the kinds of gods that cultures evolve.

Sillitoe, P. 1998. *An Introduction to the Anthropology of Melanesia: Culture and Tradition*. Cambridge University Press, Cambridge. To expand your view of cultural differences, note the great diversity of cultures in Melanesia and the differences (e.g., direct manipulation of gods; symbolic trade networks) and similarities (male genital mutilation; rites of passage) when compared with our own.

Stark, R. 1996. *The Rise of Christianity: A Sociologist Reconsiders History*. Princeton University Press, Princeton, NJ. How Christianity won out over other Oriental mystery religions in the Roman Empire by promoting "Christian" behavior—essential reading if you are interested in the evolution of religion.

Wilson, E. O., and W. L. Brown. 1953. The subspecies concept and its taxonomic application. *Systematic Zoology* 2:97–111. Classic paper on the biological unreality of races.

Chapter 7: The Ups and Downs of Populations

Connelly, M. 2008. *Unnatural Selection: Population Control and the Struggle to Remake Humanity*. Harvard University Press, Cambridge, MA. A detailed source with a good discussion of early motives and programs.

Daily, G. C., A. H. Ehrlich, and P. R. Ehrlich. 1994. Optimum human population size. *Population and Environment* 15:469–75. A rough cut at determining what might be an optimal human population size.

Daily, G. C., and P. R. Ehrlich. 1992. Population, sustainability, and Earth's carrying capacity. *BioScience* 42:761–71. A discussion of how many people Earth might support.

Donohue, J. J., and S. D. Levitt. 2001. The impact of legalized abortion on crime. *Quarterly Journal of Economics* 116:379–420. Indicates that availability of abortion has accounted for as much as 50 percent of the observed drop in crime rates.

Ehrlich, P. R., and A. H. Ehrlich. 2006. Enough already. *New Scientist* 191:46–50. An overview of the population aging "problem."

Population Reference Bureau. 2008. *2008 World Population Data Sheet*. Population Reference Bureau, Washington, DC. Invaluable source on human population statistics—published annually, and available on the Web.

United Nations. 2007. *World Population Prospects: The 2006 Revision*. United Nations, New York. Comprehensive statistics on the world's populations, revised every two years and available on the Web.

United Nations Population Fund (UNFPA). 2007. *State of the World Population, 2006*. UNFPA, New York. An annual report; available on the Web.

Chapter 8: History as Cultural Evolution

Atkinson, R. 2002. *An Army at Dawn: The War in North Africa, 1942–1943*. Henry Holt, New York. Shows the impact of cultural microevolution—poor preparation on the part of the U.S. military could greatly affect military outcomes.

Barnett, C. 1983. *The Desert Generals*. 2nd ed. Cassell and Company, London. Another wonderful source on microevolution in the military.

Basalla, G. 1988. *The Evolution of Technology*. Cambridge University Press, Cambridge. A brilliant and innovative overview on which we based much of our treatment.

Braudel, F. 1993. *A History of Civilizations*. Penguin Books, New York. An outstanding example from the Annales school.

Crosby, A. 1986. *Ecological Imperialism: The Biological Expansion of Europe, 900–1900*. Cambridge University Press, Cambridge. The role of disease, parasites, and other ecological factors in letting Europeans take over most of the world.

Fracchia, J., and R. C. Lewontin. 1999. Does culture evolve? *History and Theory* 38:52–78. A thought-provoking article that convincingly demonstrates that overall Darwinian evolution is a poor model for cultural evolution. The authors claim that history shouldn't be seen as a story of cultural evolution—on this point we disagree. Well worth reading.

Jannetta, A. 2007. *The Vaccinators: Smallpox, Medical Knowledge, and the "Opening" of Japan*. Stanford University Press, Stanford, CA. A wonderful history book, describing part of the conquest of one of humanity's most important diseases, which also could serve as a textbook on cultural evolution.

Kay, J. H. 1997. *Asphalt Nation: How the Automobile Took Over America and How We Can Take It Back*. Crown, New York. Analysis of what may have been America's biggest domestic organizational mistake.

McPherson, J. M. 2007. *This Mighty Scourge: Perspectives on the Civil War*. Oxford University Press, Oxford. A wonderful read, the first chapter of which deals with the "slavery as the cause of the war" issue.

Tuchman, B. W. 1984. *The March of Folly: From Troy to Vietnam*. Alfred A. Knopf, New York. Four wonderful examples of folly (and cultural stickiness)—defined in this case as governments blundering down ill-chosen paths against the advice of knowledgeable people.

Chapter 9: Cycles of Life (and Death)

Baskin, Y. 2005. *Under Ground: How Creatures of Mud and Dirt Shape Our World*. Island Press, Shearwater Books, Washington, DC. A beautiful look at an unseen world.

Chapin, F. S., III, H. A. Mooney, and P. A. Matson. 2004. *Principles of Terrestrial Ecosystem Ecology*. Springer, Berlin. Authoritative and clearly written.

Cohen, J. E., and D. Tilman. 1996. Biosphere 2 and biodiversity: The lessons so far. *Science* 274:1150–51. Source on the attempt to replicate Earth on a small scale.

Ehrlich, P. R. 1991. Coevolution and its applicability to the Gaia hypothesis. In *Scientists on Gaia*, edited by S. H. Schneider and P. J. Boston, 19–22. MIT Press, Cambridge, MA.

Odling-Smee, F. J., K. N. Laland, and M. W. Feldman. 2003. *Niche Construction: The Neglected Process in Evolution*. Princeton University Press, Princeton, NJ. A comprehensive look.

Schneider, S. H. 1997. *Laboratory Earth: The Planetary Gamble We Can't Afford to Lose*. Basic Books, New York. Authoritative.

Vitousek, P. M., J. D. Aber, R. Howarth, G. Likens, P. A. Matson, W. Schlesinger, D. Schindler, and D. Tilman. 1997. *Human Alteration of the Nitrogen Cycle*. Ecological Society of America, Washington, DC. Excellent summary.

Chapter 10: Ecosystems and Human Domination of Earth

Beattie, A. J., and P. R. Ehrlich. 2001. *Wild Solutions: How Biodiversity Is Money in the Bank*. Yale University Press, New Haven, CT. A popular overview of the various benefits (especially novel products) that are supplied to humanity by ecosystems, and a simple description of how ecosystems work.

Brauman, K. A., G. C. Daily, T. K. Duarte, and H. A. Mooney. 2007. The nature and value of ecosystem services: An overview highlighting hydrologic services. *Annual Review of Environment and Resources* 32:67–98. A fine recent overview.

Ehrlich, P. R., and J. Holdren. 1971. Impact of population growth. *Science* 171:1212–17. Foundational work on the $I = PAT$ analysis.

Ehrlich, P. R., and H. A. Mooney. 1983. Extinction, substitution, and ecosystem services. *BioScience* 33:248–54. Reviews the possibilities of maintaining the services artificially—and shows they are very limited.

Holdren, J. P., and P. R. Ehrlich. 1974. Human population and the global environment. *American Scientist* 62:282–92. Basic paper on ecosystem services.

Jackson, J. B. C. 2008. Ecological extinction and evolution in the brave new ocean. *Proceedings of the National Academy of Sciences USA* (in press). A very depressing but important report.

Odling-Smee, F. J., K. N. Laland, and M. W. Feldman. 2003. *Niche Construction: The Neglected Process in Evolution*. Princeton University Press, Princeton, NJ. A comprehensive look at this phenomenon, which is far from restricted to *Homo sapiens*.

Chapter 11: Consumption and Its Costs

Colborn, T., D. Dumanoski, and J. P. Myers. 1996. *Our Stolen Future*. Dutton, New York. Classic source on hormone-mimicking chemicals.

Daily, G. C., and P. R. Ehrlich. 1996. Impacts of development and global change on the epidemiological environment. *Environment and Development Economics*

1:309–44. Detail and references on the increasing threats of infectious diseases.

Davis, M. 2001. *Late Victorian Holocausts: El Niño Famines and the Making of the Third World*. Verso, New York. A controversial book on the role of the rich in generating poverty.

Gleick, P. 2007. *The World's Water 2006–2007*. Island Press, Washington, DC. The fifth in a series of comprehensive biennial reports that are a basic source on humanity's water supply.

Harte, John. 2007. Human population as a dynamic factor in environmental degradation. *Population and Environment* 28:223–236. Details disproportionate impacts of population growth.

Krueger, A. B. 2004. Inequality: Too much of a good thing. In *Inequality in America: What Role for Human Capital Policies?*, edited by J. J. Heckman and A. B. Krueger, 1–75. MIT Press, Cambridge, MA.

Laurance, W. F., and C. A. Peres, eds. 2006. *Emerging Threats to Tropical Forests*. University of Chicago Press, Chicago. The most up-to-date discussion of the loss of Earth's greatest reservoir of terrestrial biodiversity.

Liu, J., G. Daily, P. R. Ehrlich, and G. Luck. 2003. Effects of household dynamics on resource consumption and biodiversity. *Nature* 421:530–33. Basic paper on demographics of household size and environment.

Myers, J. P., R. T. Zoeller, and F. S. von Saal. 2008. A clash of scientific disciplines, with important implications for public health. Submitted to *Science*. Points out that some hormone-mimicking chemicals may be more dangerous in low doses than high doses.

Myers, N., and J. Kent. 2004. *The New Consumers: The Influence of Affluence on the Environment*. Island Press, Washington, DC. The key source.

Smith, A. 1974 (1759). *The Theory of Moral Sentiments*. Clarendon Press, Oxford. An important predecessor of *The Wealth of Nations*, in which Smith makes clear that once basic necessities are cared for, one should turn away from economic goals.

Smith, K. R. 2006. Health impacts of household fuelwood use in developing countries. *Unasylva* 224, no. 57:1–4. A superb brief treatment that shows one major way the poor suffer health effects from their poverty.

Chapter 12: A New Imperative

Bierregaard, R. O., Jr., T. E. Lovejoy, V. Kapos, A. A. D. Santos, and R. W. Hutchings. 1992. The biological dynamics of tropical rainforest fragments: A prospective comparison of fragments and continuous forest. *BioScience* 42:859–66. An early description of the Biological Dynamics of Forest Fragments Project, when it was known as the Critical Size of Ecosystems Project.

Colborn, T., D. Dumanoski, and J. P. Myers. 1996. *Our Stolen Future*. Dutton, New York. Classic source on hormone-mimicking chemicals, which, for example, *may* be lowering human sperm counts as well as disrupting the development of human children and other organisms.

Ehrlich, P. R., A. H. Ehrlich, and G. C. Daily. 1995. *The Stork and the Plow: The Equity Answer to the Human Dilemma*. Putnam, New York. More on agriculture and the food-population balance.

Flannery, T. 1994. *The Future Eaters: An Ecological History of the Australasian Lands and People*. Grove Press, New York. Detailed exploration of how humans conquered Australia and its megafauna.

Kelly, B. C., M. G. Ikonomou, J. D. Blair, A. E. Morin, and F. A. P. C. Gobas. 2007. Food web–specific biomagnification of persistent organic pollutants. *Science* 317:236–38. Latest on the complexities of the accumulation of persistent organic pollutants (POPs)—see also a more popular description on pp. 182–83 of the same *Science* issue.

Laurance, W. F., et al. 2006. Rapid decay of tree-community composition in Amazonian forest fragments. *Proceedings of the National Academy of Sciences USA* 103:19010–14. Fragmentation hurts.

Lobell, D. D., and C. B. Field. 2007. Global scale climate–crop yield relationships and the impacts of recent warming. *Environmental Research Letters* 2:014002. Very bad news for humanity's food supply if warming continues.

Loreau, M., S. Naeem, P. Inchausti, J. Bengtsson, J. P. Grime, A. Hector, D. U. Hooper, et al. 2001. Biodiversity and ecosystem functioning: Current knowledge and future challenges. *Science* 294:804–8. A fine review.

Pauly, D., V. Christensen, J. Dalsgaard, R. Froese, and F. Torres Jr. 1998. Fishing down marine food webs. *Science* 279:860–63. The fate of fisheries.

Pollan, M. 2006. *The Omnivore's Dilemma: A Natural History of Four Meals.* Penguin Books, New York. Essential source on the dark side of industrial agriculture. Highly recommended.

Pringle, R. M. 2005. The origins of the Nile perch in Lake Victoria. *BioScience* 55:780–87. Birth of an invasive species.

Sterner, T., M. Troell, J. Vincent, S. Aniyar, S. Barrett, W. Brock, S. Carpenter, et al. 2006. Quick fixes for the environment: Part of the solution or part of the problem? *Environment* 48:19–27. Overview of tactics for solving environmental problems.

Stuart, S., J. S. Chanson, N. A. Cox, B. E. Young, A. S. L. Rodrigues, D. L. Fishman, and R. W. Waller. 2004. Status and trends of amphibian declines and extinctions worldwide. *Science* 306:1783–86.

Tilman, D., P. B. Reich, J. Knops, D. Wedin, T. Mielke, and C. Lehman. 2001. Diversity and productivity in a long-term grassland experiment. *Science* 294:843–45. An example of a fine field experiment on a key topic related to agriculture.

Vitousek, P. M. 1994. Beyond global warming: Ecology and global change. *Ecology* 75:1861–76. A pioneering summary—still well worth reading; has references to many key papers.

Vitousek, P. M., P. R. Ehrlich, A. H. Ehrlich, and P. A. Matson. 1986. Human appropriation of the products of photosynthesis. *BioScience* 36:368–73. Shows how scientists calculate such global numbers and why they are rough approximations.

Wackernagel, M., N. B. Schulz, D. Deumling, A. C. Linares, M. Jenkins, V. Kapos, C. Monfreda, et al. 2002. Tracking the ecological overshoot of the human economy. *Proceedings of the National Academy of Sciences USA* 99:9266–71. Humanity's footprint.

Worm, B., E. B. Barbier, N. Beaumont, D. J. Emmett, C. Folke, B. S. Halpern, J. B. C. Jackson, et al. 2006. Impacts of biodiversity loss on ocean ecosystem services. *Science* 314:787–90.

Chapter 13: Altering the Global Atmosphere

Ayres, J. G. 2006. *Air Pollution and Health*. Imperial College Press, London. An overview of the more conventional issues related to using the atmosphere as a sewer.

Canadell, J. G., C. Le Quéré, M. R. Raupach, C. B. Field, E. T. Buitenhuis, P. Ciais, T. J. Conway, et al. 2007. Contributions to accelerating atmospheric CO_2 growth from economic activity, carbon intensity, and efficiency of natural sinks. *Proceedings of the National Academy of Sciences USA* 104:18866–70. Frightening paper showing that the growth rate of CO_2 in the atmosphere is increasing rapidly.

Dukes, J. S., N. R. Chiariello, E. E. Cleland, L. A. Moore, M. R. Shaw, S. Thayer, T. Tobeck, H. A. Mooney, and C. B. Field. 2005. Responses of grassland production to single and multiple global environmental changes. *PLoS Biology* 3:1829–37. A classic example of biological research attempting to sort out impacts of atmospheric global change.

Ehrlich, P. R., J. Harte, M. A. Harwell, P. H. Raven, C. Sagan, G. M. Woodwell, et al. 1983. Long-term biological consequences of nuclear war. *Science* 222: 1293–1300.

Field, C. B., and M. R. Raupach, eds. 2004. *The Global Carbon Cycle: Integrating Humans, Climate, and the Natural World*. Island Press, Washington, DC. Many interesting contributions.

Gelbspan, R. 1997. *The Heat Is On: The High Stakes Battle over Earth's Threatened Climate*. Addison-Wesley, Reading, MA. A fine introduction to the politics of global heating and the successful efforts of paid industry operatives to delay action on this crucial issue.

Goulder, L. H., and B. M. Nadreau. 2002. International approaches to reducing greenhouse gases. In *Climate Change Policy*, edited by S. H. Schneider, A.

Rosencranz, and J. O. Niles, 115–49. Island Press, Washington, DC. Goulder is a leader in thinking about cap-and-trade and tax shifting.

Hames, R. S., K. V. Rosenberg, J. D. Lowe, S. E. Barker, and A. A. Dhondt. 2002. Adverse effects of acid rain on the distribution of the wood thrush *Hylocichla mustelina* in North America. *Proceedings of the National Academy of Sciences USA* 99:11235–40. A recent indicator of why acid precipitation should be carefully monitored.

Harte, J. 1988. *Consider a Spherical Cow: A Course in Environmental Problem Solving.* University Science Books, Sausalito, CA. A wonderful introduction to quantitative approaches to solving the human predicament. If you like this one, there is a second *Cow*, with more adventures, same author, published in 2001.

Holdren, J. P. 2002. Beyond the Moscow Treaty. Testimony before Foreign Relations Committee of the United States Senate, Hearings on Treaty on Strategic Offensive Reductions, 12 September.

Johnson, C. 2007. *Nemesis: The Last Days of the American Republic.* Metropolitan Books, New York. See chapter 6 for the pollution of space.

Kintisch, E. 2007. Scientists say continued warming warrants closer look at drastic fixes. *Science* 318:1054–55. Examining large-scale, dangerous experiments to slow warming.

Lobell, D. D., and C. B. Field. 2007. Global scale climate–crop yield relationships and the impacts of recent warming. *Environmental Research Letters* 2:014002, doi:10.1088/1748-9326/2/1/014002. Climate change is already hurting crop yields.

Pacala, S., and R. Socolow. 2004. Stabilization wedges: Solving the climate problem for the next fifty years with current technologies. *Science* 305:968–72. The portfolio approach to the difficult problem of damping the disruptive effects of climate change.

Parmesan, C. 2006. Ecological and evolutionary responses to recent climate change. *Annual Review of Ecology, Evolution, and Systematics* 37:637–69. Overview of changes in distribution of organisms already thought to be responses to climate change.

Root, T. L., D. P. MacMynowski, M. D. Mastandrea, and S. S. Schneider. 2005. Human-modified temperatures induce species changes: Joint attribution. *Proceedings of the National Academy of Sciences USA* 102:7465–69. A key paper tying human actions to changing distributions.

Scheffer, M., and S. R. Carpenter. 2003. Catastrophic regime shifts in ecosystems: Linking theory to observation. *Trends in Ecology and Evolution* 18:648–56. A fine overview of the kinds of phenomena that make ecologists nervous about the human future.

Schneider, S. H. 2007. Climate Change. http://stephenschneider.stanford.edu/. Best Web site on global climate disruption.

Victor, D. G., and D. Cullenward. 2007. Making carbon markets work. *Scientific American* (24 September), http://www.sciam.com/article.cfm?id=making-carbon-markets-work. Excellent, in-depth analysis.

Chapter 14: Energy: Are We Running Out of It?

Brown, L. R. 2006. Rescuing a planet under stress. *The Futurist* 40 (July–August). Contains an optimistic and we hope realistic view of the possibilities of wind power.

Casten, T. R. 1998. *Turning Off the Heat*. Prometheus, Amherst, NY. Analysis of energy systems and how they could be made much more efficient.

Fargione, J., J. Hill, D. Tilman, S. Polasky, and P. Hawthorne. 2008. Land clearing and the biofuel carbon debt. *Science* 319:1235–1238 (February). Concludes that whether or not biofuels reduce the flux of CO_2 depends on how they are produced. Converting rain forests, peatlands, savannas, or grasslands to produce biofuels in Brazil, Southeast Asia, and the United States creates a "carbon debt" by releasing 17 to 800 times more CO_2 than the annual greenhouse gas reductions from the resulting displacement of fossil fuels. In contrast, biofuels made from waste biomass or from biomass grown on abandoned agricultural lands as perennial plants offer immediate and sustained reductions in greenhouse gas releases.

Geller, H. 2003. *Energy Revolution: Policies for a Sustainable Future*. Island Press, Washington, DC.

Hill, J., E. Nelson, D. Tilman, S. Polasky, and D. Tiffany. 2006. Environmental, economic, and energetic costs and benefits of biodiesel and ethanol biofuels. *Proceedings of the National Academy of Sciences USA* 103:11206–10. An authoritative source on biofuels.

Holdren, J. P. 1991. Population and the energy problem. *Population and Environment* 12:231–55. A short, classic paper that, with adjustments in the numbers for the passage of time, still gives a brilliant overview of the basis of the Holdren scenario.

Holdren, J. P. 2006. The energy innovation imperative: Addressing oil dependence, climate change, and other 21st century energy challenges. *Innovations: Technology/Governance/Globalization* (Spring): 3–23. A fine overview by today's top energy-environment analyst.

Victor, D. G., and D. Cullenward. 2007. Can we stop global warming? The only practical approach is to pursue technologies that burn coal more cleanly. *Boston Review* 32, no. 1 (January–February). An excellent summary of the coal situation.

Chapter 15: Saving Our Natural Capital

Aronson, J., S. J. Milton, and J. N. Blignaut, eds. 2007. *Restoring Natural Capital: Science, Business, and Practice*. Island Press, Washington, DC. The best recent book on restoration ecology.

Baskin, Y. 2002. *A Plague of Rats and Rubbervines: The Growing Threat of Species Invasions*. Island Press, Shearwater Books, Washington, DC. A fine non-technical summary.

Beissinger, S. R., and D. R. McCullough, eds. 2002. *Population Viability Analysis*. University of Chicago Press, Chicago. Leading scientists discuss how to estimate extinction risk and how to reduce that risk.

Boggs, C., C. Holdren, I. Kulahci, T. Bonebrake, B. Inouye, J. Fay, A. McMillan, et al. 2006. Delayed population explosion of an introduced butterfly. *Journal of Animal Ecology* 75:466–75. Recent paper on a classic transplant experiment, which shows how difficult it is to achieve success.

Brosi, B. J., G. C. Daily, and P. R. Ehrlich. 2007. Bee community shifts with landscape context in a tropical countryside. *Ecological Applications* 17:418–30. Studying the dynamics of a pollinator community in a disturbed landscape.

Ceballos, G., and P. Ehrlich. 2006. Global mammal distributions, biodiversity hotspots, and conservation. *Proceedings of the National Academy of Sciences USA* 103:19374–78. Looks closely at the nature of hotspots.

Courchamp, F., E. Angulo, P. Rivalan, R. J. Hall, L. Signoret, L. Bull, and L. Meinard. 2006. Rarity value and species extinction: The anthropogenic Allee effect. *PLoS Biology* 4:e415. The Allee effect is a tendency for reproductive success to decline as populations grow smaller, through such mechanisms as increasing difficulty of finding mates.

Daily, G. C., P. R. Ehrlich, and A. Sanchez-Azofeifa. 2001. Countryside biogeography: Utilization of human-dominated habitats by the avifauna of southern Costa Rica. *Ecological Applications* 11:1–13. Demonstrates that "forest birds" can utilize agricultural countryside in some circumstances.

Daily, G. C., and K. Ellison. 2002. *The New Economy of Nature: The Quest to Make Conservation Profitable*. Island Press, Shearwater Books, Washington, DC. Pioneers a new, and newly successful, approach to aligning financial and conservation incentives.

Daly, H. E., and J. Farley. 2004. *Ecological Economics: Principles and Applications*. Island Press, Washington, DC. See especially chapter 21 on direct regulation and marketable permits.

Damschen, E. L., N. M. Haddad, J. L. Orrock, J. J. Tewksbury, and D. J. Levey. 2006. Corridors increase plant species richness at large scales. *Science* 313: 1284–86. Recent addition to a series of papers that demonstrates the Haddad group's approach to corridor research.

Donlan, J. C., J. Berger, C. E. Bock, J. H. Bock, D. A. Burney, J. A. Estes, D. Foreman, et al. 2006. Pleistocene rewilding: An optimistic agenda for twenty-first century conservation. *American Naturalist* 168:660–81. Overview of the most ambitious "reserve" approach.

Ehrlich, P. R., and A. H. Ehrlich. 1981. *Extinction: The Causes and Consequences of the Disappearance of Species*. Random House, New York. See especially chapter 6, on esthetic and ethical reasons for preserving biodiversity, and chapter 5, on ecosystem services.

Ehrlich, P. R., and I. Hanski, eds. 2004. *On the Wings of Checkerspots: A Model System for Population Biology*. Oxford University Press, Oxford. Shows, among many other things, the complexity of determining conservation strategies for a single butterfly species.

Goble, D. D., J. M. Scott, and F. W. Davis. 2005. *Endangered Species Act at Thirty: Renewing the Conservation Promise*. Island Press, Washington, DC.

Hilty, J. A., W. Z. Lidicker Jr., and A. M. Merenlender. 2006. *Corridor Ecology: The Science and Practice of Linking Landscapes for Biodiversity Conservation*. Island Press, Washington, DC. A fine overview.

Holl, K. D. 1996. The effect of coal surface mine reclamation on diurnal lepidopteran conservation. *Journal of Applied Ecology* 33:225–36. An example of nitty-gritty work in restoration ecology.

Hughes, J. B., G. C. Daily, and P. R. Ehrlich. 1997. Population diversity: Its extent and extinction. *Science* 278:689–92. Key paper on the importance of populations.

Janzen, D. H. 2000. Costa Rica's Area de Conservación Guanacaste: A long march to survival through non-damaging biodevelopment. *Biodiversity* 1:7–20. A success story put together by one of Earth's leading tropical biologists.

Kareiva, P., and M. Marvier. 2003. Conserving biodiversity coldspots. *American Scientist* 91:344–51. An important article showing the need to modify the hotspot approach to conservation.

Lindenmayer, D. B., and R. J. Hobbs, eds. 2007. *Managing and Designing Landscapes for Conservation*. Blackwell, London. Shows the great complexity of issues facing managers.

MacArthur, R. H., and E. O. Wilson. 1967. *The Theory of Island Biogeography*. Princeton University Press, Princeton, NJ. This book summarizes the original theory; their pathbreaking 1963 paper in the journal *Evolution* was titled "An equilibrium theory of insular zoogeography." Either one will let you see how a simple mathematical theory can be developed and help scientists to understand a complex issue.

McLachlan, J. S., J. J. Hellmann, and M. W. Schwartz. 2007. A framework for debate of assisted migration in an era of climate change. *Conservation Biology* 21:297–302, doi:10.1111/j.1523-1739.2007.00676.x. An overview on the transplantation debate.

Pimm, S. L., and R. A. Askins. 1995. Forest losses predict bird extinctions in eastern North America. *Proceedings of the National Academy of Sciences USA* 92:9343–47. One pioneering paper from the now-vast literature on extinctions that shows the predictability of the process.

Quammen, D. 1997. *The Song of the Dodo: Island Biogeography in an Age of Extinctions*. Scribner, New York. A well-written popular exposition.

Ricketts, T. H. 2001. The matrix matters: Effective isolation in fragmented landscapes. *American Naturalist* 158:87–99. Shows why pure island biogeographic theory does not work so well with habitat islands.

Ricketts, T. H., G. C. Daily, P. R. Ehrlich, and C. D. Michener. 2004. Economic value of tropical forest to coffee production. *Proceedings of the National Academy of Sciences USA* 101:12579–82. How leaving some potential farmland in forest can increase agricultural productivity.

Roberts, C. 2007. *Unnatural History of the Sea*. Island Press, Shearwater Books, Washington, DC. The latest word on the rape of the seas.

Sekercioglu, C. H., S. R. Loarie, V. Ruiz-Gutierrez, F. Oviedo Brenes, G. C. Daily, and P. R. Ehrlich. 2007. Persistence of forest birds in tropical countryside. *Conservation Biology* 21:482–94. Shows that mixed agricultural countryside has conservation value.

Sodhi, N., B. W. Brook, and C. J. A. Bradshaw. 2007. *Tropical Conservation Biology*. Blackwell, Oxford. A first-rate book on conservation in the tropics, where the need is greatest.

Soulé, M. E. 1999. An unflinching vision: Networks of people for networks of wildlands. *Wildlands* 9:38–46. Ambitious restoration ecology.

Sterner, T. 2003. *Policy Interests for Environmental and Natural Resource Management*. Resources for the Future, Washington, DC. Excellent, detailed source on how to apply economics to policy issues in resource management.

Vellend, M., L. J. Harmon, J. L. Lockwood, M. M. Mayfield, A. R. Hughes, J. P. Wares, and D. F. Sax. 2007. Effects of exotic species on evolutionary diversification. *Trends in Ecology and Evolution* 22:481–88. Discusses some positive effects of invasive species on biodiversity.

Whittaker, R. J., and J. M. Fernández-Palacios. 2007. *Island Biogeography: Ecology, Evolution, and Conservation*. 2nd ed. Oxford University Press, Oxford. A fine modern treatment.

Wilcove, D. S. 1999. *The Condor's Shadow: The Loss and Recovery of Wildlife in America*. W. H. Freeman, New York. A very readable overview.

Wuethrich, B. 2007. Reconstructing Brazil's Atlantic rainforest. *Science* 315:1070–72. Shows how complex restoration can be.

Chapter 16: Governance: Tackling Unanticipated Consequences

Barber, B. R. 1995. *Jihad vs. McWorld*. Ballantine Books, New York. Provocative discussion of globalism versus tribalism.

Barrett, S. 2003. *Environment and Statecraft: The Strategy of Environmental Treaty-Making*. Oxford University Press, New York. A discussion of how to make treaties more effective from a viewpoint of economics and game theory.

Barrett, S. 2007. *Why Cooperate: The Incentive to Supply Global Public Goods*. Oxford University Press, Oxford. A brilliant book that shows how international cooperation could be achieved on such crucial issues as controlling greenhouse gas emissions and eliminating nuclear proliferation.

Dworkin, R. 2000. Free speech and the dimensions of democracy. In *If Buckley Fell: A First Amendment Blueprint for Regulating Money in Politics*, edited by E. J. Rosenkranz, 63–102. Century Foundation Press, New York. A definitive look at the consequences of the *Buckley v. Valeo* Supreme Court decision, which made it unconstitutional to cap the amount of money that could be spent on election campaigns.

Ehrlich, P. R. 2006. Environmental science input to public policy. *Social Research* 73:915–48. A close look at the influence of environmental science on the political system.

Gazzaniga, M. S. 2005. *The Ethical Brain*. Dana Press, New York. Controversial views on many of today's tough ethical issues.

Goulder, L. H., and R. N. Stavins. 2002. An eye on the future. *Nature* 419:673–4. A closer look at the mysteries of discounting.

Jianguo, L., T. Dietz, S. R. Carpenter, M. Alberti, C. Folke, et al. 2007. Complexity of coupled human and natural systems. *Science* 317:1513–16. Excellent overview with lots of examples.

Johnson, C. 2007. *Nemesis: The Last Days of the American Republic*. Metropolitan Books, New York. A provocative summary of the record of the American empire with thorough references, raising the issue of whether the nation can avoid becoming a dictatorship as the military and executive continually accumulate power.

Klare, M. T. 2001. *Resource Wars: The New Landscape of Global Conflict*. Metropolitan Books, New York. A classic work and the source of much of the material on resource wars in this chapter.

Klare, M. T. 2004. *Blood and Oil: The Dangers and Consequences of America's Growing Dependency on Imported Petroleum*. Metropolitan Books, New York. The leading resource war analyst's views; see especially his comments on the 2003 war in Iraq.

Milanovic, B. 2005. *Worlds Apart: Measuring International and Global Inequity*. Princeton University Press, Princeton, NJ. An excellent close look at inequity in a globalizing world.

Mukherjee, V., and G. Gupta. 2006. Of guns and trees: Impact of terrorism on forest conservation. *Environment and Development Economics* 11:221–33.

Ostrom, E., M. A. Janssen, and J. M. Anderies. 2007. Going beyond panaceas. *Proceedings of the National Academy of Sciences USA* 104:15176–78. Warning against settling on simple solutions to the problems of governance of social-ecological systems.

Perrow, C. 2007. *The Next Catastrophe: Reducing Our Vulnerabilities to Natural, Industrial, and Terrorist Disasters*. Princeton University Press, Princeton, NJ. An important source on how to increase the resilience of our complex societies.

Sassen, S. 2006. *Territory, Authority, Rights*. Princeton University Press, Princeton, NJ. An interesting analysis of globalization, well worth reading, though it lacks adequate discussion of the population-resource-environment issues that will soon be the dominant factors in discussions of globalization.

Sunstein, C. R. 2007. *Republic.com 2.0*. Princeton University Press, Princeton, NJ. Important analysis of possible influence of the Internet on politics.

Tainter, J. A. 1988. *The Collapse of Complex Societies*. Cambridge University Press, Cambridge. Classic work on collapse.

Epilogue

Ehrlich, P. R., G. Wolff, G. C. Daily, J. B. Hughes, S. Daily, M. Dalton, and L. Goulder. 1999. Knowledge and the environment. *Ecological Economics* 30:267–84. A close look at issues concerning the "information revolution."

Oak Ridge National Laboratory. 1968. *Nuclear Energy Centers, Industrial and Agro-industrial Complexes*. Summary Report, ORNL-4291. One of the Pollyannaish proposals for feeding everyone, of a sort common forty years ago. There are still almost a billion very hungry people today.

Ornstein, R., and P. Ehrlich. 1989. *New World/New Mind: Moving toward Conscious Evolution*. Doubleday, New York. This is the source on slow change and the need for directed cultural evolution of our perceptual systems.

Acknowledgments

Our research group, now Stanford University's Center for Conservation Biology (CCB), for almost four decades has specialized in trying to keep the "big picture" in focus, not just looking at artificially isolated problems and how they might be solved if society poured all its resources into the solution of one while neglecting other related problems. Our debt to the many group members who have discussed these issues with us over the years is enormous.

Some of the most outstanding scholars dedicated to understanding the human condition have reviewed our earlier writings on the subject (especially *Human Natures* or *One with Nineveh*), selected portions of this manuscript, or this one in its entirety, or have put much time into discussing issues with us. They include John Allman (Division of Biology, California Institute of Technology); Kenneth Arrow and Lawrence Goulder (Department of Economics, Stanford University); Steven Beissinger (Department of Environmental Science, Policy & Management, University of California, Berkeley); Derek Bickerton (Department of Linguistics, University of Hawai'i, Honolulu); Carol Boggs, Kate Brauman, Berry Brosi, Gretchen Daily, Marcus Feldman, Elizabeth Hadly, Susan McConnell, Harold Mooney, Robert Pringle, Deborah Rogers, Terry Root, Joan Roughgarden, Robert Sapolsky, Stephen Schneider, Ward Watt, and Charles Yanofsky (Department of Biological Sciences, Stanford University); Margaret M. Breinholt (Office of the General Counsel, U.S. Department of Agriculture); Nona Chiariello (Research Coordinator, Jasper Ridge Biological Preserve, Stanford University); Danny Cullenward (Program on Energy and Sustainable Development, Stanford University); Lisa Daniel (Department of Economics, University of Maryland); Timothy Daniel (NERA, Inc.); Partha Dasgupta (Faculty of Economics, University of Cambridge); Jared Diamond (Department of Geography, University of California, Los Angeles); Walter Falcon, Donald Kennedy, and Rosamond Naylor (Woods Institute for the Environment, Stanford University); Sylvia Fallon (Natural Resources Defense Council); Christopher B. Field (Director, Department of Global Ecology, Carnegie Institution); Carl Folke, Karl-Göran Mäler, Tasos Xepapadeas, and

Aart de Zeeuw (Beijer Institute of Ecological Economics, Stockholm); and Peter and Rosemary Grant (Department of Ecology and Evolutionary Biology, Princeton University).

Corey S. Goodman (Department of Molecular and Cell Biology, University of California, Berkeley); Geoffrey Heal (Graduate School of Business, Columbia University); Thomas Heller (School of Law, Stanford University); Jessica Hellmann (Department of Biological Sciences, University of Notre Dame); Ann Holdren (anthropologist, Monterey, California); John P. Holdren (John F. Kennedy School of Government, Harvard University; Woods Hole Research Center); Jean-Jacques Hublin (CNRS IRESCO, Paris); David Inouye (Department of Biology, University of Maryland); Nina Jablonski (Department of Anthropology, Pennsylvania State University); Allen W. Johnson (Department of Anthropology, University of California, Los Angeles); Leslie Kaufman (Department of Biology, Boston University); Patrick Kirch (Department of Anthropology, University of California, Berkeley); Richard Klein (Department of Anthropology, Stanford University); Jon Krosnick (Departments of Communication and Political Science, Stanford University); Simon A. Levin (Department of Ecology and Evolutionary Biology, Princeton University); Thomas E. Lovejoy (The Heinz Center); Jennifer B. H. Martiny (Department of Ecology and Evolutionary Biology, University of California, Irvine); Charles D. Michener (Department of Entomology, University of Kansas); Robert Ornstein (Institute for the Study of Human Knowledge); Sally Otto (Department of Zoology, University of British Columbia); Graham Pyke (Macquarie University, New South Wales, Australia); Taylor Ricketts (World Wildlife Fund); Merritt Ruhlen (linguist, Palo Alto, California); James Salzman (School of Law, Duke University); Kirk Smith (Department of Environmental Health Sciences, University of California, Berkeley); Robert R. Sokal (Department of Ecology and Evolution, State University of New York, Stony Brook); Michael Soulé (Paonia, Colorado); Donald Symons (Department of Anthropology, University of California, Santa Barbara); Timothy Wirth (United Nations Foundation); Wren Wirth (Winslow Foundation); and Edward Zackery (Ed Zackery Expeditions). We are indebted to all of them, as well as to many colleagues at the Beijer Institute of Ecological Economics of the Royal Swedish Academy of Sciences, whose stimulating discussions over seventeen years have changed our thinking on many issues. We also owe much to the members and correspondents of the Sierra Club's Global Warming and Energy Committee, whose e-mail discussions have added considerably to our understanding of these rapidly evolving issues over the past several years.

In particular, Steve Beissinger, Margaret Breinholt, Berry Brosi, Danny Cullenward, Gretchen Daily, Tim and Lisa Daniel, Partha Dasgupta, Marc Feldman, Chris Field, Larry Goulder, John Harte, John Holdren, Dave Inouye,

Richard Klein, Tom Lovejoy, Hal Mooney, Rob Pringle, Graham Pyke, Robert Sapolsky, Kirk Smith, Tim and Wren Wirth, and Charley Yanofsky have put special effort into this manuscript, and we are deeply grateful. Ed Zackery was instrumental in clearing up theological issues in relation to the FSM, and we are in his debt for helping us with that complexity and with our fieldwork.

The help of all these colleagues on this manuscript, on earlier ones, or both has been invaluable, and many of their general views and specific suggestions have been incorporated. Needless to say, however, any errors are ours and ours alone. We can no longer say that we'll miss late-night conversations with our friend Jonathan Cobb after the book is published. He's still our brilliant editor at Island Press, but he's become even more our friend, and so the conversations go on, book or no. In this case, once again, his contributions have been invaluable—no, monumental. Pat Harris, who also struggled with *Human Natures* and *One with Nineveh*, copyedited this manuscript with her usual enormous patience and great skill. As before, it was a great pleasure to interact with her.

Our colleague Richard Klein kindly allowed us to base figures 3-3 and 4-1 on ones he has prepared for the third edition of his classic text *The Human Career*. We are also indebted to Kelvin Pollard, senior demographer at the Population Reference Bureau in Washington, DC, for producing the fine population profiles of three nations in chapter 7, and to the Population Reference Bureau for their excellent work and support over many decades. Darryl Wheye allowed us to reproduce her wonderful *Archaeopteryx* drawing, and Peter and Rosemary Grant gave permission to use their photographs in the Galápagos (Darwin's) finch story in chapter 1. John Harte got us Scott Saleska's wonderful picture of his global heating experiment (it's even better in color). Rick Stanley kindly took some photos for us on our educational (and thoroughly enjoyable) field trip to Brazil. And Judy and John Waller, as they did for *Human Natures*, prepared many of the drawings with great skill. It was again fun and rewarding to work with them.

Peggy Vas Dias of Stanford's Center for Conservation Biology was, as always, a tremendous help with the diversity of chores that accompany the creation of a book. Jill Otto and the rest of the staff of the Falconer Biology Library once more proved that the proximity of that wonderful and obliging facility is one of the best things about the wonderful department we've enjoyed as a home for nearly fifty years. And, not surprisingly, Pat Browne and Steve Masley as usual handled photocopying chores perfectly. Once again (it must be habit-forming), we've enjoyed publishing with Island Press, a non-profit company still struggling to roll back the tide of environmental ignorance and destruction.

We remain deeply in debt to the many people who have supported our work over the years, in particular Peter and Helen Bing, Larry Condon, Stanley and

Marion Herzstein, Walter and Karen Loewenstern, and Wren Wirth. Finally, we wish to acknowledge a special debt to five close colleagues and dear friends who have passed away: Loy Bilderback, Dick Holm, LuEsther Mertz, John Montgomery, and John Thomas. All were dedicated to making the world a better place, and their intellectual contributions and moral and direct support have been invaluable to us.

Index

Note: Page numbers in *italics* refer to illustrations

About the Authors

PAUL R. EHRLICH is Bing Professor of Population Studies, Department of Biological Sciences, Stanford University and a Beijer Fellow. An expert in the fields of evolution, ecology, and human biology, Ehrlich has devoted his career to the study of such topics as dynamics of insect populations, coevolution of plants and herbivores, and human cultural evolution. His fieldwork has carried him to all continents, from the Arctic and the Antarctic to the tropics, and from high mountains to the ocean floor. Professor Ehrlich has written more than 900 scientific papers and popular articles as well as over 40 books, including *The Population Bomb, The Process of Evolution, The Science of Ecology, The Birder's Handbook, New World/New Mind, A World of Wounds, Human Natures*, and *Wild Solutions*.

Among his many scientific honors, Ehrlich is a Fellow of the American Academy of Arts and Sciences, an honorary member of the British Ecological Society, and a member of the United States National Academy of Sciences and the American Philosophical Society. He received the Crafoord Prize in Population Biology and the Conservation of Biological Diversity, an explicit substitute for the Nobel Prize in fields of science for which the latter is not given. Ehrlich has also received a MacArthur Fellowship, the Volvo Environment Prize, the International Center for Tropical Ecology's World Ecology Medal, the International Ecology Institute's ECI Prize, the Dr. A. H. Heineken Prize for Environmental Sciences, the Blue Planet Prize, and the Eminent Ecologist Award of the Ecological Society of America. He was the first recipient of the Roger Tory Peterson Memorial Medal.

ANNE H. EHRLICH is Senior Research Scientist in the Department of Biological Sciences of Stanford University as well as Policy Coordinator of Stanford's Center for Conservation Biology. She is a Fellow of the American Academy of Arts and Sciences and the Californian Academy of Sciences. Besides having conducted research on butterflies and reef fish, she has been active in working with a number of environmental organizations, including the Sierra Club. She was deeply involved in issues such as environmental impacts of nuclear war as well as interactions between population, resources, and the environment. She has served as a government advisor and on boards of directors of several organizations and small foundations.

Anne was co-editor of *Hidden Dangers*, and Anne and Paul together have authored a series of books, including *Ecoscience, Extinction, Earth, The Population Explosion, Healing the Planet, The Stork and the Plow, Betrayal of Science and Reason*, and *One with Nineveh*.

Anne and Paul together have jointly received the Sasakawa Environment Prize of the United Nations Environment Programme (UNEP), the Heinz Award in the Environment, the American Humanist Association's Distinguished Service Award, the Nuclear Age Peace Foundation's Distinguished Peace Leadership Award, and the Tyler Prize for Environmental Achievement.

About the Center for Conservation Biology

In 1984, Paul R. Ehrlich founded Stanford University's Center for Conservation Biology (CCB) to develop the science of conservation biology and to help devise ways and means to protect Earth's life-support systems. Ehrlich serves as President of the CCB.

Under direction of Professor Gretchen C. Daily, the Center for Conservation Biology designs experiments to address specific and general questions in conservation biology, especially in the new fields of countryside biogeography and conservation finance. It also conducts research on broad-scale policy issues, including human cultural evolution, overconsumption, environmental deterioration, and ecological economics. Among its major goals are to communicate the results of this scientific and policy research to conservation biologists, planners, non-governmental organizations, decision makers, and the public; to educate students and professionals; and to foster collaboration with social scientists and conservation groups around the world.

The Center for Conservation Biology is supported by donations and grants from individuals, private foundations, and corporations.